4.4.78

PRACTICAL ARITHMETIC

PRACTICAL ARITHMETIC

The Third "R"

CAROL L. JOHNSON

HARVEY E. REYNOLDS

LOYD V. WILCOX

Golden West College

PRENTICE-HALL, INC., *Englewood Cliffs, New Jersey 07632*

Library of Congress Cataloging in Publication Data

Johnson, Carol L (date)
Practical arithmetic.

Includes index.
1. Arithmetic—1961- I. Reynolds, Harvey E., (date) joint author. II. Wilcox, Loyd V., (date) joint author. III. Title.
QA107.J63 513 76-48918
ISBN 0-13-689273-6

10 9 8 7 6 5 4 3 2 1

Printed in the United States of America

Prentice-Hall International, Inc., *London*
Prentice-Hall of Australia Pty. Limited, *Sydney*
Prentice-Hall of Canada, Ltd., *Toronto*
Prentice-Hall of India Private Limited, *New Delhi*
Prentice-Hall of Japan, Inc., *Tokyo*
Prentice-Hall of Southeast Asia Pte. Ltd., *Singapore*
Whitehall Books Limited, *Wellington, New Zealand*

Contents

PREFACE IX

INTRODUCTION 1

Unit One WHOLE NUMBERS AND WHOLE NUMBER OPERATIONS 3

LESSON

1 The Whole Numbers *4*

2 Addition and Subtraction of Whole Numbers *8*

3 Multiplication and Division of Whole Numbers *17*

4 Powers and Roots of Whole Numbers *31*

5 Order of Operations and Rules of Divisibility *39*

Drill Exercises *45*
Self-test *47*
Answers *50*

Unit Two FRACTIONAL NUMBERS 57

LESSON

6 Prime Numbers and Prime Factorizations *58*

7 Common Multiples and Common Factors *62*

8 An Introduction to Fractions *68*

*9 Word Problems *80*

Drill Exercises *86*
Self-test *88*
Answers *92*

*An asterisk preceding a lesson number indicates that the lesson is optional.

Unit Three FRACTIONAL NUMBER OPERATIONS 99

LESSON

10 Addition and Subtraction of Fractions *100*

11 Division of Fractions *108*

12 Mixed Number Operations *115*

*13 Fractional Parts and Probability *127*

Drill Exercises *138*
Self-test *142*
Answers *146*

Unit Four AN INTRODUCTION TO ALGEBRA 153

LESSON

14 Mathematical Phrases and Sentences *154*

15 Solving Equations *161*

16 Proportions *169*

17 Formulas *175*

Drill Exercises *181*
Self-test *184*
Answers *187*

Unit Five DECIMALS AND DECIMAL OPERATIONS 191

LESSON

18 Decimal Numeration *192*

19 Decimal Operations *199*

20 Decimal Division *208*

*21 Additional Decimal Topics *224*

Drill Exercises *237*
Self-test *240*
Answers *243*

Unit Six PERCENT 249

LESSON

22 The Meaning of Percent *250*

23 Solving Percent Problems *263*

*24 Applications of Percent *270*

Drill Exercises *280*
Self-test *283*
Answers *286*

Unit Seven MEASUREMENT 291

LESSON

25 An Introduction to Measurement *292*

26 Metric Measures *304*

27 Other Metric Measures *319*

*28 Applications of Measurements *335*

Drill Exercises *356*
Self-test *360*
Answers *363*

Unit Eight USING NUMERICAL INFORMATION 369

LESSON

*29 Descriptive Statistics *370*

*30 Graphs and Tables *389*

Drill Exercises *413*
Self-test *420*
Answers *425*

APPENDIX 431

TABLE

1 Addition and Multiplication Combinations *432*

2 Squares and Square Roots *433*

3 English System Weights and Measures *434*

4 Metric Prefixes *435*

5 Metric System Weights and Measures *435*

6 Metric-English Equivalents *437*

7 Measures of Time *437*

8 Geometric Figures *438*

9 Percent, Decimal, and Common Fraction Equivalents *440*

10 Miscellaneous Formulas *441*

INDEX 443

Preface

"Why study arithmetic? I don't need to know fractions to balance my checkbook." The primary purpose of this book is to provide the individual in need of a refresher and renewal course in basic arithmetic skills an answer to this question by providing insight into the reasons "why" everyone can benefit from the knowledge of the areas of mathematics that affect their lives daily. Drills, algorithms, and theory must be included as some part of any course in arithmetic, but we have tried to motivate the needs for the necessary skills by including topics that are relevant to the young adult completing a course in refresher mathematics as well as to the mature adult who finds it necessary to study arithmetic.

Methods of problem solving are introduced at the end of the second unit. From this point on, each unit ends with a lesson relating the material to applied topics that can be understood with only the skills that have been acquired in this and preceding units. The applied topics and all other subjects are introduced, whenever possible, with a problem to be solved, followed by the rationale for and the method of solution. "Rules and rigor" have not taken precedence over intuition and familiarity of basic notions as the key to understanding the material.

This book has also been designed so that it can provide the individual who needs a review of arithmetic with a positive experience so that they may overcome any preconceived notions about their inability to understand mathematical concepts. In order to make this material as understandable as possible, clearly stated performance objectives are given at the beginning of each lesson within a unit. These objectives are designed to let the user of this book know exactly what is to be learned in a lesson; what skills and concepts should be mastered before going on to other topics.

Pre- and post-tests matched to the objectives for each lesson, as well as practice exercises within the lesson, extensive drill exercise problems, and a unit self-test allow the person using this text to fit the contents to his or her own individual needs.

This book could not have been written without the thousands of students at Golden West College whose experiences and comments over the last five years have guided its evolution. We are indebted to the excellent (and patient) tutors, particularly Roy Pinkerton, who have worked with these students and have passed invaluable information on to us. Our thanks is extended to Jerry Karl and Ellen Church for their comments and criticisms and to John Wadhams who helped to initiate this undertaking. Special acknowledgement is made of our debt to Nancy Halford of Rio Hondo College for her constructive criticisms and corrections, her time spent in proofreading, and for her analysis of the objectives and the related self-test material. Finally, there would have been no book without the assistance of Dena Reynolds and Paul Steen and without the support and patience of Ginnie Wilcox, Leta Reynolds, and Ken Johnson.

CAROL L. JOHNSON
HARVEY E. REYNOLDS
LOYD V. WILCOX

PRACTICAL ARITHMETIC

Introduction

This book probably differs a great deal from other arithmetic texts that you have seen or used. You should take a moment to read through this introduction on how to use the book so that you can get the most out of the material from the very beginning of your studies.

On the first page of each lesson you will see a list of objectives and vocabulary terms. Do read carefully through this material before you start to study the lesson that follows. The purpose of the objectives is to let you know just what is important and what should be emphasized in each lesson. The vocabulary list gives the new terms that will be introduced and defined within the lesson. It also provides you with a ready reference to where a term first appears in case you find you'd like to review a definition.

A pre-test also precedes each lesson. You should work the problems in the pre-test and correct your answers after studying the objectives. The purpose of the pre-test is to help you identify what you should take particular care with when you study the material in the lesson. For example, the first objective in Lesson 2, Unit 1 states that you should be able to determine the sum of two or more whole numbers; the first pre-test question tests this objective by asking you to add four whole numbers. If after correcting the pre-test you are reasonably sure that you can add whole numbers, you may wish to only skim the lesson material that deals with addition of whole numbers. (This related lesson material is identified by the frame numbers that follow the objective.) If, however, you feel that you can't add whole numbers, you should work carefully through the related lesson material. The pre-test allows you to decide where and how to spend your time while you work through each lesson.

Within the lesson itself you will be asked to fill in some information in certain frames and to complete practice exercises in other frames. Even if you decide that you only need to skim certain lesson material, do stop and follow the directions in these frames. At the end of the lesson there is a post-test that covers exactly the same material and objectives as the lesson pre-test. (You shouldn't worry if you can't answer a pre-test question correctly, but you should take the time to review the related lesson material if you can't answer a post-test question.)

After all the lessons in a unit have been completed, you should do the unit drill exercise problems before you attempt the unit self-test. The self-tests are your final check of how well you know the material within the unit as a whole. The self-test gives you the opportunity to discover any weaknesses that you may still have and to review before being tested in class.

Finally, give the material and yourself a fair chance by not going into any lesson intimidated by its title! Past experience has shown that many students form self-fulfilling prophecies about their success in a mathematics course by having preconceived notions about the difficulty of the subject. We are confident that you can succeed. You, too, should have that same confidence in yourself and in your abilities.

UNIT ONE

Whole Numbers and Whole Number Operations

Lesson 1	THE WHOLE NUMBERS
Lesson 2	ADDITION AND SUBTRACTION OF WHOLE NUMBERS
Lesson 3	MULTIPLICATION AND DIVISION OF WHOLE NUMBERS
Lesson 4	POWERS AND ROOTS OF WHOLE NUMBERS
Lesson 5	ORDER OF OPERATIONS AND RULES OF DIVISIBILITY

LESSON 1
THE WHOLE NUMBERS

Objectives:

1. Write a whole number in word form. (1–7)
2. Write a whole number in expanded notation. (1–7)
3. Determine the name of the place-value position of specified digits in a given whole number. (8–11)

Vocabulary:

Numeration System	Numeral
Digit	Expanded Notation
Whole Number	Place Value

PRE-TEST

1. Write each of the following numbers in word form:
   (a) 513 ____________________
   (b) 7,028 ____________________
2. Write each of the following numbers in expanded notation:
   (a) 235 ____________________
   (b) 5,709 ____________________
3. Determine the name of the place-value position of each of the underlined digits:
   (a) 44,4$\underline{4}$4 ____________________
   (b) 7,3$\underline{0}$8,010 ____________________

A *numeration system* is any systematic method for naming and counting with numbers. The written form of a number is called a *numeral*. Our numeration system is called a positional system because the position of each digit in a numeral has assigned to it exactly one *place value*. All of the numerals in our familar system are formed from the ten digits: (1)

0, 1, 2, 3, 4, 5, 6, 7, 8, and 9

The *whole numbers* are the counting numbers and zero. (2)

0, 1, 2, 3, 4, 5, 6, 7, 8, 9, 10, 11, 12,
13, 14, 15, 16, 17, 18, 19, and so on

A whole number is written exactly as it is read. (3)

For example,

43	is read or written as	**forty-three**
703	is read or written as	**seven hundred three**
8,888	is read or written as	**eight thousand eight hundred eighty-eight**

Notice that the word "and" does not appear in the written forms of 703 and 8,888. Being able to read a number correctly helps in determining the meaning of the number. The numbers between 21 and 99 are hyphenated (-) when they are written in word form.

The meaning of a number: (4)

43	means	**40 + 3 or 4 tens + 3 ones**
703	means	**700 + 3 or 7 hundreds + 0 tens + 3 ones**
8,888	means	**8,000 + 800 + 80 + 8 or 8 thousands + 8 hundreds + 8 tens + 8 ones.**

The meaning of a number can also be shown using *expanded notation.* (5)

For example, since 43 means 40 + 3, it can also be written as

$$(4 \times 10) + (3 \times 1)$$

The parentheses are used to show that the multiplications, 4 × 10 and 3 × 1, are completed before adding. So, 4 × 10 is 40 and 3 × 1 is 3; the sum of these two numbers is 43.

703	**means**	(7 × 100) + (0 × 10) + (3 × 1)
8,888	**means**	(8 × 1,000) + (8 × 100) + (8 × 10) + (8 × 1)

Writing whole numbers in word form and expanded notation: (6)

Complete the following as shown:

A. 10,056
Word form: ten thousand fifty-six
Expanded notation: (1 × 10,000) + (0 × 1,000) + (0 × 100) + (5 × 10) + (6 × 1)

B. 4,040
Word form: four ______________ forty

Expanded notation: (△ × 1,000) + (⬡ × 100) + (△ × 10) + (⬡ × 1)

C. 25,067
Word from: twenty-five ______________ sixty-seven

Expanded notation: (2 × □) + (5 × 1,000) + (0 × ○) + (6 × 10) + (7 × ▽)

Practice Exercises ⑦

Write each of the following numbers in word form and in expanded notation:

		Word Form	Expanded Notation
1.	321	______________	______________
2.	4,024	______________	______________
3.	77,777	______________	______________
4.	66,355	______________	______________

Expanded notation is useful in determining the place value of a digit within a numeral. Consider Example C in Frame 6; what is the place value of the digit 2 in the numeral 25,067? ⑧

The word form, twenty-five thousand sixty-seven, gives little information about the place value of the digit 2. From the expanded notation,

$$(2 \times 10{,}000) + (5 \times 1{,}000) + (0 \times 100) + (6 \times 10) + (7 \times 1)$$

it is possible to see that the place value of the 2 is *ten-thousands*; there are 2 ten-thousands in 25,067.

The place value of a digit can also be determined by naming the place-value positions of all the digits in the numeral:

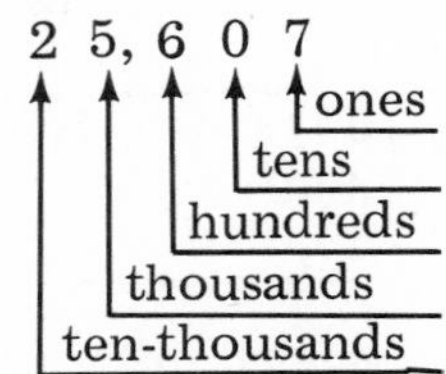

The Place-Value Names: ⑨

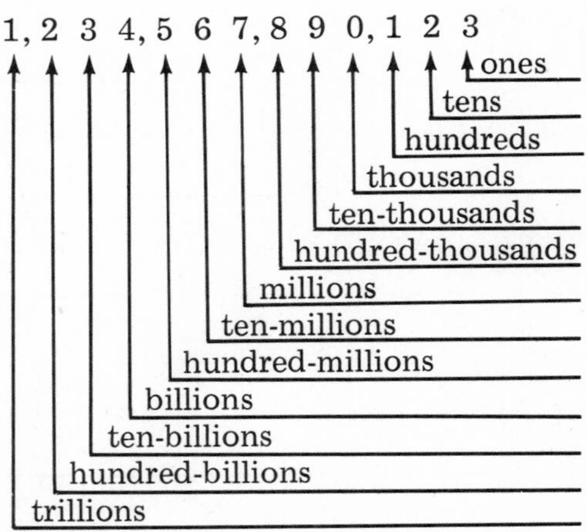

Naming the place value of specified digits in a numeral: (10)

Consider the following example: The population of the United States is approximately 270,925,000.

This number is read: two hundred seventy million, nine hundred twenty-five thousand.

Complete the following:

A. The place value of the digit 5 is ________________ .

B. The place value of the digit 7 is ________________ .

C. The place value of the digit 0 that appears between the 7 and the 9 is ____________ .

Practice Exercises (11)

Write the place-value name for the position of each underlined digit.

1. 32,4<u>5</u>9 ______________________________
2. 1<u>1</u>8,113 ______________________________
3. 10,<u>0</u>08 ______________________________
4. 7<u>8</u>4,324,000 ______________________________
5. 9,00<u>0</u>,000,421 ______________________________

POST-TEST

1. Write each of the following numbers in word form:

   (a) 5,007______________________________

   (b) 234,031______________________________

2. Write each of the following numbers in expanded notation:

   (a) 40,384 ______________________________

   (b) 745,999______________________________

3. Determine the name of the place-value position of each of the underlined digits:

   (a) 38,59<u>2</u> ______________________________

   (b) <u>4</u>99,003 ______________________________

LESSON 2
ADDITION AND SUBTRACTION OF WHOLE NUMBERS

Objectives:

1. Determine the sum of two or more whole numbers, showing any carried digits. (1–13)
2. Determine the difference between two whole numbers, showing any borrowed digits. (14–21)
3. Write the missing numbers to correctly complete inverse operations statements for addition and subtraction. (22–25)

Vocabulary:

Addition	Subtraction
Addend	Minuend
Sum	Subtrahend
Carrying	Difference
Grouping	Borrowing
Reordering	Inverse Operations

PRE-TEST

1. Determine each sum, showing any carried digits.

(a)
```
   472
 3,808
    95
+  670
------
```

(b)
```
 3,747
   992
 4,018
+  345
------
```

2. Determine each difference, showing any borrowed digits.

(a)
```
 58,014
-39,408
-------
```

(b)
```
 90,041
-69,853
-------
```

3. Write the missing number to correctly complete each inverse operations statement.

(a) (743 + ____) – 231 = 743

(b) (____ + 19) – 19 = 33

The Addition Operation: ①

Addition is an operation that replaces the process of counting.

The symbol for addition is +, the plus sign.

3 plus 5 is written **3 + 5**

The numbers that are added are called *addends* and the answer is called the *sum.*

$$3 + 5 = 8 \leftarrow \text{sum}$$

addends ↑ ↑ (3 and 5)

$$\begin{array}{r} 3 \\ +5 \\ \hline 8 \end{array}$$

3, +5 ← addends; 8 ← sum

Addition Combinations: 

Complete the following simple addition combinations by filling in each of the blank spaces with the sum of the numbers that appear in the same row and column. For example, notice that the number 3 has been written in one of the spaces. Three is the sum of the number 2 (the number at the top of the same column) and the number 1 (the number at the left of the same row).

+	0	1	2	3	4	5	6	7	8	9
0										
1			3							
2										
3										
4										
5										
6										
7										
8										
9										

A complete table of addition combinations can be found in the Appendix. This table can be used to check the answers in this frame.

Notice that in filling in the table in the preceding frame there is no difference in the sums if the row number is added to the column number or if the column number is added to the row number. That is, ③

$$2 + 1 = 3 \quad \text{and} \quad 1 + 2 = 3$$

Knowing that the *reordering* of the addends does not affect the sum of two numbers reduces the number of addition combinations that must be learned. Since 2 + 1 and 1 + 2 both equal the same sum,

$$2 + 1 = 1 + 2$$

This is an example of the Commutative Property of Addition. The Commutative Property of Addition simply states that any two numbers can be added in either order without affecting the sum.

■ + ● = ● + ■

Another useful property of addition is the Associative Property. The Associative Property of Addition is a statement of the fact that three or more addends can be *grouped* in any manner without changing the sum. The order of the addition is indicated by the use of parentheses. For this reason, the Associative Property may be referred to as the *grouping* property of addition. (4)

Consider the following example: **1 + 4 + 3**

In order to determine the sum, it is necessary to first add two of the addends. Using parentheses to show which two addends are to be added first, we get

$$(1 + 4) + 3 = 5 + 3 \qquad 1 + (4 + 3) = 1 + 7$$
$$= 8 \qquad\qquad = 8$$

In either case, the answer is 8, so it is possible to say that

$$(1 + 4) + 3 = 1 + (4 + 3)$$

Or for any three addends,

(■ + ●) + ▲ = ■ + (● + ▲)

Another useful property to remember is that zero added to any number is just that number. (5)

214 + 0 = 214 and **0 + 17 = 17**

Or for any number, ■

0 + ■ = ■ and 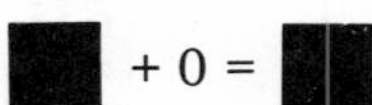

The addition operation: 417 + 322 (6)

"How"

In order to determine the sum of these two numbers, the digits of like place value are added. The problem is first written so that the digits that are to be added are aligned:

$$\begin{array}{r} 417 \\ +322 \\ \hline \end{array}$$

Then, adding from right to left,

$$\begin{array}{r} 417 \\ +322 \\ \hline 9 \end{array}$$

The sum of the digits in the ones' place-value position, 7 and 2, is written in the one's position in the answer.

$$\begin{array}{r} 417 \\ +322 \\ \hline 39 \end{array}$$

The sum of the digits in the tens' position, 2 and 1, is written in the same place-value position in the answer.

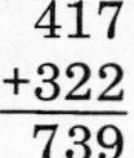

$$\begin{array}{r} 417 \\ +322 \\ \hline 739 \end{array}$$

The sum of the digits in the hundreds' position, 4 and 3, is written in the same place-value position in the answer.

"Why"

$$\begin{array}{r} 417 \\ +322 \\ \hline 739 \end{array} \quad \rightarrow \quad \begin{array}{rccccc} & \text{hundreds} & & \text{tens} & & \text{ones} \\ & (4 \times 100) & + & (1 \times 10) & + & (7 \times 1) \\ \text{add} & (3 \times 100) & + & (2 \times 10) & + & (2 \times 1) \\ \hline & (7 \times 100) & + & (3 \times 10) & + & (9 \times 1) \end{array}$$

The addition operation: 357 + 85 ⑦

"How"

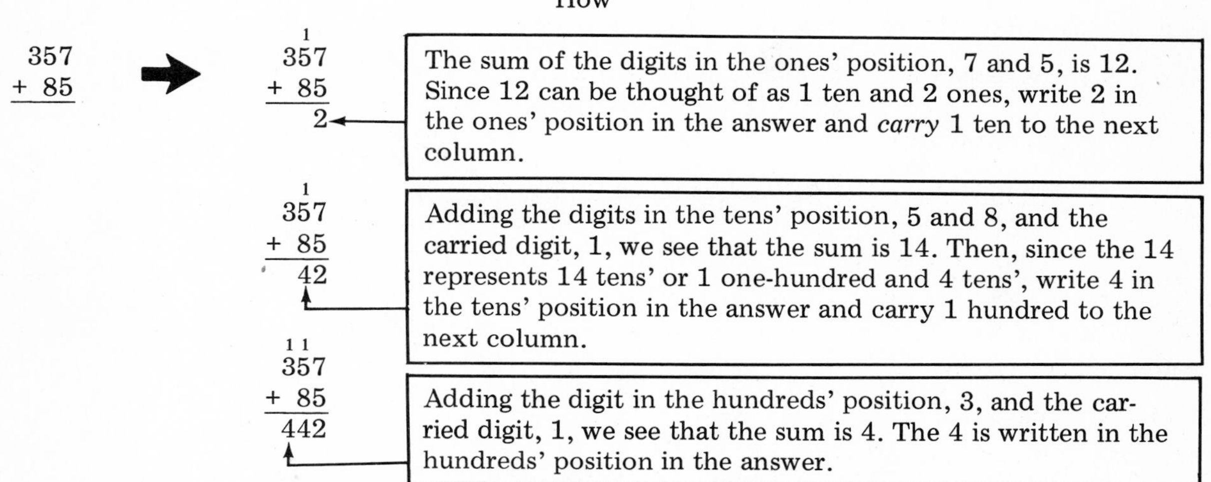

$$\begin{array}{r} 357 \\ +\ 85 \\ \hline \end{array} \rightarrow \begin{array}{r} {}^{1} \\ 357 \\ +\ 85 \\ \hline 2 \end{array}$$

The sum of the digits in the ones' position, 7 and 5, is 12. Since 12 can be thought of as 1 ten and 2 ones, write 2 in the ones' position in the answer and *carry* 1 ten to the next column.

$$\begin{array}{r} {}^{1} \\ 357 \\ +\ 85 \\ \hline 42 \end{array}$$

Adding the digits in the tens' position, 5 and 8, and the carried digit, 1, we see that the sum is 14. Then, since the 14 represents 14 tens' or 1 one-hundred and 4 tens', write 4 in the tens' position in the answer and carry 1 hundred to the next column.

$$\begin{array}{r} {}^{1\,1} \\ 357 \\ +\ 85 \\ \hline 442 \end{array}$$

Adding the digit in the hundreds' position, 3, and the carried digit, 1, we see that the sum is 4. The 4 is written in the hundreds' position in the answer.

"Why"

$$\begin{array}{r} 357 \\ +\ 85 \\ \hline 442 \end{array} \rightarrow \begin{array}{r} 300 + 50 + 7 \\ + 80 + 5 \\ \hline 300 + 130 + 12 \end{array} \rightarrow \begin{array}{r} (3 \times 100) + (5 \times 10) + (7 \times 1) \\ + (8 \times 10) + (5 \times 1) \\ \hline (3 \times 100) + (13 \times 10) + (12 \times 1) \end{array}$$

1st carry

$$(3 \times 100) + (14 \times 10) + (2 \times 1)$$

2nd carry

$$(4 \times 100) + (4 \times 10) + (2 \times 1)$$

Carrying is used whenever the sum of the digits in the same place-value position is greater than 9 since 9 is the digit with the greatest value that can be written.

Carrying in an addition operation: ⑧

$$\begin{array}{r} 427 \\ 31 \\ 9{,}674 \\ +\quad 9 \\ \hline \end{array} \rightarrow \begin{array}{r} {}^{1\,1\,2} \\ 427 \\ 31 \\ 9{,}674 \\ +\quad 9 \\ \hline \underline{1}0{,}141 \end{array} \leftarrow \text{carried digits}$$

The underlined digit in the sum is in the ten-thousands position although none of the addends in this problem has a digit with this place value.

Checking the sum of an addition operation: (9)
An addition can be checked by reversing the order of addition. That is, if the place-value columns are first added from top to bottom, the sum can be checked by re-adding each column from bottom to top.

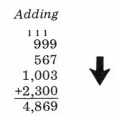

For example, the sum of 6,598; 376; and 4,059:

Add and check: 999 + 567 + 1,003 + 2,300 (10)

Add and check: 9,538 + 1,427 (11)

Complete the following by writing in the correct carry digits in the addition and by checking the sum:

Adding

```
      Δ
   9,5 3 8
+  1,4 2 7
 1 0,9 6 5
```

Checking

```
 [       ]
   9,5 3 8
+  1,4 2 7
 1 0,9 6 5
```

Adding and checking the sum: (12)

Complete each of the following by writing in the missing digits in each sum. Check each answer by reversing the order of addition.

A.

```
   1 1 1 2
   4 1,2 4 8
     1,0 0 5
   2 7,9 9 9
+  1 6,5 1 4
   8 _,7 _ 6
```

B.

```
     1 2     1
   1 2 5,7 2 8
     3 0,7 4 1
+  4 1 5,6 2 8
   _ 7 _,_ 9 _
```

Practice Exercises (13)

Find the indicated sums, showing all carried digits. Check each sum by reversing the order of addition.

1.	2.	3.
165 394 276 +505	146 92 981 + 17	268 74 93 +192

4.	5.	6.
740,765 617,500 879,282 456,123 +123,456	43,433 801,801 564,792 32,462 + 21,442	181,753 90,000 327,462 1,572,486 + 438,966

The Subtraction Operation:

(14)

The symbol for *subtraction* is –, the minus sign.

7 subtract 3 is written **7 – 3**

The parts of a subtraction problem have special names:

7 – 3 = 4 ← difference
minuend ↑ ↑ subtrahend

7 ← minuend
–3 ← subtrahend
4 ← difference

The subtraction operation: 8,352 – 5,142

(15)

"How"

In order to determine the difference, the digits of like place value are subtracted. The problem is first written so that the digits of like place value are aligned:

8,352
–5,142

Then, subtracting from right to left,

8,352
–5,142
0 ← The difference between the digits in the ones' position, 2 – 2, is written in the ones' position in the answer.

8,352
–5,142
10 ← The difference between the digits in the tens' position, 5 – 4, is written in the same place-value position in the answer.

8,352
–5,142
210 ← The difference between the digits in the hundreds' position, 3 – 1, is written in the same place-value position in the answer.

8,352
–5,142
3,210 ← The difference between the digits in the thousands' position, 8 – 5, is written in the same place-value position in the answer.

"Why"

		thousands		hundreds		tens		ones
8,352		$(8 \times 1{,}000)$	+	(3×100)	+	(5×10)	+	(2×1)
−5,142	subtract	$(5 \times 1{,}000)$	+	(1×100)	+	(4×10)	+	(2×1)
3,210		$(3 \times 1{,}000)$	+	(2×100)	+	(1×10)	+	(0×1)

Checking the difference in a subtraction problem: (16)

The difference in a subtraction problem can be checked by addition. The difference and the subtrahend are added. The sum should equal the minuend.

The subtraction operation: 42 − 23 (17)

"How"

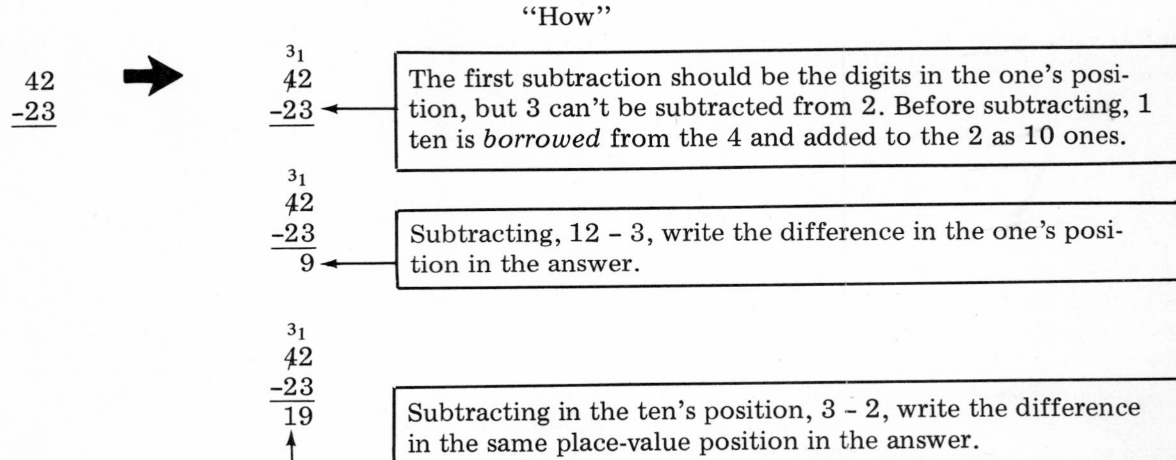

42
−23
The first subtraction should be the digits in the one's position, but 3 can't be subtracted from 2. Before subtracting, 1 ten is *borrowed* from the 4 and added to the 2 as 10 ones.
Subtracting, 12 − 3, write the difference in the one's position in the answer.
Subtracting in the ten's position, 3 − 2, write the difference in the same place-value position in the answer.

"Why"

42		40 + 2		30 + 12
−23	subtract	20 + 3	subtract	20 + 3
				10 + 9

Borrowing is used whenever a larger digit is to be subtracted from a smaller digit. Borrowing is the opposite of carrying.

Repeated borrowing in a subtraction: 302 − 89 

Subtracting

$\overset{2}{\not3}\,{}^{1}0\,2$	→	$\overset{2}{\not3}\,\overset{9}{{}^{1}\not0}\,{}^{1}2$
− 8 9		− 8 9
		2 1 3

Checking

213
\+ 89
302

Since there are 0 tens in 302, it is first necessary to borrow 1 hundred from 300 and add it to the 0 as 10 tens. It is then possible to borrow 1 ten which is added to the 2 as 10 ones.

Repeated borrowing in a subtraction: 436 − 297 ⑲

Subtracting	*Checking*
3 12 1 4̸ 3̸ 6 − 2 9 7 1 3 9	297 +139 436

Subtracting and checking the difference: 

Complete the following by writing in the missing digits in the differences. Check each subtraction by adding the subtrahend and the difference.

A. 1, 0 0 7
− 7 0 8
2 __ 9

Check: 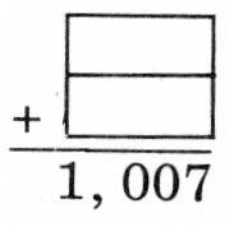
+ □
1, 007

B. 9, 7 1 2
− 2, 6 5 1
7, __ __ 1

Check:
□
+ □
9, 712

Practice Exercises 

Find the indicated differences, showing all borrowed digits. Check each difference by using addition.

1. 53 − 21
2. 777 − 578
3. 4,302 − 1,670
4. 439 − 249
5. 14,888 − 10,999
6. 5,040 − 4,053
7. 384,107 − 126,439
8. 700,000 − 390,909
9. 555,666 − 483,987

Inverse operations are pairs of operations that "undo" each other. 

Some examples of inverse operations:

Dressing and undressing;
Packing and unpacking a suitcase;
Going up in an elevator and then back down;
Adding 6 to a number and subtracting 6 from the sum.

Addition and subtraction are inverse operations: 

A subtraction "undoes" an addition:

$(17 + 6) - 6 = 17$

(▲ + ■) − ■ = ▲

An addition "undoes" a subtraction:

$(99 - 75) + 75 = 99$

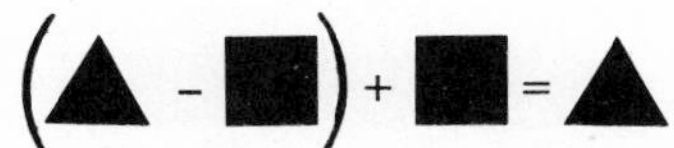

Completing an inverse operations statement: (24)

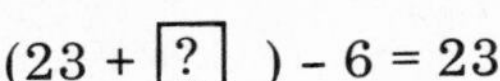
(23 + [?]) - 6 = 23

In this example the statement is true if [?] is replaced by 6.

([?] - 2) + 2 = 13

In this example the statement is true if [?] is replaced by 13.

Practice Exercises

Write the missing numbers to correctly complete each inverse operations statement.

1. (53 - 12) + 12 = ____
2. ____ = (200 + 73) - 73
3. (27 - 7) + ____ = 27
4. (74 + 19) - ____ = 74
5. (____ + 3) - 3 = 100
6. (42 + ____) - ____ = 42

POST-TEST

1. Determine each sum, showing any carried digits.

(a)
```
  274
8,083
   59
+ 760
-----
```

(b)
```
7,473
  299
8,014
+ 543
-----
```

2. Determine each difference, showing any borrowed digits.

(a)
```
 80,493
-41,085
-------
```

(b)
```
 35,986
-14,997
-------
```

3. Write the missing number to correctly complete each inverse operations statement.

(a) (621 + 241) - ____ = 621

(b) (____ - 44) + 44 = 93

LESSON 3
MULTIPLICATION AND DIVISION OF WHOLE NUMBERS

Objectives:

1. Write the missing numbers to correctly complete a Distributive Property statement. (9–12)
2. Determine the product of two whole numbers, showing any carried digits. (13–20)
3. Determine the quotient and the remainder when a whole number is divided by another whole number. (21–33)
4. Write the missing numbers to correctly complete inverse operations statements for multiplication and division. (36–38)

Vocabulary:

Multiplication	Quotient
Factor	Divisor
Product	Dividend
Distributive Property	Remainder
Division	

PRE-TEST

1. Write the missing numbers to correctly complete each Distributive Property statement.

   (a) 5(8 + 3) = (5 × ___) + (5 × 3)

   (b) 26(700 + 30 + 2) = (___ × 700) + (___ × 30) + (___ × 2)

2. Determine the product of each of the following, showing any carried digits:

   (a) $\begin{array}{r} 938 \\ \underline{\times 407} \end{array}$ (b) $\begin{array}{r} 3{,}061 \\ \underline{\times \quad 739} \end{array}$

3. Determine the quotient and the remainder for each of the following:

   (a) 1,437 ÷ 58   (b) 7,204 ÷ 203

4. Write the missing number to correctly complete each inverse operations statement.

   (a) (216 ÷ ___) × 17 = 216

   (b) (432 × 13) ÷ 13 = ___

How many eggs are there in a box that has 3 rows if there are 6 eggs in each row? (1)

One way to answer this question is to use addition:

$$\underbrace{6+6+6}_{\text{3 rows}} = 18$$

This question can also be answered by using *multiplication:*

$$3 \times 6 = 18 \qquad \text{means} \qquad \underbrace{6+6+6}_{\text{3 addends}} = 18$$

Multiplication can be thought of as repeated addition.

How many points are possible on a 20-question examination if each question is worth 4 points? (2)

Using addition:

$$\underbrace{4+4+4+\ldots+4}_{\text{20 questions}} = 80$$

The notation " . . . " means that 16 addends of 4 have not been written. In general, this notation is used to indicate that certain information that follows in the pattern shown has been omitted.

Using multiplication:

$$20 \times 4 = 80 \qquad \text{means} \qquad \underbrace{4+4+4+\ldots+4}_{\text{20 addends}} = 80$$

The Multiplication Operation: (3)

One symbol for multiplication is ×.

2 times 5 is written 2 × 5

The numbers which are multiplied are called *factors* and the answer is called the *product.*

$2 \times 5 = 10$ ← product; 2 and 5 ← factors

$$\begin{array}{r} 5 \\ \times 2 \\ \hline 10 \end{array}$$

(5 and 2 ← factors; 10 ← product)

The multiplication 2 times 5 can also be indicated by any of the following:

$2 \cdot 5 \qquad (2)(5) \qquad 2(5) \qquad (2)5 \qquad 2(4+1)$

Multiplication Combinations: 4

Complete the following simple multiplication combinations by filling in each of the blank spaces with the product of the numbers that appear in the same row and column. For example, notice that the number 6 is written in one of the spaces. Six is the product of 3 (the number that appears at the top of the same column) and 2 (the number at the left of the same row).

X	0	1	2	3	4	5	6	7	8	9
0										
1										
2				6						
3										
4										
5										
6										
7										
8										
9										

A complete table of multiplication combinations can be found in the Appendix. Use the table to check the answers in this frame.

Notice that in filling in the table in the preceding frame there is no difference in the products 5
if the row number is multiplied by the column number or if the column number is multiplied by the row number. That is,

$$2 \times 3 = 6 \qquad\qquad 3 \times 2 = 6$$

Knowing that the *reordering* of the factors does not affect the product of two numbers reduces the number of multiplication combinations that must be learned. Since 2×3 and 3×2 both equal 6,

$$2 \times 3 = 3 \times 2$$

This is an example of the Commutative Property of Multiplication. The Commutative Property of Multiplication states that any two numbers can be multiplied in either order without affecting the product. (Compare this property with the Commutative Property of Addition in Frame 3, Lesson 2.)

$$\blacksquare \times \blacktriangle = \blacktriangle \times \blacksquare$$

Just as there is a Commutative Property for both addition and multiplication, so is there an 6
Associative Property for Multiplication comparable to the Associative Property of Addition.

Consider the following example: $2 \times 5 \times 3$

$$(2 \times 5) \times 3 = 10 \times 3 = 30 \qquad\qquad 2 \times (5 \times 3) = 2 \times 15 = 30$$

No matter which two factors are multiplied first, the product is the same. The Associative Property of Multiplication is a statement of the fact that three or more factors can be grouped in any manner without changing the product.

$$(2 \times 5) \times 3 = 2 \times (5 \times 3)$$

Or, for any three factors,

$$(■ \times ▲) \times ● = ■ \times (▲ \times ●)$$

"Multiplication by One": ⑦

A property of multiplication that will be used repeatedly in this course is "Multiplication by One": the product of any number and one is that same number.

$(23)(1) = 23$ and $(1)(17) = 17$

Or for any number, ■,

$■ \times 1 = ■$ and $1 \times ■ = ■$

It will also be useful to remember that the product of any number and zero is always zero. 

$16 \times 0 = 0$ and $0 \times 413 = 0$

For any number, ■,

$■ \times 0 = 0$ and $0 \times ■ = 0$

One further property relates the operations of addition and multiplication: the *Distributive Property.* ⑨

For example, $2(5 + 4)$

$$2(5 + 4) = (2 \times 5) + (2 \times 4)$$
$$2(9) = 10 + 8$$
$$18 = 18$$

For any numbers,

$$■ \times (▲ + ●) = (■ \times ▲) + (■ \times ●)$$

The Distributive Property can also relate the operations of subtraction and multiplication. (10)

For example, 3(5 – 2)

$$
\begin{aligned}
3(5-2) &= (3 \times 5) - (3 \times 2) \\
3(3) &= 15 - 6 \\
9 &= 9
\end{aligned}
$$

Or, in general,

$$\blacksquare \times (\blacktriangle - \bullet) = (\blacksquare \times \blacktriangle) - (\blacksquare \times \bullet)$$

The Distributive Property can be used to simplify computations. (11)

Consider the product of 8 and 17:

A.	8 × 17 = 8 × (10 + 7)	Rewriting 17 as 10 + 7
	= (8 × 10) + (8 × 7)	Using the Distributive Property
	= 80 + 56	Multiplying
	= 136	Adding the products
B.	8 × 17 = 8 × (20 – 3)	Rewriting 17 as 20 – 3
	= (8 × 20) – (8 × 3)	Using the Distributive Property
	= 160 – 24	Multiplying
	= 136	Subtracting the products

Practice Exercises (12)

Write the missing numbers to correctly complete each of the following:

1. 5 × (4 + 7) = (5 × ____) + (5 × ____)
2. 6 × 18 = 6 × (____ + 8)
3. 4(59) = 4(50 + ____) = (4 × ____) + (4 × ____)
4. 8 × 999 = 8(900 + 90 + 9)
   = (____ × 900) + (____ × 90) + (____ × 9)
5. 153 × 17 = (100 + ____ + 3) × 17
6. 33 × 456 = 33 × (400 + 50 + ____)
7. 6(8 – 1) = (____ × 8) – (____ × 1)
8. 9 × 38 = 9 × (40 – 2)
   = (9 × ____) – (9 × ____)

Determine the product of 3 and 13. (13)

Using the Distributive Property:

$$3 \times 13 = 3 \times (10 + 3)$$
$$= (3 \times 10) + (3 \times 3)$$
$$= 30 + 9$$
$$= 39$$

This problem can also be worked as follows:

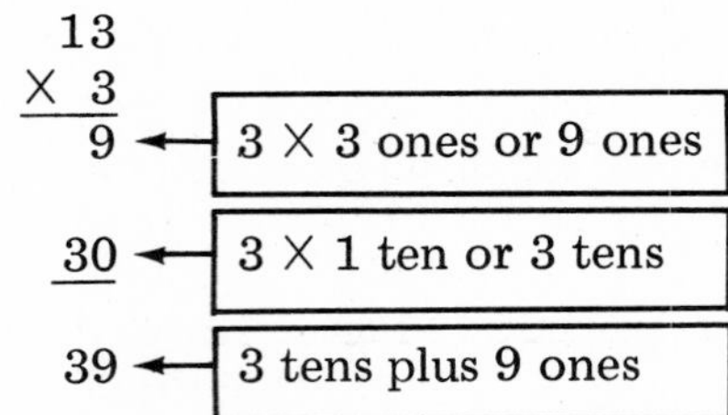

Notice in the answer that the digit in the one's position is the product of 3 and the number in the ones' position in the number 13 and that the digit in the tens' position is the product of 3 and the number in the ten's position. The following shortcut method is usually used for all multiplication problems:

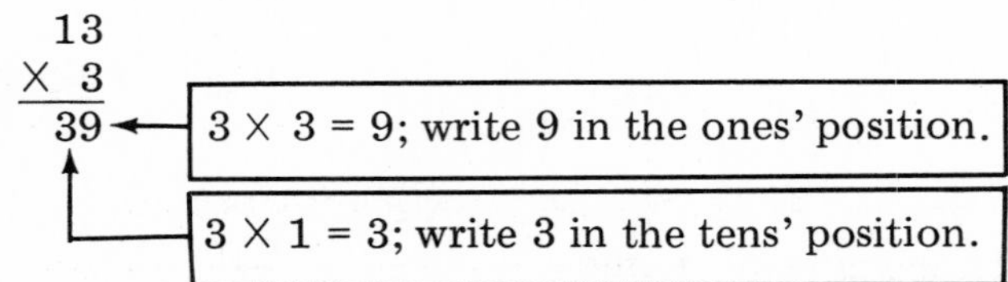

13
X 3
39
3 X 3 = 9; write 9 in the ones' position.
3 X 1 = 3; write 3 in the tens' position.

Determine the product of 4 and 637: (14)

"How"

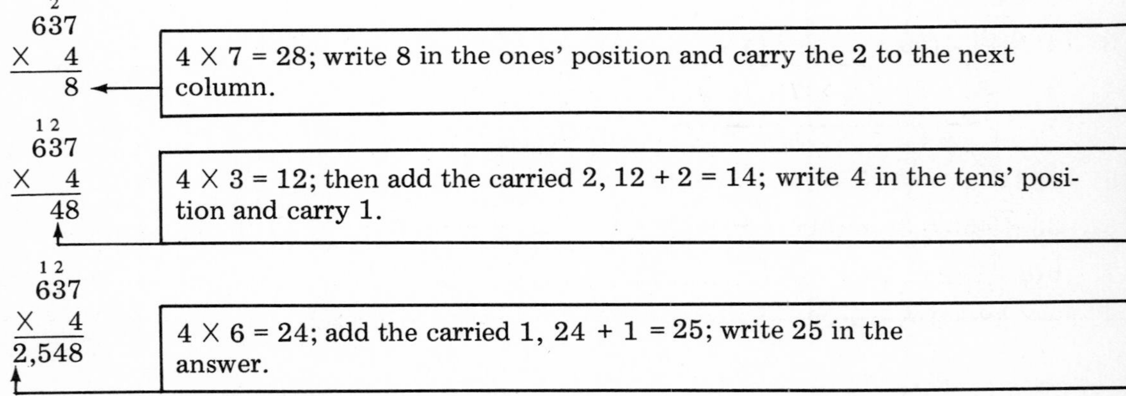

2
637
X 4
8
4 X 7 = 28; write 8 in the ones' position and carry the 2 to the next column.
1 2
637
X 4
48
4 X 3 = 12; then add the carried 2, 12 + 2 = 14; write 4 in the tens' position and carry 1.
1 2
637
X 4
2,548
4 X 6 = 24; add the carried 1, 24 + 1 = 25; write 25 in the answer.

"Why"

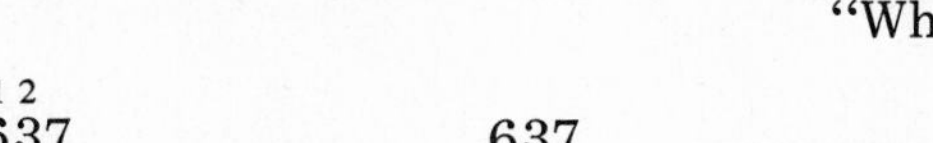

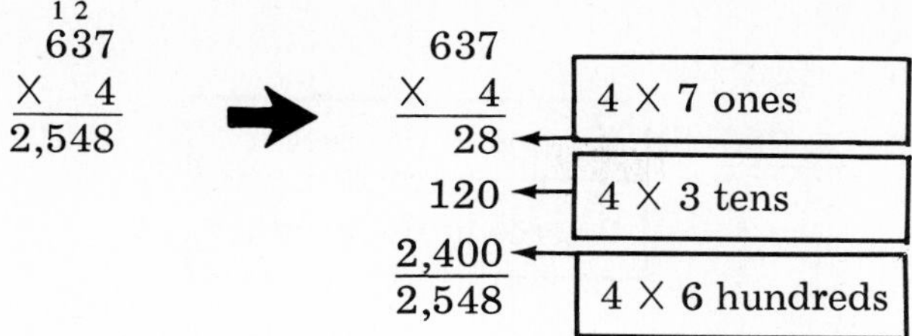

Practice Exercises

Multiply each of the following, showing any carried digit.

1. 589 X 7
2. 4,036 X 8
3. 710,502 X 9

Multiply: 20 X 17

Using the Distributive Property:

$$
\begin{aligned}
20 \times 17 &= 20 \times (10 + 7) \\
&= (20 \times 10) + (20 \times 7) \\
&= 200 + 140 \\
&= 340
\end{aligned}
$$

If 17 is multiplied by 2, the product is 34. Compare this with the product when 17 is multiplied by 20. When multiplying by a whole number that ends in one or more zeros, the following method can be used:

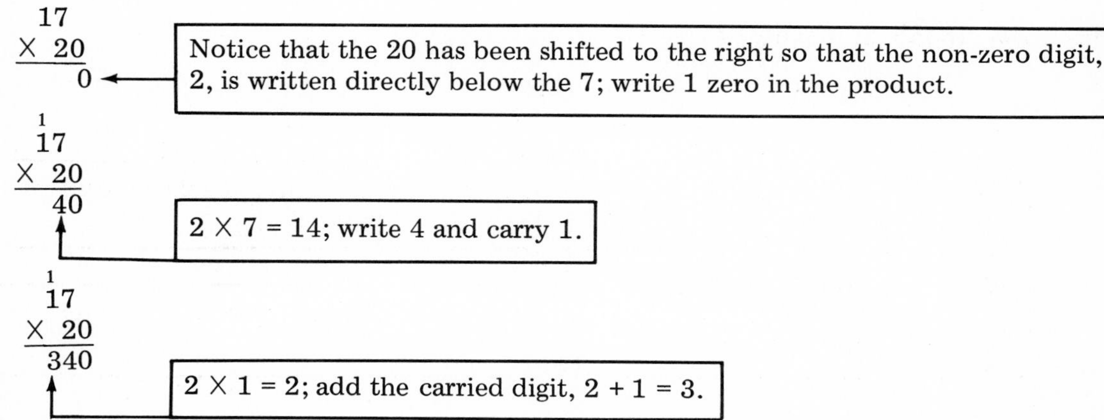

17
X 20
0
Notice that the 20 has been shifted to the right so that the non-zero digit, 2, is written directly below the 7; write 1 zero in the product.
1
17
X 20
40
2 X 7 = 14; write 4 and carry 1.
1
17
X 20
340
2 X 1 = 2; add the carried digit, 2 + 1 = 3.

Multiplying by a number that ends in one or more zeros: (17)

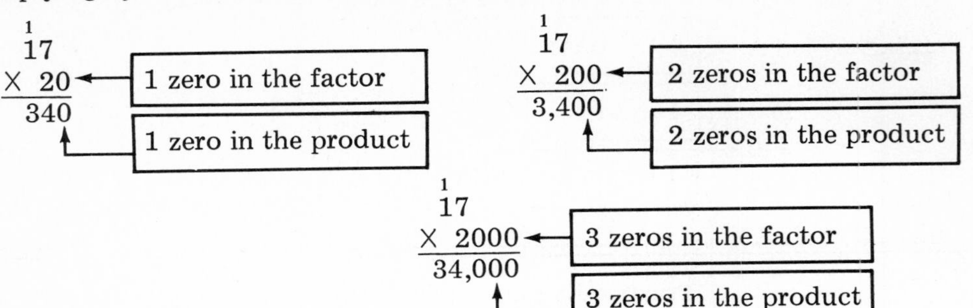

Find the product of 172 and 45: (18)

Using the Distributive Property, it can be shown that

$$\begin{array}{r} 172 \\ \times\ 45 \\ \hline \end{array} \quad \text{is equivalent to} \quad \begin{array}{r} \scriptstyle 2 \\ 172 \\ \times\ \ 40 \\ \hline 6{,}880 \end{array} + \begin{array}{r} \scriptstyle 3\,1 \\ 172 \\ \times\ \ 5 \\ \hline 860 \end{array} = 7{,}740$$

The numbers, 6,880 and 860, are called the *partial products* of the multiplication 172 X 45. This problem is usually worked as follows:

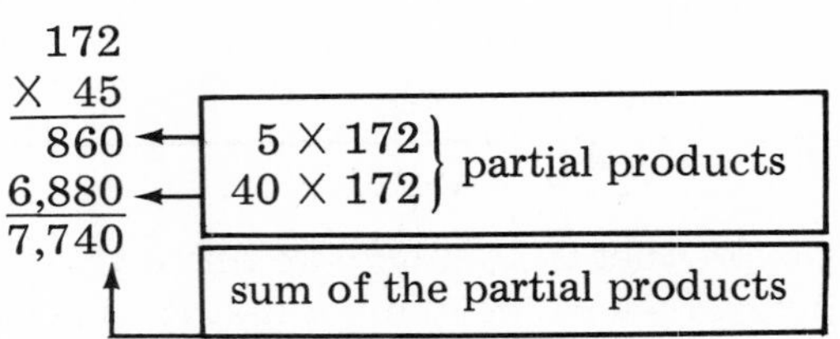

Multiply: 239 X 747 (19)

"How"

$$\begin{array}{r} 747 \\ \times 239 \\ \hline 6{,}723 \\ 22{,}410 \\ 149{,}400 \\ \hline 178{,}533 \end{array} \left.\vphantom{\begin{array}{r} 6 \\ 2 \\ 1 \end{array}}\right\} \boxed{\text{partial products}}$$

"Why"

$$\begin{array}{r} 747 \\ \times 239 \\ \hline \end{array} \quad \rightarrow \quad \begin{array}{r} 747 \\ \times\ \ 9 \\ \hline 6{,}723 \end{array} + \begin{array}{r} 747 \\ \times\ 30 \\ \hline 22{,}410 \end{array} + \begin{array}{r} 747 \\ \times 200 \\ \hline 149{,}400 \end{array} = 178{,}533$$

partial products

Practice Exercises

Find each product, showing all carried digits.

1. 637 × 34	2. 715 × 403	3. 998 × 889
4. 471 × 107	5. 808 × 303	6. 4,723 × 46
7. 8,034 × 701	8. 6,633 × 5,353	9. 2,637 × 902

The Division Operation:

Two symbols used for division are ÷ and $\overline{)\quad}$

12 divided by 4 is written **12 ÷ 4 or 4 $\overline{)12}$**

The parts of a division problem have special names:

12 ÷ 4 = 3 ← quotient; 12 is the dividend, 4 is the divisor

divisor → 4 $\overline{)12}$ ← dividend, with 3 ← quotient above

Multiplication and division are related in the following way:

3 × 4 = 12 means **4 + 4 + 4 = 12** (3 addends)

Multiplication is a repeated addition. Division can be thought of as a repeated subtraction.

12 ÷ 4 = 3 means 12 − 4 = 8; 8 − 4 = 4; 4 − 4 = 0

The division 12 ÷ 4 = 3 means that 4 can be subtracted from 12 three times.

The relationship between multiplication and division.

Comparing the product of 3 × 4 and the quotient when 12 is divided by 3 or by 4:

3 × 4 = 12 **12 ÷ 3 = 4** **12 ÷ 4 = 3**

When a product is divided by either of its factors, the quotient is the other factor.

Division can be used to answer questions such as 

$$3 \times \boxed{?} = 51$$

This question could also be answered by subtracting 3 repeatedly from 51, but it is much easier to determine the missing number by finding the quotient.

$$51 \div 3 = \boxed{?} \qquad \text{or} \qquad 3\overline{)51}\ \ (\text{quotient } \boxed{?})$$

Determine the quotient: 51 ÷ 3 

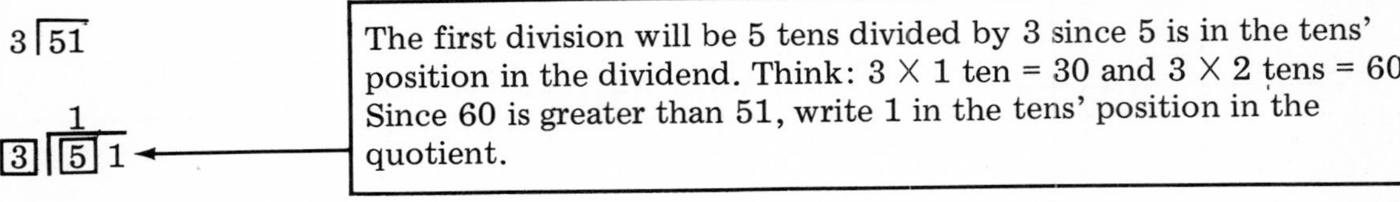

3 ⟌51
1
3 ⟌5 1
The first division will be 5 tens divided by 3 since 5 is in the tens' position in the dividend. Think: 3 X 1 ten = 30 and 3 X 2 tens = 60. Since 60 is greater than 51, write 1 in the tens' position in the quotient.

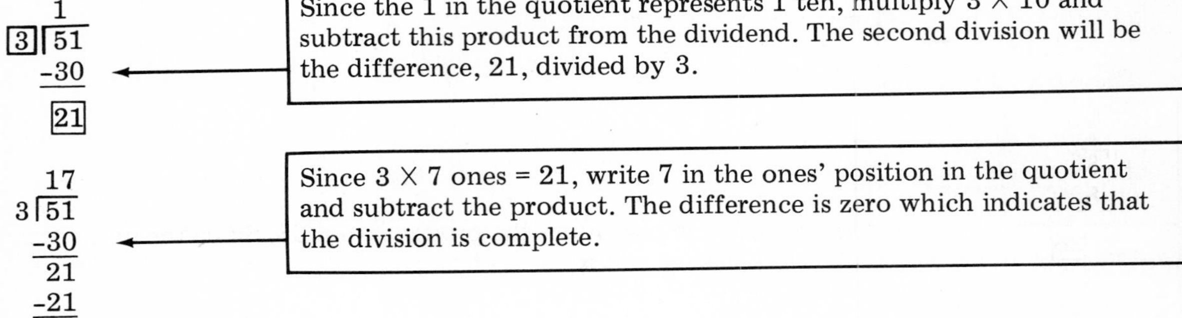

1
3 ⟌51
-30
21
Since the 1 in the quotient represents 1 ten, multiply 3 X 10 and subtract this product from the dividend. The second division will be the difference, 21, divided by 3.
17
3 ⟌51
-30
21
-21
0
Since 3 X 7 ones = 21, write 7 in the ones' position in the quotient and subtract the product. The difference is zero which indicates that the division is complete.

Since 51 ÷ 3 = 17, the missing number in $3 \times \boxed{?} = 51$ is 17. Notice that the multiplication of the quotient, 17, by the divisor, 3, results in a product that is equal to the dividend. Multiplication can be used to check the quotients in division problems such as this example.

Determine the quotient: 204 ÷ 7 

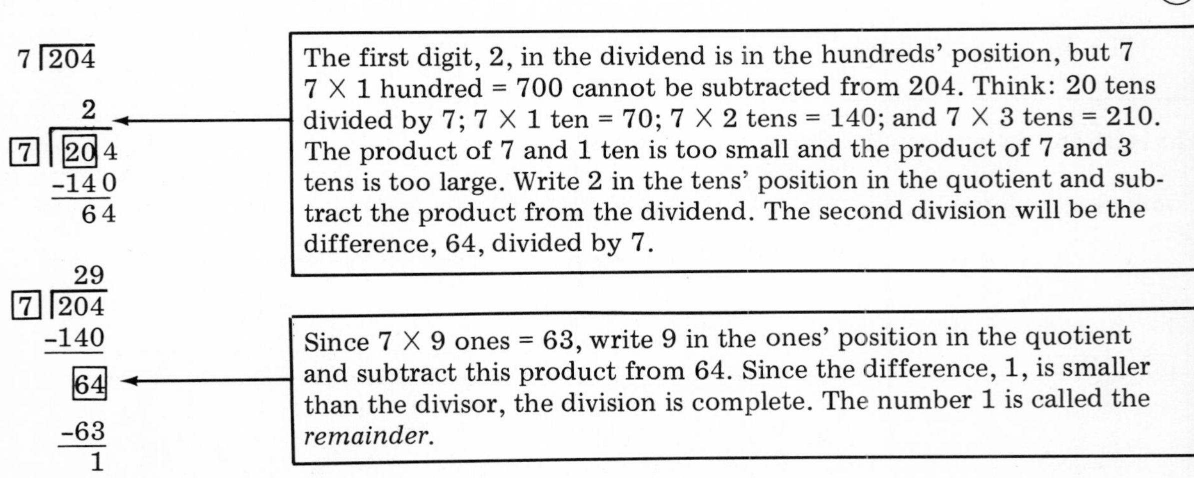

7 ⟌204
2
7 ⟌20 4
-14 0
6 4
The first digit, 2, in the dividend is in the hundreds' position, but 7 7 X 1 hundred = 700 cannot be subtracted from 204. Think: 20 tens divided by 7; 7 X 1 ten = 70; 7 X 2 tens = 140; and 7 X 3 tens = 210. The product of 7 and 1 ten is too small and the product of 7 and 3 tens is too large. Write 2 in the tens' position in the quotient and subtract the product from the dividend. The second division will be the difference, 64, divided by 7.
29
7 ⟌204
-140
64
-63
1
Since 7 X 9 ones = 63, write 9 in the ones' position in the quotient and subtract this product from 64. Since the difference, 1, is smaller than the divisor, the division is complete. The number 1 is called the remainder.

Non-zero remainders are written with the quotient as shown below.

$$7\overline{)204}\quad 29 \text{ R } 1$$

To check the answer, multiply the quotient by the divisor and then add the remainder to this product.

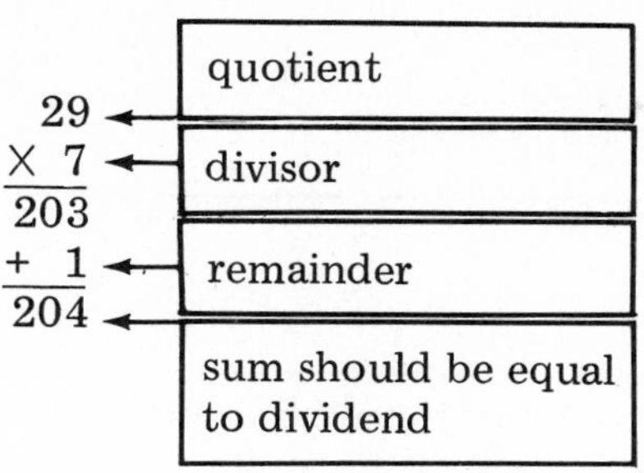

Divide: 836 ÷ 4 27

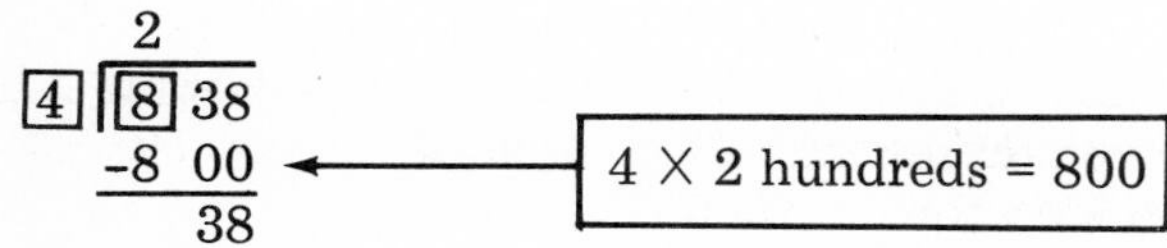

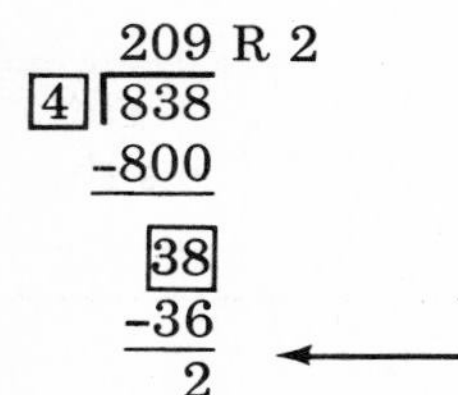

Since 4 × 1 ten = 40 is greater than 38, write 0 in the quotient in the tens' position. Then, 4 × 9 ones = 36; write 9 in the ones' position in the quotient and subtract the product from 38. Since the difference, 2, is less than the divisor, the division is complete. Write the remainder with the quotient.

Check: (209 × 4) + 2 = 836 + 2 = 838

"Long" division can also be done using the following method: 28

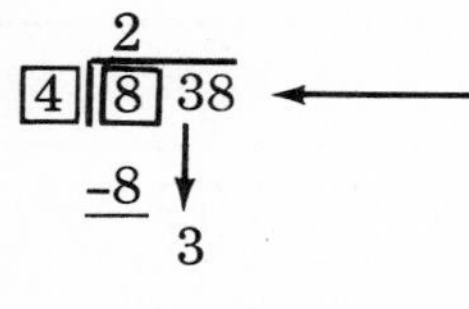

First step: Think 4 "goes into" 8 two times; write 2 in the quotient directly above the 8. Multiply 2 × 4 and subtract this product from 8. "Bring down" the next digit in the dividend, 3.

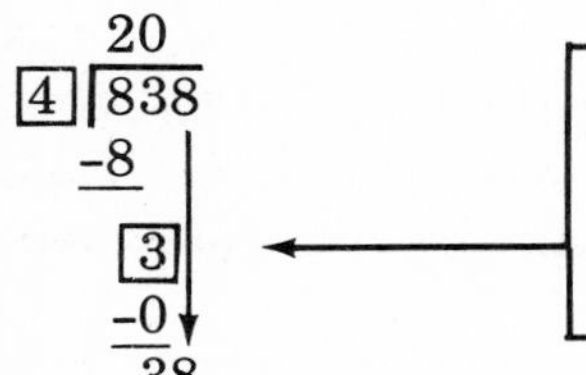

Second step: Think 4 "goes into" 3 zero times; write 0 in the quotient directly above the 3. Multiply 0 × 4 and subtract the product from 3. "Bring down" the next digit, 8, and write it to the right of the difference, 3.

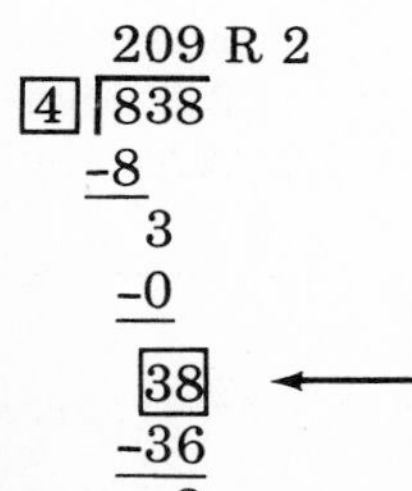

Third step: Think 4 "goes into" 38 nine times; write 9 in the quotient directly above the 8. Multiply 9 × 4 and subtract the product from 38. The difference is 2. Since there are no further digits that can be "brought down," the division is complete. Write the difference, 2, as the remainder.

Practice Exercises

29

Calculate each quotient and remainder. Check each result by using multiplication and addition.

1. $7\overline{)983}$
2. $5\overline{)4{,}263}$
3. $4\overline{)30{,}300}$
4. $8\overline{)911{,}377}$

Divide: 343 ÷ 15 

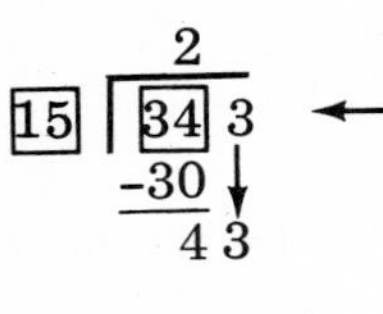

First division: Since 15 won't go into 3, think 15 "goes into" 34 two times; write 2 in the quotient directly above the 4. Multiply 2 × 15 and subtract the product from 34. Bring down the 3.

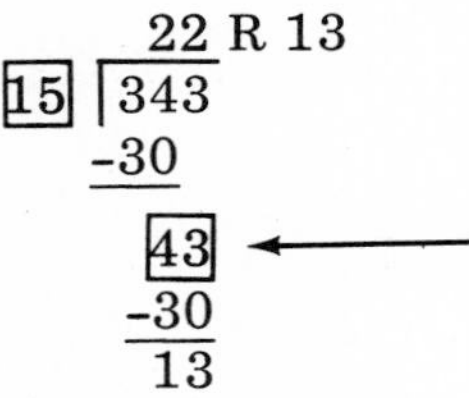

Second division: Think 15 "goes into" 43 two times (since two 15's is 30 and three 15's is 45; 45 can't be subtracted from 43); write 2 in the quotient above the 3 and subtract the product from 43. The remainder, 13, is written with the quotient.

Divide: 2,875 ÷ 37 

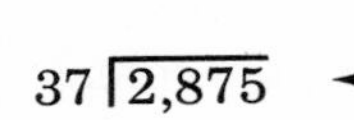

First division: Since 37 won't go into either 2 or 28, the first division will be 37 into 287. With a divisor such as 37, it is useful to think that 37 is almost 40. Think 287 ÷ 40 to estimate the digit that will be placed in the quotient.

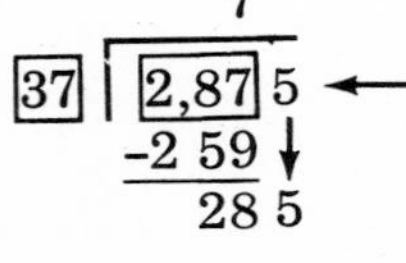

Think: 40 × 7 = 280; use 7 as a first guess. Then, 37 × 7 = 259. Subtracting 259 from 287, the difference is 28. Since 28 is less than 37, 7 is the correct number to be placed in the quotient (directly above the 7 in the dividend). Bring down the 5.

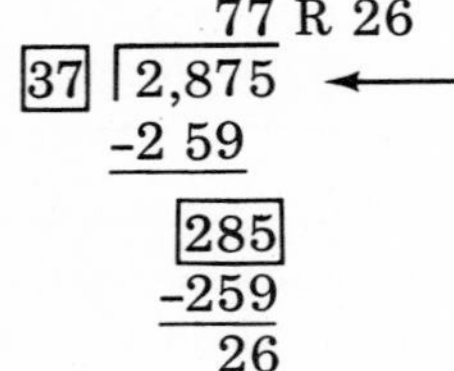

Second division: 285 ÷ 37; write another 7 in the quotient directly above the 5. Subtracting the product, 7 × 37, the remainder is 26.

Determine the quotient: 53,327 ÷ 176 

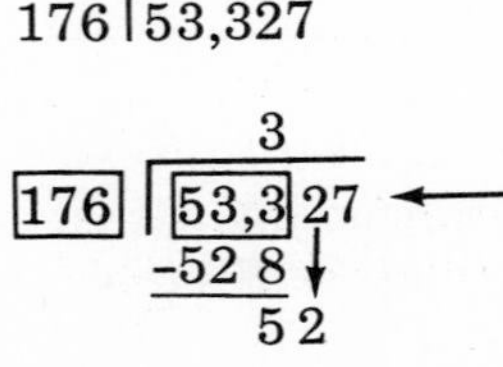

First division: 533 ÷ 176. Estimate the first digit to be written in the quotient by thinking 530 ÷ 180. Then, 2 × 180 = 360 and 3 × 180 = 540. Try 2; 2 × 176 = 352 and 533 − 352 = 181, a number that is greater than 176; the estimate is incorrect. Next try 3; 3 × 176 = 528 and 533 − 528 = 5; write 3 in the quotient and subtract the product of 3 and 176 from 533. Bring down the next digit, 2.

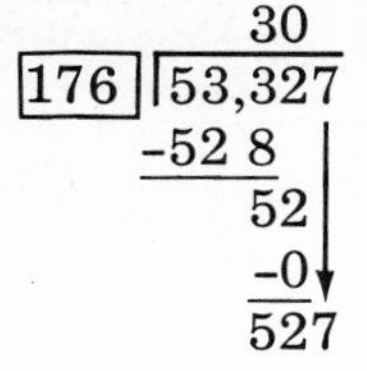

Second division: 176 won't go into 52; write 0 in the quotient and subtract the product of 0 and 176 from 52. Bring down the 7.

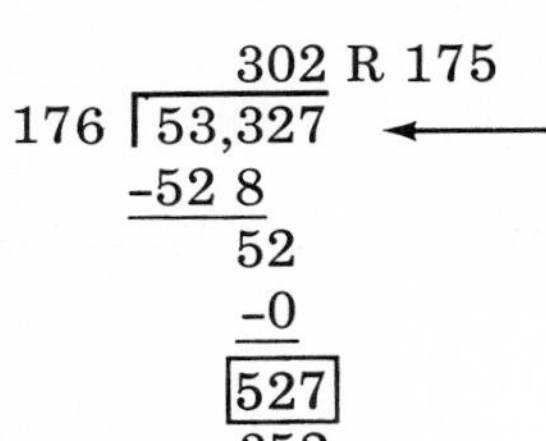

Third division: It is apparent from the first division that 176 goes into 527 two times; write 2 in the quotient and subtract the product of 2 and 176 from 527. Write the remainder, 175, with the quotient.

Practice Exercises (33)

Divide and check each of the following:

1. 58 ⟌3,777
2. 169 ⟌958,920
3. 231 ⟌6,583
4. 93 ⟌90,393
5. 82 ⟌88,886
6. 418 ⟌748,209
7. 1,003 ⟌53,217
8. 714 ⟌407,140
9. 5,302 ⟌424,198
10. 922 ⟌122,222

There are several important division properties that will be used in this text: (34)

Any non-zero number divided by itself is 1.

$$13 \div 13 = 1$$

 = 1

Any number divided by 1 is just that same number.

$$68 \div 1 = 68$$

 ÷ 1 = ■

Zero divided by any non-zero number is 0.

$$0 \div 118 = 0$$

0 ÷  = 0

Notice in the preceding frame that two of these properties restrict the divisor to non-zero numbers. What happens if a whole number is divided by 0? (35)

$$11 \div 0 = \boxed{?}$$

Since division indicates a repeated subtraction, how many times can 0 be subtracted from 11? It is possible to subtract 0 from 11 an endless number of times. For this reason, division by 0 is not possible. Another way to think of this is

If $11 \div 0 = \boxed{?}$, then $\boxed{?} \times 0 = 11$

But since the product of any number and 0 is always 0, clearly there is no number [?] that will result in a product of 11 when multiplied by 0.

Multiplication and division are inverse operations. (36)

A division "undoes" a multiplication:

(16 X 2) ÷ 2 = 16

A multiplication "undoes" a division:

(16 ÷ 2) X 2 = 16

The relationships of the fundamental operations of arithmetic: 

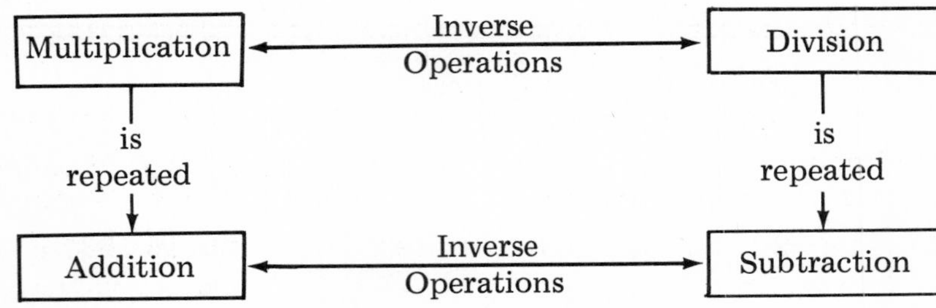

Practice Exercises 

Write the missing number to correctly complete each inverse operations statement.

1. (5 X 7) ÷ 7 = □
2. (33 X □) ÷ 61 = 33
3. (□ ÷ 14) X 14 = 129
4. (392 ÷ □) X 49 = 392

POST-TEST

1. Write the missing numbers to correctly complete each distributive property statement.
   (a) 13(7 + 8) = (____ X 7) + (____ X 8)
   (b) ____ (300 + 70 + 1) = (9 X 300) + (9 X 70) + (9 X 1)
2. Determine the product of each of the following, showing any carried digits:
   (a) 704 X839
   (b) 1,603 X 937
3. Determine the quotient and the remainder for each division operation.
   (a) 7,431 ÷ 85
   (b) 3,704 ÷ 307
4. Write the missing number to correctly complete each inverse operations statement.
   (a) (444 X 18) ÷ ____ = 444
   (b) (____ ÷ 6) X 6 = 702

LESSON 4
POWERS AND ROOTS OF WHOLE NUMBERS

Objectives:

1. Rewrite the repeated multiplication of a whole number using an exponent. (1–6)
2. Write a whole number which is a power of 2, 3, or 10 in exponential form. (1–6)
3. Determine a specified power of a given whole number base. (1–6)
4. Write a whole number in expanded notation using the exponential forms of the powers of ten. (7–9)
5. Determine the square root of a given square number. (10–14)
6. Determine the approximate value of the square root of a non-square number by placing the root between two consecutive whole numbers. (15–21)
7. Determine the cube root of a given cube number. (22–25)
8. Determine the approximate value of the cube root of a non-cube number by placing the root between two consecutive whole numbers. (26–29)

Vocabulary:

Exponent
Base
Power
Square Number
Cube Number
Square Root
Cube Root

PRE-TEST

1. Rewrite the following repeated multiplication using an exponent:

   $$7 \times 7 \times 7 \times 7 \times 7$$

2. Write 81, a power of 3, in exponential form.
3. Find the 4th power of 5.
4. Write 30,428 in expanded notation using the exponential forms of the powers of 10.
5. Determine the square root of 81.
6. Determine the approximate value of $\sqrt{53}$ by placing it between two consecutive whole numbers.
7. Determine the cube root of 64.
8. Determine the approximate value of $\sqrt[3]{300}$ by placing it between two consecutive whole numbers.

How many great-grandparents does each person have? (1)

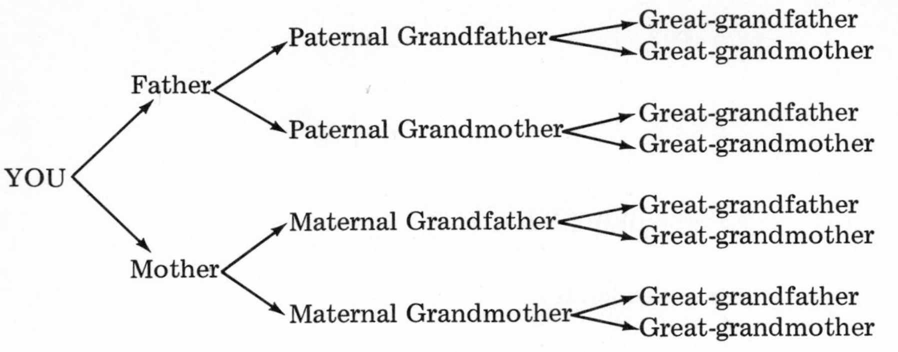

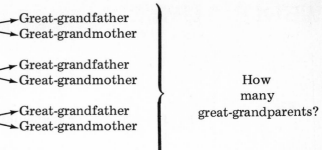

This problem can also be solved using repeated multiplication:

$2 \times 2 \times 2 = 8$ Great-grandparents

The triad tree is a botanical oddity. Each triad tree has 3 trunks, each trunk has 3 branches, each branch has 3 stems, and each stem has 3 leaves. How many leaves does a triad tree have? (2)

Using repeated multiplication,

$3 \times 3 \times 3 \times 3 = 81$ Leaves

An *exponent* is used to indicate a repeated multiplication. (3)

$$2^3 \quad \text{means} \quad \underbrace{2 \times 2 \times 2}_{\text{3 factors}}$$

$$3^4 \quad \text{means} \quad \underbrace{3 \times 3 \times 3 \times 3}_{\text{4 factors}}$$

The exponent tells how many times the *base* is to be used as a factor. (4)

In 2^3, 2 is the base and 3 is the exponent.

In 3^4, 3 is the base and 4 is the exponent.

The exponential form of a number: (5)

3^4 is the exponential form of the number 81.

exponent
↓
$3^4 = 81$
↑ ↑
base power

81 is said to be the 4th *power* of 3, since $3 \times 3 \times 3 \times 3 = 81$.

Practice Exercises

6

Rewrite the following repeated multiplications using exponents:

1. 7 × 7 × 7 × 7 × 7
2. 21 × 21 × 21
3. 5 × 5
4. 10 × 10 × 10 × 10 × 10 × 10 × 10 × 10 × 10

Write each of the following numbers in exponential form using the base 2 and the appropriate exponent. (Hint: First rewrite each number as a repeated multiplication.)

5. 8
6. 32
7. 256
8. 16

Find the indicated power for each of the following:

9. 3rd power of 7
10. 7th power of 10
11. 4th power of 4
12. 5th power of 2

The exponential forms of the powers of 10:

$$10^0 = 1$$
$$10^1 = 10$$
$$10^2 = 100$$
$$10^3 = 1{,}000$$
$$10^4 = 10{,}000$$
$$10^5 = 100{,}000$$
$$10^6 = 1{,}000{,}000$$
$$10^7 = 10{,}000{,}000$$
$$10^8 = 100{,}000{,}000$$
$$10^9 = 1{,}000{,}000{,}000$$

$$\vdots \qquad \vdots$$

Note that the 1st power of 10 has 1 zero, the 2nd power has 2 zeros, and so on.

Writing a number in expanded notation using the exponential forms of the powers of 10: (8)

$$368 = (3 \times 100) + (6 \times 10) + (8 \times 1)$$
$$= (3 \times 10^2) + (6 \times 10^1) + (8 \times 10^0)$$

$$573{,}094 = (5 \times 100{,}000) + (7 \times 10{,}000) + (3 \times 1{,}000) + (0 \times 100) + (9 \times 10) + (4 \times 1)$$
$$= (5 \times 10^5) + (7 \times 10^4) + (3 \times 10^3) + (0 \times 10^2) + (9 \times 10^1) + (4 \times 10^0)$$

Practice Exercises (9)

Write each of the following numbers in expanded notation using the exponential forms of the powers of 10.

1. 7,403
2. 1
3. 555
4. 8,002,509

There are special names for some exponential forms: (10)

$6^2 = 36$ 6^2 is read **"six squared"; 36** is a **square number**

When a number is raised to the 3rd power, the base is said to be *cubed* and the power is called a *cube number.*

$4^3 = 64$ 4^3 is read **"four cubed"; 64** is a **cube number**

The Square Number Pattern: (11)

What whole number has been squared if the power is 36? (12)

$$\boxed{?}^2 = 36$$

The most efficient way to find this number is to introduce a new operation: *square root.*

Square Root: (13)

The symbol used for square root is $\sqrt{}$.
The square root of 36 is written $\sqrt{36}$.

$$\sqrt{36} = 6 \text{ since } 6^2 = 36$$

Squaring and square root are inverse operations.

$$5^2 = 25$$
$$5 = \sqrt{25}$$

The square root of any square number is a whole number.

The Square Root Pattern: 

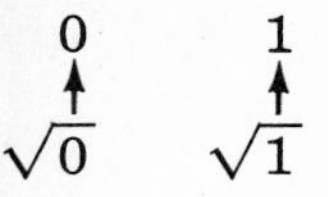 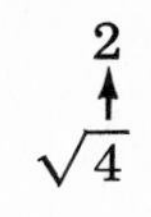 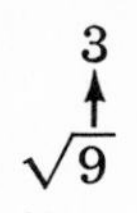

0	1	2	3	4	5	6	7	8	9	10 ...
↑	↑	↑	↑	↑	↑	↑	↑	↑	↑	↑
$\sqrt{0}$	$\sqrt{1}$	$\sqrt{4}$	$\sqrt{9}$	$\sqrt{16}$	$\sqrt{25}$	$\sqrt{36}$	$\sqrt{49}$	$\sqrt{64}$	$\sqrt{81}$	$\sqrt{100}$ ...

The square roots of non-square numbers are not whole numbers. 

For example, the square root of 3, written $\sqrt{3}$, is the only number that when squared is equal to 3.

$$\boxed{?}^2 = 3$$
$$(\sqrt{3})^2 = 3$$

For all numbers greater than zero:

$$(\sqrt{\blacktriangle})^2 = \blacktriangle \quad \text{and} \quad \sqrt{\blacktriangle^2} = \blacktriangle$$

Approximations may be used for the square roots of non-square numbers.

An approximate value for the square root of a non-square number can be determined by placing the square root between two consecutive whole numbers. 

For example, $\sqrt{3}$ is between what two consecutive whole numbers?

Looking at the Square Root Pattern:

0	1	2	3	4 ...
↑	↑	↑	↑	↑
$\sqrt{0}$	$\sqrt{1}$	$\sqrt{4}$	$\sqrt{9}$	$\sqrt{16}$ ...

Since 3 is between the square numbers 1 and 4, $\sqrt{3}$ is between $\sqrt{1}$ and $\sqrt{4}$.

0	1	$\sqrt{3}$	2 ...
↑	↑	↑	↑
$\sqrt{0}$	$\sqrt{1}$	$\sqrt{3}$	$\sqrt{4}$ ...

Since $\sqrt{3}$ is between $\sqrt{1} = 1$ and $\sqrt{4} = 2$, $\sqrt{3}$ is between the two consecutive whole numbers 1 and 2.

The Square Root Pattern could be written as follows: 

0	1	$\sqrt{2}$	$\sqrt{3}$	2	$\sqrt{5}$	$\sqrt{6}$	$\sqrt{7}$	$\sqrt{8}$	3	$\sqrt{10}$	$\sqrt{11}$	$\sqrt{12}$ ...
↑	↑	↑	↑	↑	↑	↑	↑	↑	↑	↑	↑	↑
$\sqrt{0}$	$\sqrt{1}$	$\sqrt{2}$	$\sqrt{3}$	$\sqrt{4}$	$\sqrt{5}$	$\sqrt{6}$	$\sqrt{7}$	$\sqrt{8}$	$\sqrt{9}$	$\sqrt{10}$	$\sqrt{11}$	$\sqrt{12}$ ...

The square roots of numbers greater than 100 can be determined exactly or can be approximated by extending the Square Root or Square Number Pattern. (18)

$$\sqrt{144} = \boxed{?}$$

Since 144 is greater than 100, the greatest square number shown in the square number pattern, $\sqrt{144}$ is greater than 10, the square root of 100. To determine the exact square root or the approximate value of $\sqrt{144}$, begin by squaring any whole number greater than 10:

$$11^2 = 121 \longleftrightarrow \sqrt{121} = 11$$

$$12^2 = 144 \longleftrightarrow \sqrt{144} = 12$$

The exact square root of 144 is 12 since 12 squared is 144.

$$\sqrt{172} = \boxed{?}$$ (19)

Since 172 is greater than 100, $\sqrt{172}$ is greater than 10. To determine the value of $\sqrt{172}$, begin by squaring any whole number greater than 10:

$12^2 = 144$ is less than 172
$13^2 = 169$ is less than 172
$14^2 = 196$ is greater than 172

Since 172 is between 169 and 196, $\sqrt{172}$ is between 13 and 14.

Determining the values of square roots of numbers greater than 100: (20)

Complete the following:

A. $\sqrt{225} = \boxed{?}$

$13^2 = 13 \times 13 = \triangle$

$14^2 = 14 \times 14 =$ ⬡

$15^2 = 15 \times 15 = \bigcirc$

$\sqrt{225} = \square$

B. $\sqrt{130} = \boxed{?}$

$11^2 = 11 \times 11 = \triangle$

$12^2 = 12 \times 12 = \bigcirc$

130 is between $\triangle$ and $\bigcirc$

$\sqrt{130}$ is between $\square$ and ⬡

Practice Exercises

Determine the exact value of each of the following, if possible, or the approximate value by placing the square root between two consecutive whole numbers:

1. $\sqrt{3}$
2. $\sqrt{49}$
3. $\sqrt{15}$
4. $\sqrt{256}$
5. $\sqrt{40}$
6. $\sqrt{152}$
7. $\sqrt{169}$
8. $\sqrt{123}$

The Cube Number Pattern: (22)

0^3	1^3	2^3	3^3	4^3	5^3	6^3	7^3	8^3	9^3	10^3
↓	↓	↓	↓	↓	↓	↓	↓	↓	↓	↓
0	1	8	27	64	125	216	343	512	729	1,000

What whole number has been cubed if the power is 216? (23)

$$\boxed{?}^3 = 216$$

The most efficient way to find this number is to introduce a new operation: cube root.

Cube Root: (24)

The symbol used for cube roots is $\sqrt[3]{}$

The cube root of 216 is written $\sqrt[3]{216}$.

$$\sqrt[3]{216} = 6 \qquad \text{since} \qquad 6^3 = 216$$

Cubing and cube root are inverse operations:

$$5^3 = 125$$
$$5 = \sqrt[3]{125}$$

The cube root of any cube number is a whole number.

Cube Root Pattern: (25)

0	1	2	3	4	5	6	7	8	9	10	. . .
↑	↑	↑	↑	↑	↑	↑	↑	↑	↑	↑	
$\sqrt[3]{0}$	$\sqrt[3]{1}$	$\sqrt[3]{8}$	$\sqrt[3]{27}$	$\sqrt[3]{64}$	$\sqrt[3]{125}$	$\sqrt[3]{216}$	$\sqrt[3]{343}$	$\sqrt[3]{512}$	$\sqrt[3]{729}$	$\sqrt[3]{1,000}$	. . .

The cube roots of non-cube numbers are not whole numbers. (26)
For example, the cube root of 3, written $\sqrt[3]{3}$, is the only number that when cubed is equal to 3.

$$\boxed{?}^3 = 3$$
$$\left(\sqrt[3]{3}\right)^3 = 3$$

For all numbers:

$$\left(\sqrt[3]{\blacktriangle}\right)^3 = \blacktriangle \quad \text{and} \quad \sqrt[3]{\blacktriangle^3} = \blacktriangle$$

The approximate value of a cube root of a non-cube number can be determined by placing the cube root between two consecutive whole numbers, just as is done to find the approximate values of square roots of non-square numbers. 

The Cube Root Pattern could be written as follows:

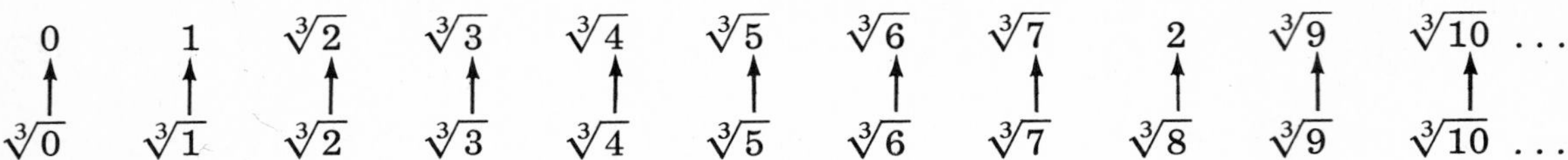

The cube roots of numbers greater than 1,000 can be determined exactly or can be approximated by "mentally" extending the Cube Root Pattern, just as is done to determine the square roots of numbers greater than 100. 

$\sqrt[3]{1{,}728} = \boxed{?}$

$11^3 = 11 \times 11 \times 11$

$= 1{,}331$

$12^3 = 12 \times 12 \times 12$

$= 1{,}728$

$\sqrt[3]{1{,}728}$ is exactly 12

$\sqrt[3]{1{,}363} = \boxed{?}$

$11^3 = 1{,}331$ is less than 1,363

$12^3 = 1{,}728$ is greater than 1,363

$\sqrt[3]{1{,}363}$ is between 11 and 12

Practice Exercises 

Determine the exact value of each of the following, if possible, or the approximate value by placing the cube root between two consecutive whole numbers:

1. $\sqrt[3]{6}$
2. $\sqrt[3]{27}$
3. $\sqrt[3]{186}$
4. $\sqrt[3]{729}$
5. $\sqrt[3]{1{,}200}$
6. $\sqrt[3]{2{,}197}$

POST-TEST

1. Rewrite the following repeated multiplication using an exponent:

   $10 \times 10 \times 10 \times 10 \times 10 \times 10$
2. Write 64, a power of 2, in exponential form.
3. Find the 5th power of 4.
4. Write 879,205 in expanded notation using the exponential forms of the powers of 10.
5. Determine the square root of 49.
6. Determine the approximate value of $\sqrt{40}$ by placing it between two consecutive whole numbers.
7. Determine the cube root of 125.
8. Determine the approximate value of $\sqrt[3]{99}$ by placing it between two consecutive whole numbers.

LESSON 5
ORDER OF OPERATIONS AND RULES OF DIVISIBILITY

Objectives:

1. Evaluate an arithmetic expression according to the order of operations. (1–10)
2. Determine if a whole number is divisible by 2, 3, 4, 5, or 9 using the Rules of Divisibility. (11–19)

Vocabulary:

Order of Operations
Parenthetical Expression
Divisible
Rules of Divisibility

PRE-TEST

1. Evaluate the following according to the order of operations.

$$12 \div 3 + 2^3 - (48 \div 4^2)$$

2. Indicate if 3,270 is divisible by 2, 3, 4, 5, or 9 by writing "yes" or "no" in the appropriate spaces.

Number	2	3	4	5	9
3,270					

Find the value of $4 \times 3 + 5$: 1

First Method

$4 \times 3 + 5$
$12 + 5$
17

Second Method

$4 \times 3 + 5$
4×8
32

Which sequence of operations gives the correct result?

The *Order of Operations* for arithmetic specifies the sequence in which computations are to be performed. The Order of Operations is necessary to obtain unique answers for computations. Without a specified order, two or more answers could result from the same evaluation problem. 

The **Order of Operations** is: (3)

Working from left to right, perform the following operations as they occur:

First: Evaluate parenthetical expressions.

Second: Evaluate powers or roots.

Third: Perform indicated multiplications or divisions.

Fourth: Perform indicated additions or subtractions.

A *parenthetical expression* is an arithmetic expression or part of an arithmetic expression that is enclosed within a set of parentheses, ( ). (4)

$(3 + 4)^2$ $\qquad\qquad$ $\sqrt{(3^2 + 4^2)}$

In the arithmetic expression $(3 + 4)^2$, the sum within the parentheses, 3 + 4, is an example of a parenthetical expression. In the second example, the parenthetical expression is $3^2 + 4^2$. Brackets, [ ], can also be used as grouping symbols.

Parenthetical expressions must be evaluated before taking a roots or raising to a power: (5)

$$(3 + 4)^2 = 7^2 = 49$$

$$\sqrt{(3^2 + 4^2)} = \sqrt{(9 + 16)} = \sqrt{25} = 5$$

Using the order of operations to evaluate 4×3^2: (6)

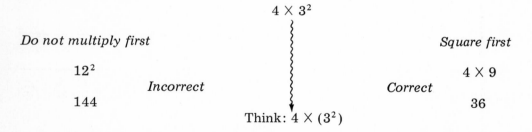

4×3^2

Do not multiply first

12^2

Incorrect

144

Square first

4×9

Correct

36

Think: $4 \times (3^2)$

Using the order of operations to evaluate $5 + 3 \times 7$: (7)

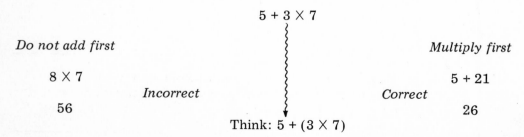

$5 + 3 \times 7$

Do not add first

8×7

Incorrect

56

Multiply first

$5 + 21$

Correct

26

Think: $5 + (3 \times 7)$

The order of operations can be used to change a number written in expanded notation to an equivalent numeral: (8)

$$\begin{aligned} 5782 &= (5 \times 10^3) + (7 \times 10^2) + (8 \times 10^1) + (2 \times 10^0) && \text{powers first} \\ &= (5 \times 1{,}000) + (7 \times 100) + (8 \times 10) + (2 \times 1) && \text{products second} \\ &= 5{,}000 + 700 + 80 + 2 && \text{sums third} \\ &= 5{,}782 \end{aligned}$$

Examples: Work step-by-step using the order of operations. (9)

A. $$\begin{aligned} 8 + 2 \times 7 - 9 \div 3 &= 8 + \underline{2 \times 7} - 9 \div 3 \\ &= 8 + 14 - \underline{9 \div 3} \\ &= \underline{8 + 14} - 3 \\ &= 22 - 3 \\ &= 19 \end{aligned}$$

B. $$\begin{aligned} 3^2 - 2^2 &= \underline{3^2} - \underline{2^2} \\ &= 9 - 4 \\ &= 5 \end{aligned}$$

C. $$\begin{aligned} 64 \div (14 - 5 \times 2)^2 - 3 &= 64 \div (14 - \underline{5 \times 2})^2 - 3 \\ &= 64 \div (\underline{14 - 10})^2 - 3 \\ &= 64 \div \underline{4^2} - 3 \\ &= \underline{64 \div 16} - 3 \\ &= 4 - 3 \\ &= 1 \end{aligned}$$

D. $$\begin{aligned} \sqrt{(8^2 + 6^2)} &= \sqrt{(\underline{8^2} + \underline{6^2})} \\ &= \sqrt{(\underline{64 + 36})} \\ &= \sqrt{100} \\ &= 10 \end{aligned}$$

Practice Exercises (10)

Evaluate each arithmetic expression according to the order of operations.

1. $100 - 8^2 \div 32$
2. $7^2 + 8 - 3^3$
3. $(6 \times 12 \div 9)^2 - 64$
4. $\sqrt{(4 + 5 \times 33)}$
5. $\sqrt{(6^2 + 8^2)}$
6. $6 \div 2 \times (7 + 11)$
7. $12 \div 3 \times 9^3$
8. $\sqrt{[9(9 - 7) \times (9 - 6) \times (9 - 3)]}$

A number is said to be *divisible* by another number if when that number is divided by the other number, the remainder is zero. (11)

For example, 12 is said to be divisible by 3 since the remainder is 0 when 12 is divided by 3.

$$\begin{array}{r} 4 \\ 3\overline{)12} \\ \underline{12} \\ 0 \end{array} \leftarrow \textbf{zero remainder}$$

In fact, 12 is divisible by 1, 2, 3, 4, 6, and 12.

While it is easy to see that 12 is divisible by 3, it is much more difficult to see that 1,327,410 is also divisible by 3. In fact, 1,327,410 is also divisible by 2, 5, and 9. The *Rules of Divisibility* can be used to determine that 2, 3, 5, and 9 are exact divisors of 1,327,410 without carrying out the necessary divisions to show that the remainders are zero. (12)

The Rules of Divisibility: (13)

Divisibility by 2:

A number is divisible by 2 if it is an even number. A number is an even number if its last digit on the right is either 0, 2, 4, 6, or 8.

56 is divisible by 2 since 56 is an even number.
51 is *not* divisible by 2 since 51 is *not* an even number.

The Rules of Divisibility: (14)

Divisibility by 3:

A number is divisible by 3 if the sum of its digits is divisible by 3.

51 is divisible by 3 since 5 + 1 = 6 and 6 is divisible by 3; 6 ÷ 3 = 2 R 0.

56 is *not* divisible by 3 since 5 + 6 = 11 and 11 is *not* divisible by 3; 11 ÷ 3 = 3 R 2.

The Rules of Divisibility: (15)

Divisibility by 4:

A number is divisible by 4 if the last 2 digits on the right represent a number that is divisible by 4.

512 is divisible by 4 since 12 is divisible by 4; 12 ÷ 4 = 3 R 0.

500 is divisible by 4 since 00 is divisible by 4; 00 ÷ 4 = 0 R 0.

361 is *not* divisible by 4 since 61 is *not* divisible by 4; 61 ÷ 4 = 15 R 1.

The Rules of Divisibility: 

Divisibility by 5:

A number is divisible by 5 if the last digit on the right is either 0 or 5.

7,15<u>5</u> is divisible by 5.

8,32<u>0</u> is divisible by 5.

9,03<u>7</u> is *not* divisible by 5.

The Rules of Divisibility: (17)

Divisibility by 9:

A number is divisible by 9 if the sum of its digits is divisible by 9.

7,155 is divisible by 9 since 7 + 1 + 5 + 5 = 18 and 18 is divisible by 9;
18 ÷ 9 = 2 R 0.

8,320 is *not* divisible by 9 since 8 + 3 + 2 + 0 = 13 and 13 is *not* divisible by 9;
13 ÷ 9 = 1 R 4.

Fill in the appropriate boxes with either "yes" or "no" by applying the Rules of Divisibility to determine some of the exact divisors of 72,423. (18)

Number	2	3	4	5	9
72,423					

72,423 is not divisible by 2 since it is not an even number.
It is divisible by 3 since 7 + 2 + 4 + 2 + 3 = 18 and 18 ÷ 3 = 6 R 0.
It is not divisible by 4 since 23 ÷ 4 = 5 R 3.
It is not divisible by 5 since the last digit on the right is neither 0 nor 5.
It is divisible by 9 since the sum of the digits, 18, is divisible by 9.

Practice Exercises (19)

Apply the Rules of Divisibility to each of the following numbers and write "yes" or "no" in the appropriate boxes.

	Number	Divisible by				
		2	3	4	5	9
1.	168,764					
2.	97,842					
3.	43,105					
4.	867,000					
5.	51,087					
6.	7,659,430					
7.	578,999					
8.	876,042					
9.	178,401					
10.	1,324,380					

POST-TEST

1. Evaluate the following according to the order of operations:

$$(3^2 - 4 + 1)^2 \div 12 \times 3$$

2. Indicate if 7,541 is divisible by 2, 3, 4, 5, or 9 by writing "yes" or "no" in the appropriate spaces.

Number	2	3	4	5	9
7,541					

DRILL EXERCISES: UNIT 1

Write each number in word form and in expanded notation:

	Expanded Notation	Word Form
1. 46	____________	____________
2. 96	____________	____________
3. 127	____________	____________
4. 386	____________	____________
5. 257	____________	____________
6. 604	____________	____________
7. 111	____________	____________
8. 500	____________	____________
9. 728	____________	____________
10. 4,444	____________	____________
11. 707,070	____________	____________
12. 211,311	____________	____________

Determine each sum, showing all carry digits. Check by reversing the order of addition.

13. 283
464
+277

14. 4,562
1,278
+2,935

15. 783
432
+269

16. 356
248
+123

17. 5,636
2,385
4,437
+2,063

18. 4,628
1,375
264
+ 24

19. 7,931
47
892
+5,001

20. 72
185
2,998
+ 341

Find each difference, showing all borrowed digits. Check by addition.

21. 382
-167

22. 465
-272

23. 62
-29

24. 43
-17

25. 1,010
- 666

26. 237
-188

27. 6,251
-2,475

28. 3,286
-2,287

Calculate each product, showing all carried digits.

29. $\begin{array}{r} 476 \\ \times\ 67 \\ \hline \end{array}$

30. $\begin{array}{r} 743 \\ \times 475 \\ \hline \end{array}$

31. $\begin{array}{r} 4{,}708 \\ \times 6{,}090 \\ \hline \end{array}$

32. $\begin{array}{r} 6{,}094 \\ \times\ 247 \\ \hline \end{array}$

33. $\begin{array}{r} 70{,}094 \\ \times\ 8{,}039 \\ \hline \end{array}$

34. $\begin{array}{r} 6{,}900 \\ \times\ 578 \\ \hline \end{array}$

35. $\begin{array}{r} 6{,}800 \\ \times\ 708 \\ \hline \end{array}$

36. $\begin{array}{r} 23{,}954 \\ \times\ 9{,}386 \\ \hline \end{array}$

37. $\begin{array}{r} 8{,}967 \\ \times 9{,}008 \\ \hline \end{array}$

Compute each quotient and remainder. Check by multiplication and addition.

38. $36\overline{)17{,}712}$

39. $97\overline{)631{,}276}$

40. $938\overline{)8{,}517{,}040}$

41. $293\overline{)164{,}080}$

42. $708\overline{)4{,}810{,}152}$

43. $978\overline{)88{,}104{,}108}$

44. $56\overline{)26{,}796}$

45. $688\overline{)55{,}701{,}340}$

Find each power.

46. 8^3

47. 2^6

48. 3^4

49. 5^4

50. 10^7

51. 17^2

Find the exact root of each square or cube number; if not a square or cube, approximate the root between a pair of consecutive whole numbers.

52. $\sqrt{64}$

53. $\sqrt[3]{27}$

54. $\sqrt{17}$

55. $\sqrt[3]{200}$

56. $\sqrt{93}$

57. $\sqrt[3]{59}$

58. $\sqrt{2}$

59. $\sqrt[3]{2}$

60. $\sqrt{10}$

Evaluate the following according to the order of operations:

61. $19 - 7 - 4$

62. $73 - (9 - 2)^2 + 4$

63. $14 \times 2 \div 7 \times 3$

64. $18 - (6 + 3) \div 3$

65. $24 + 4 \div 2 \times 6$

66. $4 \times 8 \div 2^3 \times 3$

67. $15 \div 3 \times 2^2 + 4$

68. $\sqrt{[(13^2 - 12^2) - 16]}$

69. $36 + (7 - 4)^3 \div 9$

70. $5 \times [4 + (6 \div 3)^2]^2 \div 16$

Determine whether or not 2, 3, 4, 5, or 9 are exact divisors of each of the following:

71. 673

72. 584

73. 1,482

74. 1,671

75. 2,563,780

76. 3,746,865

SELF-TEST: UNIT 1

1. Using exponential forms for the powers of ten, write the number 7,302 in expanded notation.
   (a) $(7 \times 10^3) + (3 \times 10^2) + (2 \times 10^1)$
   (b) $(7 \times 10^3) + (3 \times 10^2) + (0 \times 10^1) + (2 \times 10^0)$
   (c) $(7 \times 10^3) + (3 \times 10^2) + (0 \times 10^1) + (2)$
   (d) $(7 \times 1{,}000) + (3 \times 100) + (0 \times 10) + (2 \times 1)$
   (e) None of the above
2. Determine the sum of the following numbers: 30,215; 793; 4,006; and 67.
   (a) 34,081
   (b) 35,061
   (c) 34,981
   (d) 35,081
   (e) None of the above
3. Find the difference: 14,640 – 5,823.
   (a) 8,917
   (b) 19,907
   (c) 8,807
   (d) 19,917
   (e) None of the above
4. Calculate the product of 973 and 407.
   (a) 396,011
   (b) 45,731
   (c) 395,011
   (d) 44,731
   (e) None of the above
5. Which one of the following is correct?
   (a) $(17 - 3) + 3 = 23$
   (b) $3 \times (2 + 4) = (3 \times 2) \times (3 \times 4)$
   (c) $(4 \div 2) \times 2 = 4$
   (d) $(110 + 4) - 4 = 114$
   (e) None of the above
6. Find the quotient and remainder for 3,749 divided by 36.
   (a) 14 remainder 5
   (b) 104 remainder 15
   (c) 14 remainder 15
   (d) 104 remainder 5
   (e) None of the above
7. Find the power: 7^3
   (a) 49
   (b) 21
   (c) 343
   (d) 56
   (e) None of the above
8. Determine the approximate value of $\sqrt{61}$ by placing it between a pair of consecutive whole numbers.
   (a) 4 and 5
   (b) 5 and 6
   (c) 6 and 7
   (d) 7 and 8
   (e) None of the above

9. Evaluate according to the order of operations: $(4^2 \div 2 + 3) \times 5$
   (a) 35
   (b) 55
   (c) 23
   (d) 19
   (e) None of the above

10. The number 31,460 is divisible by
   (a) 2 and 3
   (b) 2, 3 and 5
   (c) 3 and 4
   (d) 3, 4 and 5
   (e) None of the above

11. Choose the numeral in which *both* the hundreds digit and ten-thousands digit are underlined.
   (a) $8\underline{3}2{,}05\underline{5}$
   (b) $\underline{1}29{,}2\underline{1}3$
   (c) $5\underline{4}8{,}\underline{7}90$
   (d) $3\underline{7}5{,}2\underline{6}8$
   (e) None of the above

12. The number 60,893 is written in expanded notation as:
   (a) $(60 \times 1{,}000) + (8 \times 100) + (9 \times 10) + (3 \times 1)$
   (b) $(6 \times 10{,}000) + (8 \times 100) + (9 \times 10) + (3 \times 1)$
   (c) $(6 \times 10{,}000) + (0 \times 1{,}000) + (8 \times 100) + (9 \times 10) + (3 \times 1)$
   (d) $(6 \times 1{,}000) + (0 \times 1{,}000) + (8 \times 100) + (9 \times 10) + (3 \times 1)$
   (e) None of the above

13. Write the following in exponential form: $3 \times 3 \times 3 \times 3 \times 3$
   (a) 5^3
   (b) 15
   (c) 9^3
   (d) 3^5
   (e) None of the above

14. Find the 4th power of 10.
   (a) 1,000
   (b) 10,000
   (c) 100,000
   (d) 100
   (e) None of the above

15. Determine the approximate value of $\sqrt[3]{150}$ by placing it between two consecutive whole numbers.
   (a) 2 and 3
   (b) 5 and 6
   (c) 4 and 5
   (d) 6 and 7
   (e) None of the above

16. Write the exponential form of 81, a power of 3.
   (a) 3^4
   (b) 4^3
   (c) 3^3
   (d) 9^2
   (e) None of the above

17. Write the number 2,223 in word form.
   (a) Twenty-two hundred twenty-three
   (b) Two thousand two hundred and twenty-three

(c) Twenty-two hundred and twenty-three
(d) Two thousand two hundred twenty-three
(e) None of the above

18. The square root of 121 is
    (a) 11
    (b) 14,641
    (c) Between 10 and 11
    (d) Between 11 and 12
    (e) None of the above

19. The cube root of 8 is
    (a) Between 1 and 2
    (b) 512
    (c) Between 2 and 3
    (d) 2
    (e) None of the above

20. Divide: 623,623 ÷ 89
    (a) 7,242 R 85
    (b) 7,007
    (c) 707
    (d) 77
    (e) None of the above

LESSON 1:

Pre-test

1. (a) Five hundred thirteen
   (b) Seven thousand twenty-eight
2. (a) (2 × 100) + (3 × 10) + (5 × 1)
   (b) (5 × 1,000) + (7 × 100) + (0 × 10) + (9 × 1)
3. (a) Tens
   (b) Ten-thousands

Practice Exercises

Frame 7

1. Three hundred twenty-one
   (3 × 100) + (2 × 10) + (1 × 1)
2. Four thousand twenty-four
   (4 × 1,000) + (0 × 100) + (2 × 10) + (4 × 1)
3. Seventy-seven thousand seven hundred seventy-seven
   (7 × 10,000) + (7 × 1,000) + (7 × 100) + (7 × 10) + (7 × 1)
4. Sixty-six thousand three hundred fifty-five
   (6 × 10,000) + (6 × 1,000) + (3 × 100) + (5 × 10) + (5 × 1)

Frame 11

1. Tens　　2. Ten-thousands　　3. Hundreds
4. Ten-millions　　5. Millions

Post-test

1. (a) Five thousand seven
   (b) Two hundred thirty-four thousand thirty-one
2. (a) (4 × 10,000) + (0 × 1,000) + (3 × 100) + (8 × 10) + (4 × 1)
   (b) (7 × 100,000) + (4 × 10,000) + (5 × 1,000) + (9 × 100) + (9 × 10) + (9 × 1)
3. (a) Ones
   (b) Hundred-thousands

LESSON 2:

Pre-test

1. (a) 5,045　　(b) 9,102
2. (a) 18,606　　(b) 20,188
3. (a) 231　　(b) 33

Practice Exercises

Frame 13

1. 1,340　　2. 1,236　　3. 627
4. 2,817,126　　5. 1,463,930　　6. 2,610,667

Frame 21

1. 32	2. 199	3. 2,632	4. 190
5. 3,889	6. 987	7. 257,668	8. 309,091
9. 71,679			

Frame 25

1. 53	2. 200	3. 7
4. 19	5. 100	6. Any number

Post-test

1. (a) 9,176 (b) 16,329
2. (a) 39,408 (b) 20,989
3. (a) 241 (b) 93

LESSON 3:

Pre-test

1. (a) 8 (b) 26, 26, 26
2. (a) 381,766 (b) 2,262,079
3. (a) 24 R 45 (b) 35 R 99
4. (a) 17 (b) 432

Practice Exercises

Frame 12

1. 4, 7	2. 10	3. 9, 50, 9	4. 8, 8, 8
5. 50	6. 6	7. 6, 6	8. 40, 2

Frame 15

1. 4,123	2. 32,288	3. 6,394,518

Frame 20

1. 21,658	2. 288,145	3. 887,222
4. 50,397	5. 244,824	6. 217,258
7. 5,631,834	8. 35,506,449	9. 2,378,574

Frame 29

1. 140 R 3	2. 852 R 3	3. 7,575	4. 113,922 R 1

Frame 33

1. 65 R 7	2. 5,674 R 14	3. 28 R 115	4. 971 R 90
5. 1,083 R 80	6. 1,789 R 407	7. 53 R 58	8. 570 R 160
9. 80 R 38	10. 132 R 518		

Frame 38

1. 5	2. 61	3. 129	4. 49

Post-test

1. (a) 13, 13 (b) 9
2. (a) 590,656 (b) 1,502,011
3. (a) 87 R 36 (b) 12 R 20
4. (a) 18 (b) 702

LESSON 4:

Pre-test

1. 7^5 2. 3^4 3. 625
4. $(3 \times 10^4) + (0 \times 10^3) + (4 \times 10^2) + (2 \times 10^1) + (8 \times 10^0)$ 5. 9
6. Between 7 and 8 7. 4 8. Between 6 and 7

Practice Exercises

Frame 6

1. 7^5 2. 21^3 3. 5^2 4. 10^9
5. 2^3 6. 2^5 7. 2^8 8. 2^4
9. 343 10. 10,000,000 11. 256 12. 32

Frame 9

1. $(7 \times 10^3) + (4 \times 10^2) + (0 \times 10^1) + (3 \times 10^0)$
2. (1×10^0)
3. $(5 \times 10^2) + (5 \times 10^1) + (5 \times 10^0)$
4. $(8 \times 10^6) + (0 \times 10^5) + (0 \times 10^4) + (2 \times 10^3) + (5 \times 10^2) + (0 \times 10^1) + (9 \times 10^0)$

Frame 21

1. Between 1 and 2 2. 7 3. Between 3 and 4
4. 16 5. Between 6 and 7 6. Between 12 and 13
7. 13 8. Between 11 and 12

Frame 29

1. Between 1 and 2 2. 3 3. Between 5 and 6
4. 9 5. Between 10 and 11 6. 13

Post-test

1. 10^6 2. 2^6 3. 1,024
4. $(8 \times 10^5) + (7 \times 10^4) + (9 \times 10^3) + (2 \times 10^2) + (0 \times 10^1) + (5 \times 10^0)$ 5. 7
6. Between 6 and 7 7. 5 8. Between 4 and 5

LESSON 5:

Pre-test

1. 9

2.

2	3	4	5	9
Yes	Yes	No	Yes	No

Practice Exercises

Frame 10

1. 98	2. 30	3. 0	4. 13
5. 10	6. 54	7. 2,916	8. 18

Frame 19

	2	3	4	5	9
1.	Y	N	Y	N	N
2.	Y	Y	N	N	N
3.	N	N	N	Y	N
4.	Y	Y	Y	Y	N
5.	N	Y	N	N	N
6.	Y	N	N	Y	N
7.	N	N	N	N	N
8.	Y	Y	N	N	Y
9.	N	Y	N	N	N
10.	Y	Y	Y	Y	N

Y: Yes
N: No

Post-test

1. 9

2.

Number	2	3	4	5	9
7,541	No	No	No	No	No

Answers to Drill Exercises: Unit 1

1. $(4 \times 10) + (6 \times 1)$; forty-six
2. $(9 \times 10) + (6 \times 1)$; ninety-six
3. $(1 \times 100) + (2 \times 10) + (7 \times 1)$; one hundred twenty-seven
4. $(3 \times 100) + (8 \times 10) + (6 \times 1)$; three hundred eighty-six
5. $(2 \times 100) + (5 \times 10) + (7 \times 1)$; two hundred fifty-seven
6. $(6 \times 100) + (0 \times 10) + (4 \times 1)$; six hundred four
7. $(1 \times 100) + (1 \times 10) + (1 \times 1)$; one hundred eleven
8. $(5 \times 100) + (0 \times 10) + (0 \times 1)$; five hundred
9. $(7 \times 100) + (2 \times 10) + (8 \times 1)$; seven hundred twenty-eight
10. $(4 \times 1{,}000) + (4 \times 100) + (4 \times 10) + (4 \times 1)$; four thousand four hundred forty-four
11. $(7 \times 100{,}000) + (0 \times 10{,}000) + (7 \times 1{,}000) + (0 \times 100) + (7 \times 10) + (0 \times 1)$; seven hundred seven thousand seventy
12. $(2 \times 100{,}000) + (1 \times 10{,}000) + (1 \times 1{,}000) + (3 \times 100) + (1 \times 10) + (1 \times 1)$; two hundred eleven thousand three hundred eleven
13. 1,024
14. 8,775
15. 1,484
16. 727
17. 14,521
18. 6,291
19. 13,871
20. 3,596
21. 215
22. 193
23. 33
24. 26
25. 344
26. 49
27. 3,776
28. 999
29. 31,892
30. 352,925
31. 28,671,720
32. 1,505,218
33. 563,485,666
34. 3,988,200
35. 4,814,400
36. 224,832,244
37. 80,774,736
38. 492
39. 6,508
40. 9,080
41. 560
42. 6,794
43. 90,086
44. 478 R 28
45. 80,961 R 172
46. 512
47. 64
48. 81
49. 625
50. 10,000,000
51. 289
52. 8
53. 3
54. Between 4 and 5
55. Between 5 and 6
56. Between 9 and 10
57. Between 3 and 4
58. Between 1 and 2
59. Between 1 and 2
60. Between 3 and 4
61. 8
62. 28
63. 12
64. 15
65. 36
66. 12
67. 24
68. 3
69. 39
70. 20

71. 673 is not divisible by any of the numbers 2, 3, 4, 5, or 9
72. 584 is divisible by 2 and 4
73. 1,482 is divisible by 2 and 3
74. 1,671 is divisible by 3
75. 2,563,780 is divisible by 2, 4, and 5
76. 3,746,865 is divisible by 3 and 5

Answers To Self-Test: Unit 1

1. (b)	2. (d)	3. (e)	4. (a)	5. (c)	6. (d)	7. (c)	8. (d)
9. (b)	10. (e)	11. (c)	12. (c)	13. (d)	14. (b)	15. (b)	16. (a)
17. (d)	18. (a)	19. (d)	20. (b)				

UNIT TWO

Fractional Numbers

Lesson 6	PRIME NUMBERS AND PRIME FACTORIZATIONS
Lesson 7	COMMON MULTIPLES AND COMMON FACTORS
Lesson 8	AN INTRODUCTION TO FRACTIONS
Lesson 9	WORD PROBLEMS

LESSON 6
PRIME NUMBERS AND PRIME FACTORIZATION

Objectives:

1. Classify the first fifty non-zero whole numbers as either prime, composite, or neither. (1–5)
2. Write the prime factorization for a composite number. Use exponents as needed to indicate repeated factors. (6–11)

Vocabulary:

Prime Number	Factoring
Composite Number	Prime Factorization

PRE-TEST

1. Circle each of the following that is a prime number:

   9, 15, 2, 17, 27, 31, 1, 21, 24, 41

2. Underline each of the following that is a composite number:

   13, 12, 11, 2, 4, 1, 22, 29, 33, 47

3. Using exponents as needed, write the prime factorization of 504.

The whole numbers consist of prime numbers, numbers generated by the multiplication of prime numbers, and 1 and 0. A *prime number* is a whole number greater than 1 that can only be written as a product of itself and 1. The only exact divisors of a prime number are itself and 1. (1)

The Prime Numbers between 0 and 50: (2)

2, 3, 5, 7, 11, 13, 17, 19, 23, 29, 31, 37, 41, 43, and **47**

Note that 2 is the only even prime number. Although there is no pattern of prime numbers, it can be said that all prime numbers greater than 2 are odd numbers.

The whole numbers generated by the multiplication of prime numbers are called *composite numbers*. In fact, every composite number can be written as a unique product of prime numbers. (3)

4	can be written	2×2
6	can be written	2×3

There is no other way to write 4 as the product of prime factors. It is possible to write 6 as 3×2, but this is considered to be identical to 2×3 since the factors differ only in their order.

Identifying the prime numbers between 1 and 50 using the sieve method: (4)

Starting with a list of the whole numbers from 1 to 50:

1	2	3	4	5	6	7	8	9	10
11	12	13	14	15	16	17	18	19	20
21	22	23	24	25	26	27	28	29	30
31	32	33	34	35	36	37	38	39	40
41	42	43	44	45	46	47	48	49	50

First: Cross out 1 since it is neither prime nor composite.

Second: Circle 2 and then cross out each number that can be counted by 2's:

2 , ~~4~~, ~~6~~, ~~8~~, ~~10~~, and so on.

Third: Circle 3, the first number after 2 that has not been crossed out. Cross out the numbers counting by 3's that have not already been crossed out:

3 , ~~6~~, ~~9~~, ~~12~~, and so on.

Fourth: Circle 5, the first number after 3 that has not been crossed out. Cross out the numbers counting by 5's that have not already been crossed out:

5 , ~~10~~, ~~15~~, ~~20~~, and so on.

Fifth: Circle 7, the first number after 5 that has not been crossed out. Cross out the numbers counting by 7's that have not already been crossed out:

7 , ~~14~~, ~~21~~, ~~28~~, and so on.

Sixth: Circle each of the numbers that has not been crossed out.

The circled numbers are the prime numbers between 1 and 50. The numbers greater than 2 that have been crossed out are the composite numbers between 1 and 50.

Practice Exercises (5)

Circle each of the following numbers that is a prime number, underline each of the composite numbers, and cross out any number that is neither prime nor composite:

3, 15, 9, 16, 17, 18, 19, 21,

23, 24, 11, 1, 49, 51, 2, 31,

The *prime factorization* of a composite number is the number written as the product of prime numbers. (6)

Composite Number	*Prime Factorization*
4	2×2
6	2×3
10	2×5
12	$2 \times 2 \times 3$
100	$2 \times 2 \times 5 \times 5$

What is the prime factorization of 180? (7)

To answer this question, it is necessary to learn about *factoring*, the process used to determine the prime factorization of a number.

Factoring a composite number: 180 (8)

To factor 180, choose prime divisors, using the Rules of Divisibility whenever possible, and continue dividing each successive quotient until a prime quotient is obtained.

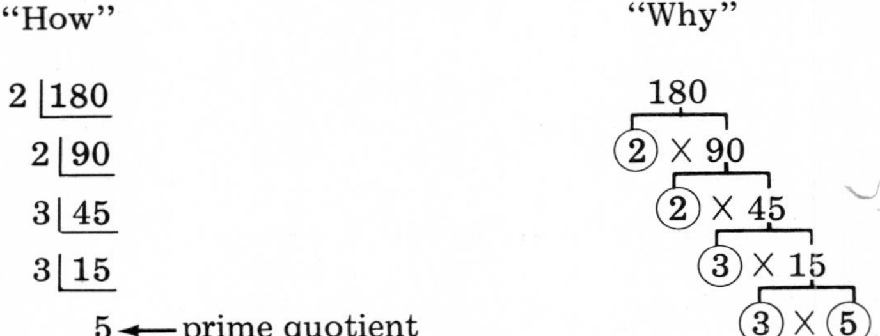

Thus, the prime factorization of 180 is $2 \times 2 \times 3 \times 3 \times 5$ or, using exponents, $2^2 \times 3^2 \times 5$. Notice that it was necessary to use the same prime number as a divisor more than one time.

Factoring a composite number: 126 (9)

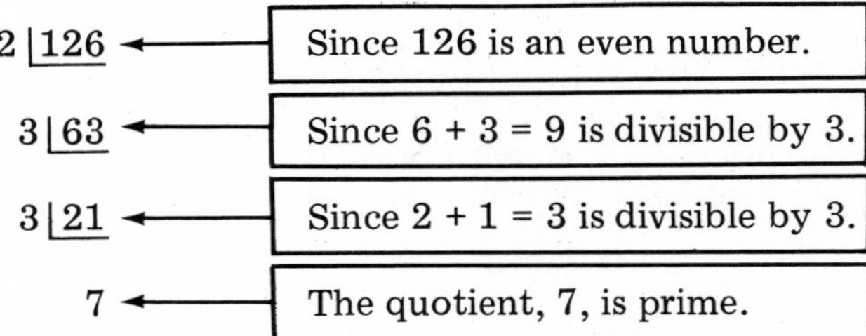

Thus, the prime factorization of 126 is $2 \times 3 \times 3 \times 7$ or $2 \times 3^2 \times 7$.

Factoring a composite number: 390 (10)

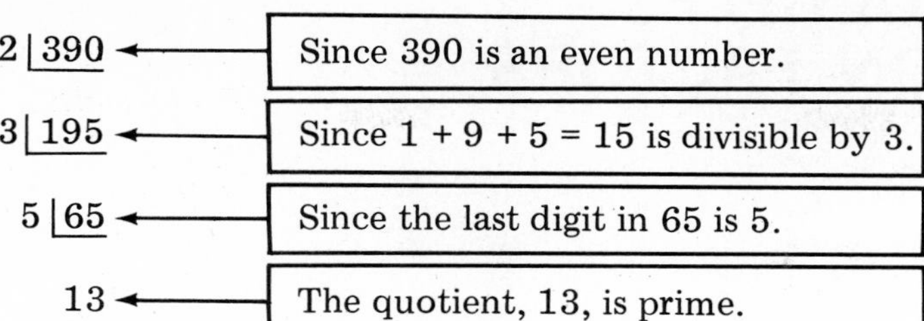

2 | 390
Since 390 is an even number.
3 | 195
Since 1 + 9 + 5 = 15 is divisible by 3.
5 | 65
Since the last digit in 65 is 5.
13
The quotient, 13, is prime.

Complete the following by writing the prime factorization of 390:____ × ____ × ____ × ____ .

Practice Exercises

(11)

Determine the prime factorizations of the following numbers, using exponents as needed to indicate repeated factors:

1. 315 2. 90 3. 84
4. 153 5. 117 6. 242

POST-TEST

1. Circle each of the following that is a prime number:

   21, 9, 15, 2, 42, 37, 29, 1, 17, 33

2. Underline each of the following that is a composite number:

   15, 21, 2, 1, 32, 37, 27, 39, 43, 46

3. Using exponents as needed, write the prime factorization of 300.

LESSON 7
COMMON MULTIPLES AND COMMON FACTORS

Objectives:

1. Determine the multiples of a whole number. (1–5)
2. Determine the Least Common Multiple (LCM) of two or more whole numbers. (6–12)
3. Determine the Greatest Common Factor (GCF) of two or more whole numbers. (13–20)

Vocabulary:

Multiples
Common Multiple
Least Common Multiple
Common Prime Factor
Common Factor
Greatest Common Factor

PRE-TEST

1. Write the first ten multiples of 7.
2. Determine the Least Common Multiple of 30 and 12.
3. Determine the Greatest Common Factor of 42 and 56.

The *multiples* of a number can be determined by counting by that number or by multiplying the number by 1, 2, 3, 4, and so on. (1)

For example, the multiples of 2 can be determined by

Counting by 2's: 2, 4, 6, 8, 10, and so on, or by

Multiplication:

$$2 \times 1 = 2$$
$$2 \times 2 = 4$$
$$2 \times 3 = 6$$
$$2 \times 4 = 8$$
$$\vdots \qquad \vdots$$

The multiples of 3: ②

Counting by 3's: 3, 6, 9, 12, 15, and so on

Multiplication: $3 \times 1 = 3$

$3 \times 2 = 6$

$3 \times 3 = 9$

⋮

The multiples of 5: ③

Complete the following:

Counting by 5's: 5, 10, ○, □, △, and so on

Multiplication: $5 \times 1 = 5$

$5 \times 2 = 10$

$5 \times 3 =$ ○

$5 \times 4 =$ □

$5 \times 5 =$ △

⋮

The multiples of 8: ④

Complete the following:

The first five multiples of 8 are 8, 16, 24, ○, and □.

Practice Exercises ⑤

Circle the multiples of 6 and underline the multiples of 7.

1, 2, 3, 6, 7, 10, 12, 14, 18, 20, 21, 22,
24, 27, 28, 30, 35, 36, 37, 40, 42, 48, 49, 50,
54, 56, 58, 60, 61, 63, 66, 69, 70, 72, 76, 77,
78, 84, 86, 89, 90, 91, 96, 98, 100, 102, 105, 112

The numbers that are multiples of two or more whole numbers are called *common multiples.* 6

For example, looking at the multiples of 2 and 3

The multiples of 2: 2, 4, (6), 8, 10, $\boxed{12}$, and so on

The multiples of 3: 3, (6), 9, $\boxed{12}$, 15, and so on

The numbers 6 and 12 are two of the many common multiples of 2 and 3. Notice that 6 is the smallest of the common multiples.

The *Least Common Multiple* of two or more numbers is the smallest of their common multiples. The Least Common Multiple, abbreviated LCM, of two or more numbers is the smallest number that is divisible by those numbers. 

For example, the LCM of 2 and 3 is 6.

Finding the Least Common Multiple, LCM, of 6, 12, and 30: 

First write the prime factorization of each of the given numbers.

$$6 = 2 \times 3$$
$$12 = 2^2 \times 3$$
$$30 = 2 \times 3 \times 5$$

Next write each base that appears in *any* of the factorizations.

2, 3, 5

Then write the exponent that denotes the *greatest* power of each of these bases that appear in any of the factorizations.

$$2^2, 3, 5$$

Finally, write the LCM as the product of these powers.

$$\text{LCM} = 2^2 \times 3 \times 5 \text{ or } 60$$

It can be verified that 60 is a common multiple of 6, 12, and 30 by dividing 60 by each of these numbers. 

$60 \div 6 = 10$ $\quad$ $60 \div 12 = 5$ $\quad$ $60 \div 30 = 2$

In fact, 60 is the smallest number that is divisible by 6, 12 and 30.

Finding the LCM of 10, 15, and 27: (10)

First: $10 = 2 \times 5$

$15 = 3 \times 5$

$27 = 3^3$

Second: 2, 3, 5

Third: 2, 3^3, 5

Fourth: LCM = $2 \times 3^3 \times 5$ or 270

Finding the LCM of 28, 70, and 105: (11)

Complete the following:

First: $28 = 2^2 \times 7$

$70 = 2 \times 5 \times 7$

$105 = 3 \times 5 \times 7$

Second: △, ○, □, ⬡

Third: $△^2$, ○, □, ⬡

Fourth: LCM = $△^2$ × ○ × □ × ⬡

Practice Exercises 

Find the Least Common Multiple for each group of numbers:

1. 26, 15, 30
2. 32, 10, 9
3. 52, 16, 27
4. 24, 54, 66
5. 20, 35, 99
6. 20, 12, 16

Any prime factors that appear in each of the prime factorizations of two or more numbers are called *common prime factors.* (13)

For example, looking at the prime factorizations of 10 and 12:

$$10 = 2 \times 5$$
$$12 = 2^2 \times 3$$

The number 2 is a common prime factor of 10 and 12. Both 10 and 12 are divisible by 2. The number 5 is not a common prime factor since 10 is divisible by 5 but 12 is not.

The *Greatest Common Factor*, GCF, of two or more numbers is the largest possible number that divides each of the given numbers exactly (the remainder is zero). The products of any of the common prime factors of two or more numbers will also divide these numbers. Any of these products are called common factors. (14)

Finding the greatest common factor, GCF, of 8, 12, and 20: (15)

First write the prime factorizations of each of the given numbers.

$$8 = 2^3$$
$$12 = 2^2 \times 3$$
$$20 = 2^2 \times 5$$

Next write each base that appears in *all* of the factorizations.

$$2$$

Then write the exponent that denotes the *least* power of each of these bases that appears in the factorizations.

$$2^2$$

Finally, write the GCF as the product of these powers.

$$\text{GCF} = 2^2 \text{ or } 4$$

Verifying that 4 is a common factor of 8, 12, and 20: (16)

$8 \div 4 = 2$ $\qquad$ $12 \div 4 = 3$ $\qquad$ $20 \div 4 = 5$

Finding the GCF of 30, 45, and 150: (17)

First: $30 = 2 \times 3 \times 5$
$45 = 3^2 \times 5$
$150 = 2 \times 3 \times 5^2$

Second: 3, 5
Third: $3^1, 5^1$
Fourth: GCF $= 3 \times 5$ or 15

Note that 3 means 3^1 and 5 means 5^1 in exponential form.

Finding the GCF of 9, 45, and 81: 

Complete the following:

First: $9 = 3^2$
$45 = 3^2 \times 5$
$81 = 3^4$

Second: △
Third: $△^2$
Fourth: GCF $= △^2$ or ⬡

Find the GCF of 25 and 36: 

$$25 = 5^2$$
$$36 = 2^2 \times 3^2$$

There is no GCF of 25 and 36 since there are no common prime factors. Note that 1 is a common divisor of 25 and 36, but it is not a common prime factor because it is not prime.

Practice Exercises 

Find the Greatest Common Factor for each group of numbers.

1. 12, 28, 48
2. 16, 56, 32
3. 50, 75, 125
4. 21, 84, 63
5. 54, 18, 90
6. 40, 60, 80

POST-TEST

1. Write the first ten multiples of 12.
2. Determine the Least Common Multiple of 32 and 42.
3. Determine the Greatest Common Factor of 102 and 170.

LESSON 8
AN INTRODUCTION TO FRACTIONS

Objectives:

1. Rewrite a fraction as a division, a ratio, a number, or a number of parts of an equally divided whole. (1–9)
2. Determine the product of two or more fractions using cancellation whenever possible. (10–22, 36–39)
3. Determine the value of a power of a fraction. (18–19)
4. Determine fractions equivalent to a given fraction using the "Multiplication by One" Property. (23–26)
5. Reduce a fraction to lowest terms by cancelling common factors in the numerator and denominator. (27–35)

Vocabulary:

Fraction	Equivalent Fractions
Numerator	Multiplication by One
Denominator	Reduced to Lowest Terms
Ratio	Cancellation

PRE-TEST

1. Rewrite each of the following fractions for the meaning named:
   (a) $\frac{3}{8}$ as a number of parts of an equally divided whole.
   (b) $\frac{7}{11}$ as a division.
   (c) $\frac{4}{5}$ as a ratio.
   (d) Nine-tenths as a number.
2. Multiply: $\frac{4}{9} \times \frac{5}{7}$
3. Determine the value of $\left(\frac{2}{5}\right)^2$.
4. Determine the equivalent fraction that results when $\frac{7}{12}$ is multiplied by $\frac{3}{3}$.
5. Reduce $\frac{36}{48}$ to lowest terms by cancelling common factors.

One morning Mrs. Bailey found that she had only 2 boxes of raisins to put into the lunches she was packing for her 3 children. What part of a box of raisins did she put into each lunch so that each child had an equal amount? (1)

This question can be answered by dividing the number of boxes of raisins by the number of children:

$$2 \div 3 = \boxed{?}$$

Since this quotient is not a whole number, it is necessary to introduce fractional numbers before this question can be answered.

A *fraction* is the written form of a fractional number. (2)

The parts of a fraction have special names:

$$\frac{2}{3} \begin{matrix} \leftarrow \text{numerator} \rightarrow \\ \leftarrow \text{denominator} \rightarrow \end{matrix} \frac{\blacksquare}{\blacktriangle}$$

The *numerator* tells "how many" or numerates.
The *denominator* names "what kind" or denominates.
The fraction, $\frac{2}{3}$, is read or written two-thirds.

The four meanings of the fraction $\frac{2}{3}$: (3)

A number	A ratio	A division	A number of parts of a whole
$\frac{2}{3}$	$2:3$	$2 \div 3$	$2 \times \frac{1}{3}$
Two-thirds	Two to three	Two divided by 3	Two parts of a whole divided into 3 equal parts

---

The number two-thirds: (4)

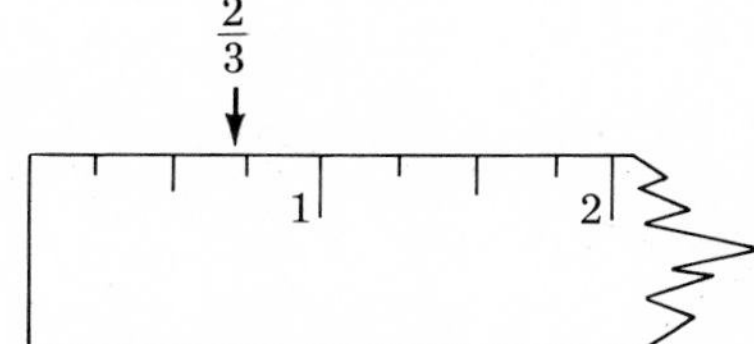

$\frac{2}{3}$ of an inch is a measurement that is less than 1 inch.

The ratio **2** to **3** : ⑤

If a committee consists of 2 men and 3 women, the ratio of men to women is 2:3 or $\frac{2}{3}$.

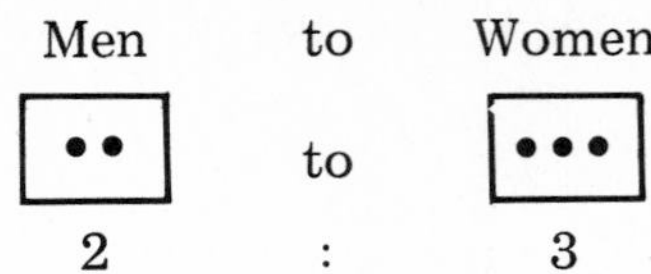

The division **2÷3** : 

To answer the question of what part of a box of raisins each of Mrs. Bailey's children had in his lunch, each equal portion was $2 \div 3$ or $\frac{2}{3}$ of a box.

The two parts of a whole divided into three parts: 

Mr. Walker's estate consisted of 3 shares of stock in a gold mine. If his favorite daughter inherited twice as much as her sister, she received $2 \times \frac{1}{3}$ or $\frac{2}{3}$ of the estate.

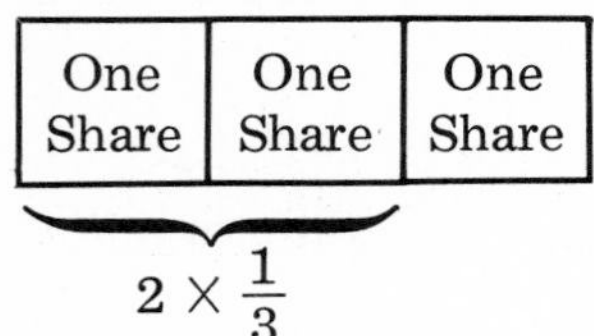

The same fraction can be written in any of the following equivalent forms: 

A number	A ratio	A division	A number of parts of a whole
			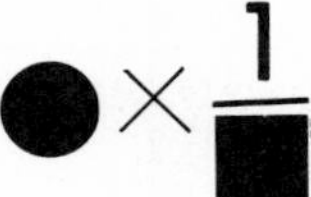

Practice Exercises 

Rewrite the following fractions for the meaning named:

1. $\frac{3}{5}$ as a ratio.
2. $\frac{2}{7}$ as a number of parts of an equally divided whole.
3. $\frac{5}{9}$ as a division.
4. One-half as a number.

Certain properties of fractions should be discussed before operations with fractions are considered. (10)

Every whole number can be written as a fraction.

For example, $5 = \frac{5}{1}$

$103 = \frac{103}{1}$

For any whole number, 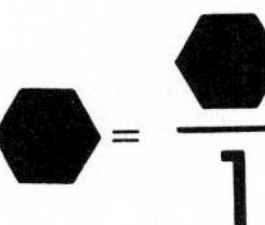

If the numerator and the denominator of a fraction are exactly the same non-zero number, the fraction is equivalent to 1.

For example, $\frac{7}{7} = 1$

For any number except 0, 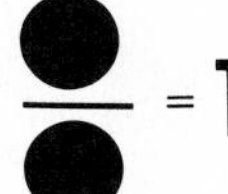

If the numerator of a fraction is 0 and the denominator is any non-zero number, the fraction is equivalent to 0.

For example, $\frac{0}{3} = 0$

For any number except 0, 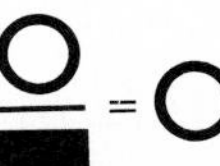

The three Segalla brothers own one-fourth interest in the Mama Mia Pizzeria. Each brother's equal share is what part of the restaurant? 

This question can be answered by determining what part of the restaurant is one-third of one-fourth?

$$\frac{1}{3} \text{ of } \frac{1}{4} = \boxed{?}$$

To answer this question, it is necessary to multiply fractions since the word "of" indicates a product.

Multiplication of fractions: $\frac{1}{3} \times \frac{1}{4}$ 

"How"

$$\frac{1}{3} \times \frac{1}{4} = \frac{1 \times 1}{3 \times 4} \text{ or } \frac{1}{12}$$

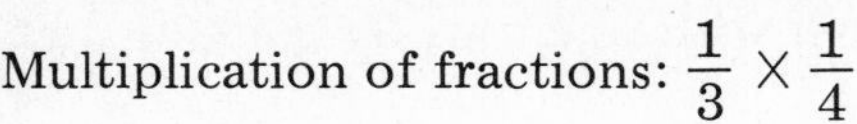

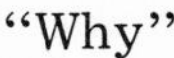

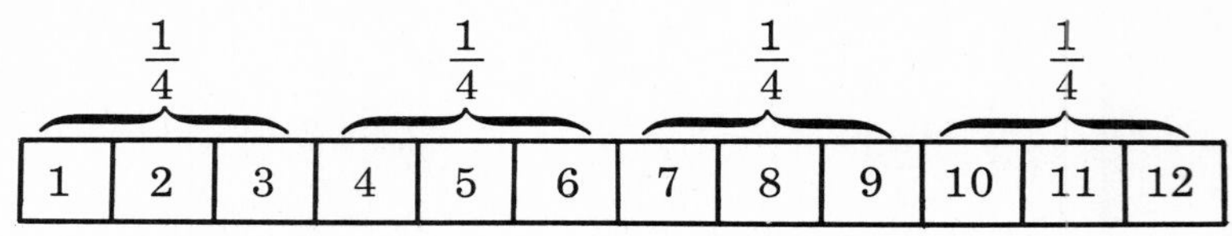

$\underbrace{\frac{1}{3} \text{ of } \frac{1}{4}}_{\frac{1}{12}}$

If each one-fourth of the business is divided into three equal portions, each portion is one-twelfth of the business.

Multiplication of fractions:

For any fractions with non-zero denominators, ▲/● × ■/⬢ = (▲ × ■)/(● × ⬢)

To multiply any fractions with non-zero denominators, form the products of the numerators and the denominators.

Multiply $\frac{3}{4} \times \frac{5}{7}$: (15)

$$\frac{3}{4} \times \frac{5}{7} = \frac{3 \times 5}{4 \times 7} = \frac{15}{28}$$

Multiply $\frac{2}{3} \times \frac{1}{7} \times \frac{5}{9}$:

When multiplying more than two fractions, the same rule applies: the numerator of the answer is the product of the numerators of all the fractions and the denominator is the product of the denominators of all the fractions.

$$\frac{2}{3} \times \frac{1}{7} \times \frac{5}{9} = \frac{2 \times 1 \times 5}{3 \times 7 \times 9} = \frac{10}{189}$$

Multiply $\frac{4}{5} \times \frac{2}{9}$: (17)

Complete the following:

$$\frac{4}{5} \times \frac{2}{9} = \frac{\square \times \bigcirc}{\triangle \times \hexagon}$$
$$= \frac{8}{45}$$

Raising a fraction to a power: $\left(\frac{1}{2}\right)^3$ (18)

Since an exponent indicates a repeated multiplication,

$$\left(\frac{1}{2}\right)^3 = \frac{1}{2} \times \frac{1}{2} \times \frac{1}{2}$$
$$= \frac{1 \times 1 \times 1}{2 \times 2 \times 2}$$
$$= \frac{1}{8}$$

Raising a fraction to a power: $\left(\frac{2}{3}\right)^2$ (19)

Complete the following:

$$\left(\frac{2}{3}\right)^2 = \frac{\triangle}{\square} \times \frac{\triangle}{\square}$$
$$= \frac{\triangle \times \triangle}{\square \times \square}$$
$$= \frac{4}{9}$$

The product of a fractional number and a whole number: $5 \times \frac{2}{7}$ (20)

$$5 \times \frac{2}{7} = \frac{5}{1} \times \frac{2}{7}$$
$$= \frac{5 \times 2}{1 \times 7} \text{ or } \frac{10}{7}$$

Multiply: $3 \times \frac{1}{4}$ 21

Complete the following:

$$3 \times \frac{1}{4} = \frac{\square}{1} \times \frac{1}{\bigcirc}$$

$$= \frac{\square \times 1}{1 \times \bigcirc}$$

$$= \frac{\square}{\bigcirc}$$

Practice Exercises

Find the indicated products of the following:

1. $\frac{2}{3} \times \frac{5}{7}$
2. $\frac{5}{2} \times 2$
3. $\frac{8}{13} \times \frac{2}{5}$
4. $\frac{3}{7} \times \frac{15}{4}$
5. $2 \times \frac{3}{5}$
6. $\frac{7}{8} \times 3$
7. $\frac{5}{9} \times \frac{7}{3}$
8. $\frac{2}{7} \times \frac{3}{8}$
9. $4 \times \frac{5}{9}$
10. $\frac{7}{9} \times \frac{1}{3}$

Equivalent fractions are fractions that name the same number.

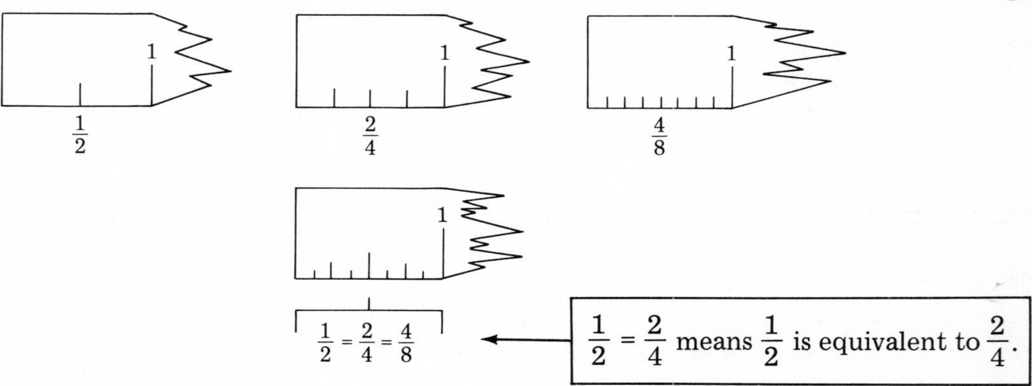

The fractions $\frac{1}{2}$, $\frac{2}{4}$, and $\frac{4}{8}$ are equivalent fractions.

Equivalent fractions can be formed by multiplying by some form of the number 1. 24

The first ten fractions equivalent to $\frac{2}{3}$:

$$\frac{2}{3} \times \frac{1}{1} = \frac{2}{3} \qquad \frac{2}{3} \times \frac{6}{6} = \frac{12}{18}$$

$$\frac{2}{3} \times \frac{2}{2} = \frac{4}{6} \qquad \frac{2}{3} \times \frac{7}{7} = \frac{14}{21}$$

$$\frac{2}{3} \times \frac{3}{3} = \frac{6}{9} \qquad \frac{2}{3} \times \frac{8}{8} = \frac{16}{24}$$

$$\frac{2}{3} \times \frac{4}{4} = \frac{8}{12} \qquad \frac{2}{3} \times \frac{9}{9} = \frac{18}{27}$$

$$\frac{2}{3} \times \frac{5}{5} = \frac{10}{15} \qquad \frac{2}{3} \times \frac{10}{10} = \frac{20}{30}$$

Since $\frac{2}{3} \times 1 = \frac{2}{3}$ and  = 1 for any non-zero number.

The "Multiplication by One" Property for Fractions: 

If a given fraction is multiplied by any form of the number 1, the resulting fraction names the same number as the given fraction; the fractions are equivalent.

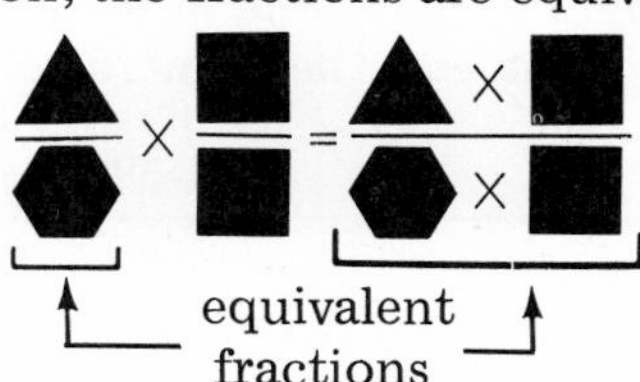

For any number ▲ and for any non-zero numbers .

Practice Exercises 

For each of the following fractions, determine the first *six* equivalent fractions using the "Multiplication by One" Property:

1. $\frac{2}{5}$ 2. $\frac{4}{3}$ 3. $\frac{5}{2}$

Are $\frac{20}{25}$, $\frac{28}{35}$, and $\frac{12}{90}$ equivalent fractions? 27

First rewrite each of the given fractions as follows and compare the resulting fractions:

$$\frac{20}{25} = \frac{2 \times 2 \times 5}{5 \times 5} = \frac{2 \times 2}{5} \times \frac{5}{5} = \frac{2 \times 2}{5} \times 1 = \frac{4}{5}$$

$$\frac{28}{35} = \frac{2 \times 2 \times 7}{5 \times 7} = \frac{2 \times 2}{5} \times \frac{7}{7} = \frac{2 \times 2}{5} \times 1 = \frac{4}{5}$$

$$\frac{12}{90} = \frac{2 \times 2 \times 3}{2 \times 3 \times 3 \times 5} = \frac{2 \times 3}{2 \times 3} \times \frac{2}{3 \times 5} = 1 \times \frac{2}{3 \times 5} = \frac{2}{15}$$

Thus, $\frac{20}{25}$ and $\frac{28}{35}$ are equivalent fractions, but $\frac{12}{90}$ is not. (28)

The process used to rewrite the fractions $\frac{20}{25}$, $\frac{28}{35}$, and $\frac{12}{90}$ is called *reduction*. The fractions $\frac{4}{5}$ and $\frac{2}{15}$ are said to be *reduced to lowest terms*. A fraction is reduced to lowest terms when there are no common factors in the prime factorizations of the numerator and the denominator.

Reducing to lowest terms: (29)

First: Write the prime factorizations of the numerator and the denominator.

Second: Use the "Multiplication by One" Property to rewrite the fractions.

Third: Write the product of the factors in the numerator and the product of the factors remaining in the denominator as the numerator and the denominator of the equivalent, reduced fraction.

Reduce $\frac{24}{30}$ to lowest terms: (30)

"Why"

$$\frac{24}{30} = \frac{2 \times 2 \times 2 \times 3}{2 \times 3 \times 5} = \frac{2 \times 2}{5} \times \frac{2}{2} \times \frac{3}{3} = \frac{2 \times 2}{5} \times 1 \times 1 = \frac{4}{5}$$

Note that 6 is the Greatest Common Factor of 24 and 30.

Reduce $\frac{24}{30}$ to lowest terms: (31)

"How"

$$\frac{24}{30} = \frac{\overset{1}{\cancel{2}} \times 2 \times 2 \times \overset{1}{\cancel{3}}}{\underset{1}{\cancel{2}} \times \underset{1}{\cancel{3}} \times 5}$$

$$= \frac{4}{5}$$

This method for reducing fractions is called *cancellation.*

Checking reduced fractions: 

In order to verify that $\frac{24}{30}$ is correctly reduced to $\frac{4}{5}$, the following property of fractions is used:

$$\frac{24}{30} = \frac{4}{5} \qquad \text{if and only if} \qquad (24)(5) = (4)(30)$$

Since $24 \times 5 = 120$ and $4 \times 30 = 120$, the fraction $\frac{24}{30}$ is correctly reduced to $\frac{4}{5}$. Note that this check does not guarantee that the fraction has been reduced to *lowest terms;* it merely shows that the fractions are equivalent.

Reduce $\frac{12}{84}$ to lowest terms: 

$$\frac{12}{84} = \frac{\overset{1}{\cancel{2}} \times \overset{1}{\cancel{2}} \times \overset{1}{\cancel{3}}}{\underset{1}{\cancel{2}} \times \underset{1}{\cancel{2}} \times \underset{1}{\cancel{3}} \times 7} = \frac{1}{7}$$

Check:

$$\frac{12}{84} \overset{?}{=} \frac{1}{7}$$

$$(12)(7) \overset{?}{=} (1)(84)$$

$$84 = 84$$

The fraction $\frac{12}{84}$ is correctly reduced to $\frac{1}{7}$. The symbol $\overset{?}{=}$ is used to show that an equivalence is being checked.

Reduce $\frac{15}{135}$ to lowest terms: 

Complete the following:

$$\frac{15}{135} = \frac{\overset{1}{\cancel{3}} \times \overset{1}{\cancel{5}}}{\underset{1}{\cancel{3}} \times 3 \times 3 \times \underset{1}{\cancel{5}}} = \frac{\bigcirc}{\square}$$

Check:

$$\frac{15}{135} \overset{?}{=} \frac{\bigcirc}{\square}$$

Practice Exercises 

Reduce each of the following fractions to lowest terms using cancellation.

1. $\frac{50}{275}$ 2. $\frac{54}{90}$ 3. $\frac{27}{72}$

4. $\frac{98}{196}$ 5. $\frac{33}{51}$ 6. $\frac{42}{56}$

Cancellation can be used to simplify the multiplication of two or more fractions. 

For example, multiplying $\frac{8}{15}$ by $\frac{9}{16}$:

$$\frac{8}{15} \times \frac{9}{16} = \frac{8 \times 9}{15 \times 16}$$

$$= \frac{\overset{1}{\cancel{2}} \times \overset{1}{\cancel{2}} \times \overset{1}{\cancel{2}} \times \overset{1}{\cancel{3}} \times 3}{\underset{1}{\cancel{3}} \times 5 \times \underset{1}{\cancel{2}} \times \underset{1}{\cancel{2}} \times \underset{1}{\cancel{2}} \times 2}$$

$$= \frac{3}{5 \times 2}$$

$$= \frac{3}{10}$$

It is not necessary to cancel only common prime factors; any common factors appearing in both the numerator and the denominator may be cancelled. 37

$$\frac{8}{15} \times \frac{9}{16} = \frac{\overset{1}{\cancel{8}} \times \overset{3}{\cancel{9}}}{\underset{5}{\cancel{15}} \times \underset{2}{\cancel{16}}}$$

← 8 is a common factor of 8 and 16; 3 is a common factor of 9 and 15.

$$= \frac{3}{5 \times 2}$$

$$= \frac{3}{10}$$

Multiply, using cancellation: $\frac{10}{33} \times \frac{9}{20} \times \frac{44}{45}$ 38

$$\frac{10}{33} \times \frac{9}{20} \times \frac{44}{45} = \frac{\overset{1}{\cancel{10}} \times \overset{1}{\cancel{9}} \times \overset{4}{\cancel{44}}}{\underset{3}{\cancel{33}} \times \underset{2}{\cancel{20}} \times \underset{5}{\cancel{45}}}$$

← Cancel 10 in 10 and 20.
Cancel 9 in 9 and 45.
Cancel 11 in 33 and 44.

$$= \frac{\overset{2}{\cancel{4}}}{3 \times \underset{1}{\cancel{2}} \times 5}$$

← Cancel 2 in 4 and 2.

$$= \frac{2}{3 \times 5}$$

$$= \frac{2}{15}$$

Practice Exercises

Multiply each of the following, using cancellation:

1. $\frac{4}{3} \times \frac{9}{10}$
2. $\frac{8}{7} \times \frac{35}{24}$
3. $\frac{2}{11} \times \frac{77}{240}$
4. $\frac{51}{34} \times \frac{26}{91}$
5. $\frac{10}{21} \times \frac{14}{15} \times \frac{18}{25}$
6. $\frac{33}{56} \times \frac{35}{44} \times \frac{16}{27}$

POST-TEST

1. Rewrite each of the following fractions for the meaning named:
   (a) $\frac{5}{12}$ as a ratio
   (b) $\frac{7}{9}$ as a number of parts of an equally divided whole
   (c) Three-fourths as a number
   (d) $\frac{2}{5}$ as a division
2. Multiply: $\frac{8}{25} \times 7$
3. Determine the value of $\left(\frac{1}{4}\right)^3$.
4. Determine the equivalent fraction that results when $\frac{3}{7}$ is multiplied by $\frac{6}{6}$.
5. Reduce $\frac{54}{63}$ to lowest terms.

LESSON 9
WORD PROBLEMS

Objective:

1. Solve a word problem by analyzing the problem, writing the complete method of solution, and checking the answer for reasonableness and accuracy. (1–11)

Vocabulary:

Analysis
Method of Solution
Check

PRE-TEST

1. Solve each of the following:
   (a) Last week, Mrs. Conner purchased 6 cans of sliced peaches that were on sale for 26 cents per can. The same can of peaches usually sells for 34 cents. How much did she save by buying the peaches while they were on sale?

   Analysis:

   Key Words

   Unknown

   Given

   Conditions

   Method of Solution:

   Check:

   Reasonableness

   Accuracy

   (b) Keith Allen purchased 2 tires at 37 dollars each, 4 shock absorbers at 9 dollars each, and a tape deck for 88 dollars. If Keith started out with 230 dollars, how much money did he have left after paying for his purchases?

Analysis:

Key Words

Unknown

Given

Conditions

Method of Solution:

Check:

Reasonableness

Accuracy

The Adams family budgets $\frac{1}{5}$ of its income for food. If the Adams family has a monthly income of 925 dollars, what is the amount of money that is budgeted for food? (1)

What is the best way to attack this or any word problem? A general strategy for working any word problem can be useful in this difficult area of arithmetic.

A general strategy for successfully solving word problems: (2)

Apply the following tactics in the order listed.

A. Analysis

1. *Read* (and reread) the problem until it is fully understood.
2. Identify the *key words* or phrases.
3. Determine what is *unknown.*
4. Determine what is *given.*
5. Establish the *conditions* that relate the unknown and the given.

B. Write a step-by-step *method of solution* leading to the answer.

C. *Check* the answer.

Applying the tactics of problem solving to a word problem: (3)

Refer to the problem in Frame 1.

First, read the problem carefully until it is fully understood.

Next identify the key words: *budgets*, *of*, *income*, and *amount*.

Determine what is unknown: the amount of their income that is budgeted for food.

What is given? $\frac{1}{5}$ of the total income is budgeted for food; the family's monthly income is 925 dollars.

What are the conditions? The amount budgeted for food is $\frac{1}{5}$ of the family's monthly income.

Write the step-by-step method of solution:

$$\text{The amount of money budgeted for food} = \frac{1}{5} \times 925 \text{ dollars}$$
$$= \frac{925}{5} \text{ dollars}$$
$$= 185 \text{ dollars}$$

Check the answer: First, is it reasonable? Second, check for accuracy in computation by multiplying 5 × 185 dollars to see if the product is 925 dollars.

The *analysis* for a word problem consists of the following steps from the general strategy: (4)

1. Read (and reread) the problem.
2. Identify the key words or phrases.
3. Determine what is unknown.
4. Determine what necessary information is given.
5. Establish the conditions that relate the unknown and the given.

The *method of solution* for a word problem is a carefully written, step-by-step calculation of the answer.

The *check* of the answer for a word problem is both a common sense test for reasonableness and an arithmetic test for accuracy.

First, is the answer reasonable?

Second, check for accuracy by using the appropriate inverse operations.

Nancy worked the following hours of overtime last week: Monday, 3 hours; Tuesday, 0 hours; Wednesday, 1 hour; Thursday, 2 hours; Friday, 0 hours. If Nancy earns 5 dollars per overtime hour, calculate her total overtime pay for the week. (7)

Analysis: Read the problem.

Key Words: total

Unknown: amount of overtime pay

Given number of hours of overtime, overtime pay of 6 dollars per hour

Conditions: The amount of overtime pay is the number of hours of overtime times the pay per hour.

Method of Solution:

Amount of overtime pay = (3 + 0 + 1 + 2 + 0) hours × 5 dollars per hour
= (6 × 5) dollars
= 30 dollars

Check: First, 30 dollars is reasonable.

Second, 30 dollars ÷ 6 = 5 dollars per hour.

Glenn has a summer job with the City of Huntington Beach that pays 2 dollars per hour. He will work 40 hours per week for 12 weeks. How much will he earn? (8)

Analysis: Read the problem.

Key Words: per hour, per week, and amount earned

Unknown: amount earned

Given: 2 dollars per hour, 40 hours per week, 12 weeks

Conditions: The total summer pay is dollars per hour times hours per week times weeks per summer.

Method of Solution:

Total summer pay = 2 dollars per hour × 40 hours per week × 12 weeks
= (80 × 12) dollars
= 960 dollars

Check: Which one of the following is reasonable for a summer pay?

96 dollars | 960 dollars | 9,600 dollars

Reverse the Method of Solution:

960 dollars divided by 2 dollars per hour = 480 hours worked per summer.
480 hours divided by 12 weeks = 40 hours per week.

Mr. and Mrs. Quan recently purchased a new home. They made a 3,500-dollar down payment and secured a loan for the difference which required monthly payments of 275 dollars for a period of 20 years. What is the total cost of the home? (9)

Analysis: Read the problem.

Key Words: down payment, difference, monthly payments, and years

Unknown: total cost

Given: 3,500-dollar down payment, 275 dollars per month, 20 years

Conditions: The total cost of home mortgage is the down payment plus 12 monthly payments per year for 20 years.

Method of Solution:

Total cost = 3,500 dollars + (275 dollars per month × 12 months per year × 20 years)
= (3,500 + 66,000) dollars
= 69,500 dollars

Check: Which of the following is reasonable for the total cost of the home?

6,950 dollars　　69,500 dollars　　695,000 dollars

Reverse the method of solution:

69,500 dollars − 3,500 dollars = 66,000 dollars for 20 years.
66,000 dollars divided by 20 years = 3,300 dollars per year.
3,300 dollars per year divided by 12 months per year = 275 dollars per month.

To successfully determine the answer for a word problem, each of the following tactics should be used: 

A. Analyze the problem.

B. Write the step-by-step method of solution leading to the answer.

C. Check the answer.

Practice Exercises 

Solve each of the following:

1. Find how much the total cost of 3 cans of soup at 18 cents each exceeds the total cost of 3 packages of dry soup mix at 11 cents each.

2. If each of 4 children in Cara's family requires a lunch for each of the 5 school days, how many lunches will be required for a 2-week period?

3. Mr. Johnson buys one-half cord of pine at 64 dollars a cord and two-thirds cord of oak at 75 dollars a cord. How much did he spend?

4. When a TV serviceman makes a house call, he charges 15 dollars for the call plus 8 dollars per hour for labor. When the TV serviceman was called to Mrs. Reynolds' home, the bill for parts, labor, and the $\frac{1}{2}$ hour call came to 24 dollars. How much did the parts cost?

POST-TEST

1. Solve each of the following:

(a) Kirk purchased 4 "Super Eagle" brand radial tires at 37 dollars each. There was a 2-dollar federal tax on each tire and an additional charge of 1 dollar per tire for balancing. What was the total price for the 4 tires?

Analysis:

Key Words

Unknown

Given

Conditions

Method of Solution:

Check:

Reasonableness

Accuracy

(b) The Wadhams wrote checks for the following amounts over the weekend: 84 dollars; 23 dollars; and 17 dollars. If their bank balance was 100 dollars on the preceding Friday, before a 45 dollar deposit was made, what amount remained in the account on Monday?

Analysis:

Key Words

Unknown

Given

Conditions

Method of Solution:

Check:

Reasonableness

Accuracy

DRILL EXERCISES: UNIT 2

Rewrite each of the following as a product of prime numbers using exponential notation:

1. 60
2. 126
3. 210
4. 252
5. 825
6. 98
7. 54
8. 143
9. 110
10. 385

Determine the least common multiple and the greatest common factor for each of the following:

11. 6, 15
12. 9, 12
13. 12, 21
14. 15, 18
15. 24, 32
16. 18, 20
17. 10, 15
18. 27, 15
19. 50, 62
20. 30, 21
21. 40, 15, 50
22. 6, 21, 28
23. 21, 30, 36
24. 28, 16, 30
25. 6, 22, 110

Multiply the following using cancellation whenever possible:

26. $\frac{4}{5} \times \frac{6}{7}$
27. $\frac{8}{11} \times \frac{2}{7}$
28. $\frac{4}{9} \times \frac{3}{28} \times \frac{6}{5}$
29. $\frac{3}{5} \times \frac{2}{11}$
30. $\frac{4}{9} \times \frac{5}{7}$
31. $\frac{3}{8} \times \frac{7}{27} \times \frac{3}{4}$
32. $\left(\frac{2}{7}\right)^2$
33. $\left(\frac{3}{5}\right)^2$
34. $\frac{5}{12} \times \frac{3}{10} \times \frac{4}{9}$
35. $\frac{1}{3} \times \frac{1}{8}$
36. $\frac{1}{6} \times \frac{1}{15}$
37. $\frac{3}{5} \times \frac{2}{9} \times \frac{25}{4}$
38. $\left(\frac{1}{5}\right)^3$
39. $\left(\frac{1}{3}\right)^3$
40. $\frac{15}{16} \times \frac{12}{20}$
41. $3 \times \frac{4}{5}$
42. $5 \times \frac{6}{7}$
43. $2 \times \frac{14}{15}$
44. $2 \times \frac{1}{7}$
45. $3 \times \frac{5}{3}$

Determine the five fractions equivalent to each of the given fractions by multiplying by $\frac{1}{1}, \frac{2}{2}, \frac{3}{3}, \frac{4}{4}$, and $\frac{5}{5}$.

46. $\frac{4}{5}$
47. $\frac{3}{7}$
48. $\frac{5}{6}$
49. $\frac{1}{11}$
50. $\frac{2}{9}$

Reduce the following fractions to lowest terms:

51. $\frac{30}{45}$
52. $\frac{21}{28}$
53. $\frac{36}{84}$
54. $\frac{30}{135}$
55. $\frac{80}{112}$
56. $\frac{24}{40}$

57. $\frac{12}{66}$

58. $\frac{125}{200}$

59. $\frac{126}{294}$

60. $\frac{189}{243}$

Solve each of the following word problems:

61. An inventory warehouseman discovers a 20-carton oversupply of bicycle tires. If there are 25 tires per carton, how many excess tires are there?

62. How many secretaries will a box of 500 sheets of paper supply if each secretary is to receive 20 sheets?

63. Cecelia purchased a car paying a 750-dollar down payment and financing the difference at 85 dollars per month for 30 months. How much more will she pay this way than if she paid the cash price of $2,570?

64. Leroy must pay his car payment, 85 dollars, and his utility bill, 23 dollars, out of this week's paycheck. If his check is for 165 dollars, how much money will he have left after paying his bills?

65. Willie Brown rented a chain saw at 7 dollars per hour, an extension ladder at 2 dollars per hour, and a set of tree climbing spurs at 3 dollars per hour for trimming his trees. What rental costs did he have for the equipment if it took him 3 hours to finish the job?

66. The Golden West College basketball team won $\frac{3}{4}$ of its games last season. If the team played 20 games, how many did it lose?

67. The McArthur family budgets $\frac{1}{10}$ of its income to its church. If the family earns $18,250 per year, how much will the church receive?

68. During the first half the college football team made the following yardage gains: 159 yards by passing and 234 yards by running but lost some yardage to 2 penalties. Calculate the net yardage gained if the penalties cost 10 yards and 15 yards.

1. Which of the following groups of numbers contain *only* prime numbers?
   (a) 1, 2, 3, 5, 7, 11, 13, 17
   (b) 1, 3, 5, 7, 9, 11, 29, 37
   (c) 2, 3, 5, 7, 11, 19, 41
   (d) 3, 5, 7, 9, 11, 23, 43
   (e) None of the above
2. Which of the following groups of numbers contain *only* composite numbers?
   (a) 1, 2, 4, 6, 8, 9, 26, 39
   (b) 1, 2, 4, 6, 26, 39
   (c) 1, 4, 6, 15, 26, 42
   (d) 4, 6, 26, 33, 39, 47
   (e) None of the above
3. The prime factorization of 198 is:
   (a) $2 \times 9 \times 11$
   (b) $2 \times 3^2 \times 11$
   (c) $2 \times 3 \times 11$
   (d) $2^2 \times 3 \times 11$
   (e) None of the above
4. The prime factorization of 756 is:
   (a) $2^2 \times 3^2 \times 7$
   (b) $2^2 \times 3^3 \times 7$
   (c) $2^2 \times 3^3$
   (d) $2^3 \times 3^2 \times 7$
   (e) None of the above
5. Which of the following groups consists of multiples of 9 *only*?
   (a) 1, 9, 18, 27, 36
   (b) 9, 27, 39, 54
   (c) 1, 18, 49, 63
   (d) 18, 27, 63, 72
   (e) None of the above
6. The least common multiple of 18 and 30 is:
   (a) 6
   (b) 30
   (c) 90
   (d) 540
   (e) None of the above
7. The least common multiple of 4, 30, and 18 is:
   (a) 180
   (b) 2
   (c) 90
   (d) 540
   (e) None of the above
8. The greatest common factor of 30 and 75 is:
   (a) 2
   (b) 3
   (c) 6
   (d) 15
   (e) None of the above

9. The greatest common factor of 18, 24, and 30 is:
   (a) 6
   (b) 12
   (c) 270
   (d) 360
   (e) None of the above

10. Which of the following is *not* an interpretation of the fraction $\frac{2}{7}$?
   (a) The number $\frac{2}{7}$
   (b) The ratio 2:7
   (c) A number of parts $2 \times \frac{1}{7}$
   (d) The division $7 \div 2$
   (e) None of the above

11. Multiply $\frac{17}{21} \times \frac{6}{34}$ and reduce if possible.
   (a) $\frac{3}{7}$
   (b) $\frac{1}{7}$
   (c) $\frac{21}{34}$
   (d) $\frac{112}{714}$
   (e) None of the above

12. Multiply $7 \times \frac{2}{35}$ and reduce if possible.
   (a) $\frac{2}{35}$
   (b) $\frac{2}{5}$
   (c) $\frac{14}{35}$
   (d) $\frac{14}{245}$
   (e) None of the above

13. Which of the following fractions with a denominator of 60 is equivalent to $\frac{2}{5}$?
   (a) $\frac{58}{60}$
   (b) $\frac{2}{60}$
   (c) $\frac{24}{60}$
   (d) $\frac{120}{300}$
   (e) None of the above

14. $\frac{90}{105}$ reduced to lowest terms is:

(a) $\frac{18}{21}$

(b) $\frac{30}{35}$

(c) $\frac{10}{35}$

(d) $\frac{6}{7}$

(e) None of the above

15. Multiply $\frac{2}{3} \times \frac{6}{7} \times \frac{1}{8}$ and reduce if possible.

(a) $\frac{0}{14}$

(b) $\frac{12}{168}$

(c) $\frac{1}{14}$

(d) 0

(e) None of the above

16. Which of the following is *not* equivalent to $\frac{3}{7}$?

(a) $\frac{27}{63}$

(b) $\frac{12}{21}$

(c) $\frac{36}{84}$

(d) $\frac{15}{35}$

(e) None of the above

17. Find: $\left(\frac{1}{3}\right)^3$

(a) $\frac{3}{9}$

(b) $\frac{3}{27}$

(c) $\frac{1}{27}$

(d) $\frac{1}{9}$

(e) None of the above

18. Nancy used one-half of a one-half pound bag of nuts in a recipe for cookies. What part of a pound of nuts did she use?

(a) $\frac{1}{16}$

(b) $\frac{1}{3}$

(c) $\frac{1}{4}$

(d) $\frac{1}{8}$

(e) None of the above

19. If a gas tank holds 24 gallons of gasoline when full, how many gallons does it contain if it is $\frac{3}{8}$ full?
(a) 10
(b) 8
(c) 9
(d) 7
(e) None of the above

20. $\frac{2}{5}$ of Joe's salary is deducted from his paycheck every month. If his monthly paycheck is 1,225 dollars, how much money does he have left after deductions?
(a) 245 dollars
(b) 490 dollars
(c) 880 dollars
(d) 735 dollars
(e) None of the above

LESSON 6:

Pre-test

1. 2, 17, 31, 41   2. 12, 4, 22, 33   3. $2^3 \times 3^2 \times 7$

Practice Exercises

Frame 5

Prime: 3, 17, 19, 23, 11, 2, 31
Composite: 15, 9, 16, 18, 21, 24, 49, 51
Neither prime nor composite: 1

Frame 11

$315 = 3^2 \times 5 \times 7$ $90 = 2 \times 3^2 \times 5$

$84 = 2^2 \times 3 \times 7$ $153 = 3^2 \times 17$

$117 = 3^2 \times 13$ $242 = 2 \times 11^2$

Post-test

1. 2, 37, 29, 17   2. 15, 21, 32, 27, 39, 46   3. $2^2 \times 3 \times 5^2$

LESSON 7:

Pre-test

1. 7, 14, 21, 28, 35, 42, 49, 56, 63, 70   2. 60   3. 14

Practice Exercises

Frame 5

Multiples of 6 = 6, 12, 18, 24, 30, 36, 42, 48, 54, 60, 66, 72, 78, 84, 90, 96, 102
Multiples of 7 = 7, 14, 21, 28, 35, 42, 49, 56, 63, 70, 77, 84, 91, 98, 105, 112

Frame 12

1. 390   2. 1,440   3. 5,616   4. 2,376   5. 13,860   6. 240

Frame 20

1. 4   2. 8   3. 25   4. 21   5. 18   6. 20

Post-test

1. 12, 24, 36, 48, 60, 72, 84, 96, 108, 120   2. 672   3. 34

LESSON 8:

Pre-test

1. (a) $3 \times \frac{1}{8}$ (b) $7 \div 11$
   (c) 4:5 (d) $\frac{9}{10}$

2. $\frac{20}{63}$ 3. $\frac{4}{25}$ 4. $\frac{7}{12} \times \frac{3}{3} = \frac{21}{36}$ 5. $\frac{3}{4}$

Practice Exercises

Frame 9

1. 3:5 2. $2 \times \frac{1}{7}$ 3. $5 \div 9$ 4. $\frac{1}{2}$

Frame 22

1. $\frac{10}{21}$ 2. $\frac{10}{2}$ or $\frac{5}{1}$ or 5 3. $\frac{16}{65}$ 4. $\frac{45}{28}$ or $1\frac{17}{28}$
5. $\frac{6}{5}$ or $1\frac{1}{5}$ 6. $\frac{21}{8}$ or $2\frac{5}{8}$ 7. $\frac{35}{27}$ or $1\frac{8}{27}$ 8. $\frac{6}{56}$ or $\frac{3}{28}$
9. $\frac{20}{9}$ or $2\frac{2}{9}$ 10. $\frac{7}{27}$

Frame 26

1.	2.	3.
$\frac{2}{5}$	$\frac{4}{3}$	$\frac{5}{2}$
$\frac{4}{10}$	$\frac{8}{6}$	$\frac{10}{4}$
$\frac{6}{15}$	$\frac{12}{9}$	$\frac{15}{6}$
$\frac{8}{20}$	$\frac{16}{12}$	$\frac{20}{8}$
$\frac{10}{25}$	$\frac{20}{15}$	$\frac{25}{10}$
$\frac{12}{30}$	$\frac{24}{18}$	$\frac{30}{12}$

Frame 35

1. $\frac{2}{11}$ 2. $\frac{3}{5}$ 3. $\frac{3}{8}$ 4. $\frac{1}{2}$ 5. $\frac{11}{17}$ 6. $\frac{3}{4}$

Frame 39

1. $\frac{6}{5}$ or $1\frac{1}{5}$ 2. $\frac{5}{3}$ or $1\frac{2}{3}$ 3. $\frac{7}{120}$ 4. $\frac{3}{7}$ 5. $\frac{8}{25}$ 6. $\frac{5}{18}$

Post-test

1. (a) 5:12 (b) $7 \times \frac{1}{9}$ (c) $\frac{3}{4}$ (d) $2 \div 5$

2. $\frac{56}{25}$ or $2\frac{6}{25}$ 3. $\frac{1}{64}$ 4. $\frac{3}{7} \times \frac{6}{6} = \frac{18}{42}$ 5. $\frac{6}{7}$

LESSON 9:

Pre-test

1. (a) Analysis:

Key Words:	cents, per can, save
Unknown:	amount saved
Given:	6 cans @ 26 cents each vs. 6 cans @ 34 cents each
Conditions:	6 × 34 cents is total cost at 34 cents per can; 6 × 26 cents is total cost at 26 cents per can

Method of Solution:

Amount saved = (6 × 34 cents) - (6 × 26 cents)
= (204 - 156) cents
= 48 cents

Check:

Reasonableness: (This check is a matter of judgment; there is no specific answer.)

Accuracy: 6 × (savings per can) = 6 × (34 - 26) cents
6 × (34 - 26) = 6 × 8 cents
= 48 cents

1. (b) Analysis:

Key Words:	cost of purchases, amount remaining
Unknown:	amount remaining
Given:	2 tires at 37 dollars each, 4 shocks at 9 dollars each, 1 tape deck at 88 dollars
Conditions:	The total cost cannot exceed 230 dollars.

Method of Solution:

Total cost = 2 × 37 dollars + 4 × 9 dollars + 88 dollars
= (74 + 36 + 88) dollars
= 198 dollars

Amount left over = 230 dollars - 198 dollars
= 32 dollars

Check:

Reasonableness: (Judgment)

Accuracy: 198 dollars + 32 dollars = 230 dollars

Practice Exercises

Frame 11

1. 21 cents   2. 40 lunches   3. 82 dollars   4. 5 dollars

Post-test

1. (a) Analysis:

Key Words:	total price
Unknown:	total price
Given:	4 tires at 37 dollars each 2 dollars tax on each of the 4 tires 1 dollar per tire on balancing
Conditions:	The total price will be the cost of the tires including tax and the cost of balancing.

Method of Solution:

Price: = (4 × 37 dollars) + (4 × 2 dollars) + (4 × 1 dollar)
= (148 + 8 + 4) dollars
= 160 dollars

Check:

Reasonableness: (Judgment)

Accuracy: 160 dollars – (148 dollars + 8 dollars + 4 dollars) = 0

1. (b) Analysis:

Key Words: amounts, balance, deposit

Unknown: amount remaining in account

Given: check for 84 dollars, 23 dollars, 17 dollars
Beginning balance of 100 dollars
deposit of 45 dollars

Conditions: The amount remaining will be the sum of the beginning balance and the deposit less the total amount of the checks.

Method of solution:

Total amount in bank = 100 dollars + 45 dollars
= 145 dollars

Total amount of all checks = 84 dollars + 23 dollars + 17 dollars
= 124 dollars

Amount remaining = 145 dollars – 124 dollars
= 21 dollars

Check:

Reasonableness: (Judgment)

Accuracy: 21 dollars + 124 dollars = 145 dollars
(balance) + (checks) = (deposit)

Answers To Drill Exercises: Unit 2

1.	$2^2 \times 3 \times 5$	2.	$2 \times 3^2 \times 7$
3.	$2 \times 3 \times 5 \times 7$	4.	$2^2 \times 3^2 \times 7$
5.	$3 \times 5^2 \times 11$	6.	2×7^2
7.	2×3^3	8.	11×13
9.	$2 \times 5 \times 11$	10.	$5 \times 7 \times 11$

	LCM	*GCF*		*LCM*	*GCF*
11.	$2 \times 3 \times 5 = 30$	3	12.	$2^2 \times 3^2 = 36$	3
13.	$2^2 \times 3 \times 7 = 84$	3	14.	$2 \times 3^2 \times 5 = 90$	3
15.	$2^5 \times 3 = 96$	$2^3 = 8$	16.	$2^2 \times 3^2 \times 5 = 180$	2
17.	$2 \times 3 \times 5 = 30$	5	18.	$3^3 \times 5 = 135$	3
19.	$2 \times 5^2 \times 31 = 1{,}550$	2	20.	$2 \times 3 \times 5 \times 7 = 210$	3
21.	$2^3 \times 3 \times 5^2 = 600$	5	22.	$2^2 \times 3 \times 7 = 84$	None
23.	$2^2 \times 3^2 \times 5 \times 7 = 1{,}260$	3	24.	$2^4 \times 3 \times 5 \times 7 = 1{,}680$	2
25.	$2 \times 3 \times 5 \times 11 = 330$	2			

26.	$\frac{24}{35}$	27.	$\frac{16}{77}$
28.	$\frac{2}{35}$	29.	$\frac{6}{55}$
30.	$\frac{20}{63}$	31.	$\frac{7}{96}$
32.	$\frac{4}{49}$	33.	$\frac{9}{25}$
34.	$\frac{1}{18}$	35.	$\frac{1}{24}$
36.	$\frac{1}{90}$	37.	$\frac{5}{6}$
38.	$\frac{1}{125}$	39.	$\frac{1}{27}$
40.	$\frac{9}{16}$	41.	$\frac{12}{5}$ or $2\frac{2}{5}$
42.	$\frac{30}{7}$ or $4\frac{2}{7}$	43.	$\frac{28}{15} = 1\frac{13}{15}$
44.	$\frac{2}{7}$	45.	$\frac{5}{1}$ or 5
46.	$\frac{4}{5}, \frac{8}{10}, \frac{12}{15}, \frac{16}{20}, \frac{20}{25}$	47.	$\frac{3}{7}, \frac{6}{14}, \frac{9}{21}, \frac{12}{28}, \frac{15}{35}$
48.	$\frac{5}{6}, \frac{10}{12}, \frac{15}{18}, \frac{20}{24}, \frac{25}{30}$	49.	$\frac{1}{11}, \frac{2}{22}, \frac{3}{33}, \frac{4}{44}, \frac{5}{55}$
50.	$\frac{2}{9}, \frac{4}{18}, \frac{6}{27}, \frac{8}{36}, \frac{10}{45}$		
51.	$\frac{2}{3}$	52.	$\frac{3}{4}$
53.	$\frac{3}{7}$	54.	$\frac{2}{9}$

55. $\frac{5}{7}$	56. $\frac{3}{5}$
57. $\frac{2}{11}$	58. $\frac{5}{8}$
59. $\frac{3}{7}$	60. $\frac{7}{9}$
61. 500 tires	62. 25 secretaries
63. 730 dollars	64. 57 dollars
65. 36 dollars	66. 5 games
67. 1,825 dollars	68. 368 yards

Answers To Self-Test: Unit 2

1. (c)	2. (e)	3. (b)	4. (b)	5. (d)	6. (c)	7. (a)
8. (d)	9. (a)	10. (d)	11. (b)	12. (b)	13. (c)	14. (d)
15. (c)	16. (b)	17. (c)	18. (c)	19. (c)	20. (d)	

UNIT THREE

Fractional Number Operations

Lesson 10	ADDITION AND SUBTRACTION OF FRACTIONS
Lesson 11	DIVISION OF FRACTIONS
Lesson 12	MIXED NUMBER OPERATIONS
Lesson 13	FRACTIONAL PARTS AND PROBABILITY

LESSON 10
ADDITION AND SUBTRACTION OF FRACTIONS

Objectives:

1. Determine the sum of two or more fractions with the same denominator. (1–9)
2. Determine the difference between two fractions with the same denominator. (6–9)
3. Determine the sum of two or more fractions with different denominators. (10–18)
4. Determine the difference between two fractions with different denominators. (19–21)

Vocabulary:

Least Common Denominator (LCD)

PRE-TEST

1. Determine the sum for each of the following, reducing all answers to lowest terms:

   (a) $\frac{7}{12} + \frac{3}{12}$ (b) $\frac{4}{15} + \frac{7}{15}$

2. Determine the difference for each of the following, reducing all answers to lowest terms:

   (a) $\frac{22}{45} - \frac{13}{45}$ (b) $\frac{3}{7} - \frac{2}{7}$

3. Determine the sum for each of the following, reducing all answers to lowest terms:

   (a) $\frac{10}{21} + \frac{6}{15}$ (b) $\frac{5}{42} + \frac{3}{20} + \frac{6}{35}$

4. Determine the difference for each of the following, reducing all answers to lowest terms:

   (a) $\frac{16}{24} - \frac{4}{15}$ (b) $\frac{15}{16} - \frac{3}{12}$

Mr. Wolberg is having his yard landscaped. One-fifth of the yard will be planted in grass and three-fifths will be planted in trees and shrubs. What part of his entire yard will be planted with either grass or trees and shrubs? 

In order to answer this question, it is necessary to add $\frac{1}{5}$ and $\frac{3}{5}$.

Adding fractions with the same denominators: ②

Adding fractions with the same denominator is not difficult. Using a rectangle divided into 5 equal parts to represent the entire yard, we can see how easy it is to add one-fifth and three-fifths.

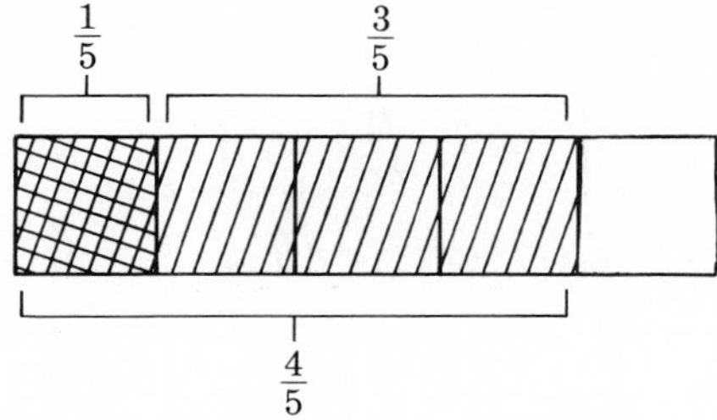

$$\frac{1}{5}+\frac{3}{5}=\frac{1+3}{5}=\frac{4}{5}$$

Adding fractions with the same denominators: ③

To add any fractions with the same denominators, simply add the numerators and keep the same denominator.

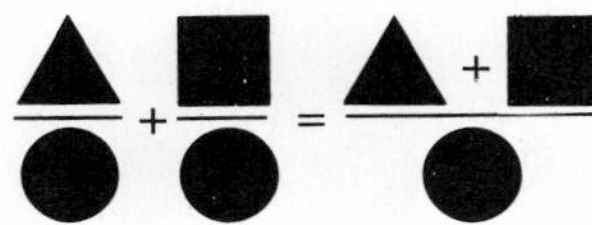

For any numbers ▲ and ■ and for any non-zero number ●.

Adding fractions with the same denominators: 

Complete the following examples as shown:

A. $\frac{2}{7}+\frac{3}{7}=\frac{2+3}{7}=\frac{5}{7}$

B. $\frac{1}{2}+\frac{3}{2}=\frac{\triangle+\square}{2}=\frac{4}{2}$ which reduces to $\frac{2}{1}$ or 2

C. $\frac{2}{13}+\frac{4}{13}+\frac{5}{13}=\frac{2+4+5}{\bigcirc}=\frac{11}{\bigcirc}$

D. $\frac{5}{19}+\frac{6}{19}+\frac{1}{19}=\frac{5+6+1}{19}=\frac{⬡}{19}$

Ramona is making an apron that calls for a total of five-sixths of a yard of material. She decides to use a contrasting fabric for the pocket and the ties. If she uses one-sixth of a yard of material for the pocket and the ties, how much yardage does she use for the rest of the apron? ⑤

In order to answer this question, it is necessary to subtract $\frac{1}{6}$ from $\frac{5}{6}$.

Subtracting fractions with the same denominators: 

Subtracting fractions with the same denominators is also easy. Using a rectangle divided into 6 equal parts to represent 1 yard of material, we see how easy it is to subtract one-sixth from five-sixths.

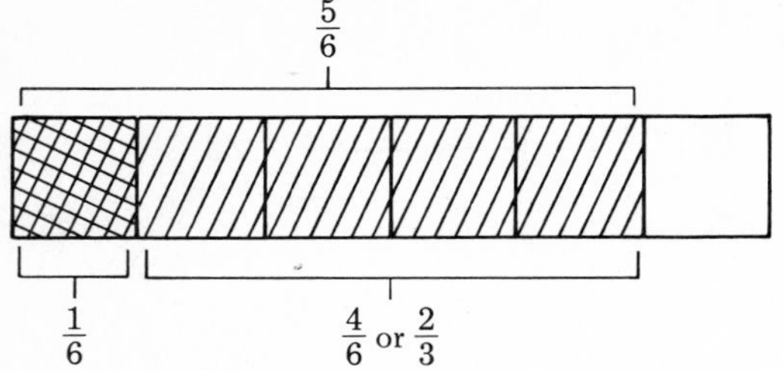

$$\frac{5}{6} - \frac{1}{6} = \frac{5-1}{6} = \frac{4}{6} \text{ or } \frac{2}{3}$$

Subtracting fractions with the same denominators: 

To subtract fractions with the same denominator, simply subtract the smaller numerator from the larger numerator and keep the same denominator.

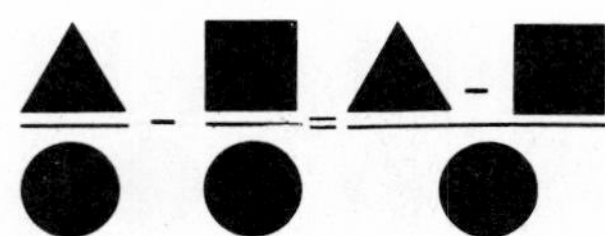

For any numbers ▲ and ■ and for any non-zero number .

Subtracting fractions with the same denominators: 

Complete the following examples as shown:

A. $\frac{4}{7} - \frac{3}{7} = \frac{4-3}{7} = \frac{1}{7}$

B. $\frac{7}{9} - \frac{2}{9} = \frac{\triangle - \square}{9} = \frac{5}{9}$

C. $\frac{5}{3} - \frac{2}{3} = \frac{5-2}{\bigcirc} = \frac{3}{3}$ or 1

D. $\frac{11}{17} - \frac{5}{17} = \frac{11-5}{17} = \frac{⬡}{17}$

Practice Exercises ⑨

Find each indicated sum or difference, reducing all answers to lowest terms.

1. $\frac{4}{7} + \frac{1}{7}$
2. $\frac{4}{9} + \frac{8}{9}$
3. $\frac{2}{11} + \frac{5}{11}$
4. $\frac{13}{2} + \frac{5}{2}$
5. $\frac{9}{7} - \frac{3}{7}$
6. $\frac{10}{13} - \frac{2}{13}$
7. $\frac{5}{3} - \frac{2}{3}$
8. $\frac{51}{53} - \frac{2}{53}$

Mrs. Burke is thinking about walking from her home to the post office down the street and then on to the supermarket. She knows it is one-third of a mile to the post office and that it is one-fifth of a mile from the post office to the market. How far will it be for her to walk back from the market to her house? (10)

In order to answer this question, it is necessary to add $\frac{1}{3}$ and $\frac{1}{5}$.

Addition of fractions with different denominators: (11)

The addition (or subtraction) of fractions which have different denominators presents a special problem.

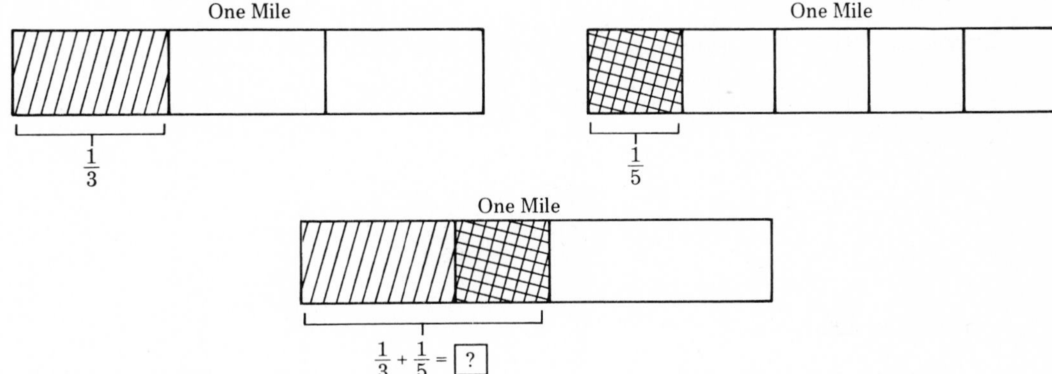

Before $\frac{1}{3}$ and $\frac{1}{5}$ can be added, they must be changed to equivalent fractions, both having the same denominator.

Addition of fractions with different denominators: (12)

The first five fractions equivalent to $\frac{1}{3}$ and to $\frac{1}{5}$ are shown below. Recall that equivalent fractions are formed by multiplying the given fraction by some form of the number one: $\frac{1}{1}, \frac{2}{2}, \frac{3}{3}$, and so on.

fractions equivalent to $\frac{1}{3}$:

$$\frac{1}{3} \times \frac{1}{1} = \frac{1}{3}$$
$$\frac{1}{3} \times \frac{2}{2} = \frac{2}{6}$$
$$\frac{1}{3} \times \frac{3}{3} = \frac{3}{9}$$
$$\frac{1}{3} \times \frac{4}{4} = \frac{4}{12}$$
$$\frac{1}{3} \times \frac{5}{5} = \boxed{\frac{5}{15}}$$

fractions equivalent to $\frac{1}{5}$:

$$\frac{1}{5} \times \frac{1}{1} = \frac{1}{5}$$
$$\frac{1}{5} \times \frac{2}{2} = \frac{2}{10}$$
$$\frac{1}{5} \times \frac{3}{3} = \boxed{\frac{3}{15}}$$
$$\frac{1}{5} \times \frac{4}{4} = \frac{4}{20}$$
$$\frac{1}{5} \times \frac{5}{5} = \frac{5}{25}$$

Note that $\frac{5}{15}$ is a fraction equivalent to $\frac{1}{3}$ and $\frac{3}{15}$ is a fraction equivalent to $\frac{1}{5}$. The fractions $\frac{5}{15}$ and $\frac{3}{15}$ can be added because they have the same denominator.

Addition of fractions with different denominators: $\frac{1}{3} + \frac{1}{5}$. (13)

$$\begin{aligned}
\frac{1}{3} + \frac{1}{5} &= \left(\frac{1}{3} \times \frac{5}{5}\right) + \left(\frac{1}{5} \times \frac{3}{3}\right)\\
&= \frac{5}{15} + \frac{3}{15}\\
&= \frac{5+3}{15}\\
&= \frac{8}{15}
\end{aligned}$$

Notice that 15 is the Least Common Multiple of 3 and 5. When two fractions have different denominators, they are first changed to equivalent fractions with the LCM as the common denominator of both before adding or subtracting. The LCM of the denominators is called the *Least Common Denominator* (LCD). (14)

Add: $\frac{3}{14} + \frac{5}{6}$ (15)

First find the LCD: $14 = 2 \times 7$, $6 = 2 \times 3$ $\longrightarrow$ LCD $= 2 \times 3 \times 7 = 42$

Then change $\frac{3}{14}$ and $\frac{5}{6}$ to equivalent fractions both having a denominator of 42 and add:

$$\frac{3}{14} + \frac{5}{6} = \left(\frac{3}{2 \times 7} \times \frac{\boxed{?}}{\boxed{?}}\right) + \left(\frac{5}{2 \times 3} \times \frac{\textcircled{?}}{\textcircled{?}}\right)$$

← Compare the factored forms of the denominators with the LCD to find what $\boxed{?}$ and $\textcircled{?}$ must be.

$$= \left(\frac{3}{2 \times 7} \times \frac{3}{3}\right) + \left(\frac{5}{2 \times 3} \times \frac{7}{7}\right)$$

← $\boxed{?}$ is 3 since $2 \times 7 \times 3 = 42$; $\textcircled{?}$ is 7 since $2 \times 3 \times 7 = 42$.

$$\begin{aligned}
&= \frac{9}{42} + \frac{35}{42}\\
&= \frac{9+35}{42}\\
&= \frac{44}{42} \text{ or } \frac{22}{21}
\end{aligned}$$

Add: $\frac{2}{15} + \frac{4}{21}$ 

Complete the following:

Finding the LCD: $15 = 3 \times 5$, $21 = 3 \times 7$ $\longrightarrow$ LCD $= 3 \times 5 \times 7 = 105$

$$\text{Adding: } \frac{2}{15} + \frac{4}{21} = \left(\frac{2}{3 \times 5} \times \frac{\square}{\square}\right) + \left(\frac{4}{3 \times 7} \times \frac{\bigcirc}{\bigcirc}\right)$$

$$= \frac{14}{105} + \frac{20}{105}$$

$$= \frac{14 + 20}{105}$$

$$= \frac{34}{105}$$

Add: $\frac{7}{30} + \frac{2}{35} + \frac{1}{15}$ (17)

Complete the following:

Finding the LCD: $30 = 2 \times 3 \times 5$, $35 = 5 \times 7$, $15 = 3 \times 5$ → LCD $= 2 \times 3 \times 5 \times 7 = 210$

$$\text{Adding: } \frac{7}{30} + \frac{2}{35} + \frac{1}{15} = \left(\frac{7}{2 \times 3 \times 5} \times \frac{7}{7}\right) + \left(\frac{2}{5 \times 7} \times \frac{2 \times 3}{2 \times 3}\right) + \left(\frac{1}{3 \times 5} \times \frac{2 \times 7}{2 \times 7}\right)$$

$$= \frac{49}{210} + \frac{12}{210} + \frac{14}{210}$$

$$= \frac{\square + \hexagon + \bigcirc}{210}$$

$$= \frac{75}{210} \text{ or } \frac{5}{14}$$

Practice Exercises

(18)

Find the sum for each of the following. Reduce all answers to lowest terms.

1. $\frac{3}{8} + \frac{3}{10}$
2. $\frac{1}{2} + \frac{5}{6} + \frac{3}{8}$
3. $\frac{1}{7} + \frac{2}{21}$
4. $\frac{4}{7} + \frac{3}{14} + \frac{1}{6}$
5. $\frac{5}{12} + \frac{3}{20}$
6. $\frac{2}{9} + \frac{3}{4} + \frac{5}{18}$
7. $\frac{3}{10} + \frac{1}{20}$
8. $\frac{3}{5} + \frac{7}{15} + \frac{3}{20}$

Subtracting fractions with different denominators: $\frac{4}{5} - \frac{2}{7}$ (19)

To subtract fractions with different denominators, the given fractions are first changed to equivalent fractions that each have the LCD as their denominators.

Finding the LCD: 5 is a prime number, 7 is a prime number → LCD = 5 × 7 = 35

Subtracting:

$$\frac{4}{5} - \frac{2}{7} = \left(\frac{4}{5} \times \frac{7}{7}\right) - \left(\frac{2}{7} \times \frac{5}{5}\right)$$
$$= \frac{28}{35} - \frac{10}{35}$$
$$= \frac{28 - 10}{35}$$
$$= \frac{18}{35}$$

Subtract: $\frac{9}{10} - \frac{5}{12}$ (20)

Complete the following:

Finding the LCD: $10 = 2 \times 5$, $12 = 2^2 \times 3$ → LCD = $2^2 \times 3 \times 5 = 60$

Subtracting:

$$\frac{9}{10} - \frac{5}{12} = \left(\frac{9}{2 \times 5} \times \frac{2 \times 3}{2 \times 3}\right) - \left(\frac{5}{2^2 \times 3} \times \frac{5}{5}\right)$$
$$= \frac{54}{\square} - \frac{25}{\square}$$
$$= \frac{\triangle - \bigcirc}{\square}$$
$$= \frac{29}{60}$$

Practice Exercises (21)

Find the difference for each of the following, reducing all answers to lowest terms.

1. $\frac{8}{9} - \frac{2}{5}$
2. $\frac{13}{15} - \frac{5}{12}$
3. $\frac{7}{30} - \frac{3}{14}$
4. $\frac{9}{7} - \frac{1}{6}$

POST-TEST

1. Determine the sum for each of the following, reducing all answers to lowest terms.

   (a) $\frac{5}{11} + \frac{7}{11}$ (b) $\frac{9}{16} + \frac{5}{16}$

2. Determine the difference for each of the following, reducing all answers to lowest terms.

   (a) $\frac{7}{9} - \frac{2}{9}$ (b) $\frac{13}{28} - \frac{5}{28}$

3. Determine the sum for each of the following, reducing all answers to lowest terms.

   (a) $\frac{11}{28} + \frac{7}{21}$ (b) $\frac{4}{9} + \frac{5}{12} + \frac{11}{27}$

4. Determine the difference for each of the following, reducing all answers to lowest terms.

   (a) $\frac{7}{12} - \frac{3}{16}$ (b) $\frac{5}{6} - \frac{5}{18}$

LESSON 11
DIVISION OF FRACTIONS

Objectives:

1. Write the reciprocal for a given fraction or whole number. (1–3)
2. Divide a whole number by a fraction. (4–12)
3. Divide a fraction by another fraction. (8–12)
4. Divide a fraction by a non-zero whole number. (11–12)
5. Rewrite a complex fraction as a division. (13–16)
6. Rewrite a division with fractions as a complex fraction. (13–16)

Vocabulary:

Complex Fraction
Reciprocal

PRE-TEST

1. Write the reciprocal for each fraction.
   (a) $\frac{2}{97}$ (b) $\frac{1}{5}$
2. Divide: $5 \div \frac{2}{3}$
3. Divide: $\frac{6}{11} \div \frac{1}{2}$.
4. Divide: $\frac{3}{7} \div 2$.
5. Rewrite each complex fraction as a division, and rewrite each division as a complex fraction.
   (a) $\frac{7}{9} \div \frac{3}{4}$ (b) $\dfrac{\frac{1}{2}}{\frac{2}{3}}$

Reciprocals are any two numbers whose product is one. Every number except 0 has a reciprocal. (1)

For example, the numbers $\frac{2}{3}$ and $\frac{3}{2}$ are reciprocals, since

$$\frac{2}{3} \times \frac{3}{2} = \frac{2 \times 3}{3 \times 2} = \frac{6}{6} \text{ or } 1$$

In fact, for all non-zero numbers, ■ and ●

$$\frac{\blacksquare}{\bullet} \times \frac{\bullet}{\blacksquare} = \frac{\blacksquare \times \bullet}{\bullet \times \blacksquare} = 1$$

This property is also true for all whole numbers, except zero, since every whole number can be written as a fraction.

The Reciprocal of a Number ②

The reciprocal of $\frac{3}{4}$ is $\frac{4}{3}$ since $\frac{3}{4} \times \frac{4}{3} = 1$.

The reciprocal of 5 is $\frac{1}{5}$ since $5 \times \frac{1}{5} = 1$.

The reciprocal of $\frac{1}{3}$ is 3 since $\frac{1}{3} \times 3 = 1$.

Practice Exercises ③

Write the reciprocal for each of the following:

1. $\frac{5}{8}$
2. $\frac{1}{9}$
3. 7
4. $\frac{7}{2}$

Recall from the definition of division of whole numbers that 8 ÷ 4 means how many times can 4 be subtracted from 8. The quotient is 2 since 4 can be subtracted from 8 two times. ④

The meaning of $4 \div \frac{1}{2}$ is exactly the same: How many times can $\frac{1}{2}$ be subtracted from 4? The quotient can be determined by counting the number of $\frac{1}{2}$'s in 4:

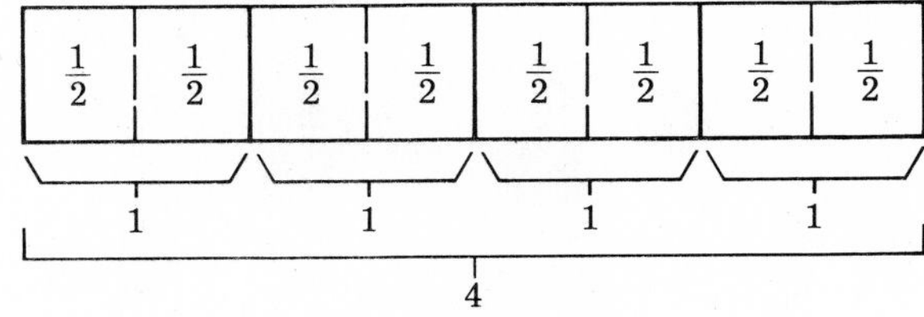

There are eight $\frac{1}{2}$'s in 4:

$$4 \div \frac{1}{2} = 8$$

Dividing a whole number by a fraction: $4 \div \frac{1}{2}$ (5)

By counting, it was shown in the last frame that $4 \div \frac{1}{2} = 8$. It is also true that $4 \times 2 = 8$. It can also be shown that $4 \div \frac{1}{2} = 4 \times 2$.

Recall that one interpretation of a fraction is that it indicates a division:

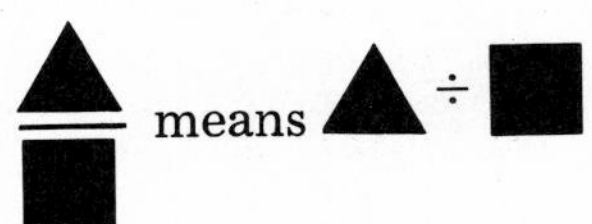

For all numbers except ■ = 0.

Since this is always true if the divisor is not 0, $4 \div \frac{1}{2}$ can be rewritten as the fraction $\frac{4}{\frac{1}{2}}$. Then

$$\frac{4}{\frac{1}{2}} = \frac{4}{\frac{1}{2}} \times \frac{2}{2}$$

$$= \frac{4 \times 2}{\frac{1}{2} \times 2}$$

$$= \frac{4 \times 2}{1}$$

$$= 4 \times 2$$

"Multiplication by One"— $\frac{2}{2}$ is chosen because 2 is the reciprocal of the divisor, $\frac{1}{2}$.

Definition of reciprocals.

Dividing a whole number by a fraction: 

It was shown in the last frame that $4 \div \frac{1}{2} = 4 \times 2$. In fact, to divide any whole number by a fraction, simply multiply the whole number by the reciprocal of the fractional divisor.

$$● \div \frac{▲}{■} = ● \times \frac{■}{▲}$$

For all numbers ● and non-zero numbers ■ and ▲.

Dividing a whole number by a fraction: 

Complete each of the following as shown:

A. $4 \div \frac{3}{5} = 4 \times \frac{5}{3} = \frac{4}{1} \times \frac{5}{3} = \frac{20}{3}$

B. $12 \div \frac{1}{3} = 12 \times \hexagon = 36$

C. $7 \div \frac{8}{5} = 7 \times \frac{\bigcirc}{\square} = \frac{7}{1} \times \frac{\bigcirc}{\square} = \frac{\triangle}{\square}$

Dividing a fraction by another fraction: $\frac{7}{5} \div \frac{2}{3}$ (8)

It has been shown that division of a whole number by a fraction is done by multiplying the whole number by the reciprocal of the divisor. Is this also true if a fraction is divided by another fraction?

$$\frac{7}{5} \div \frac{2}{3} \stackrel{?}{=} \frac{7}{5} \times \frac{2}{3}$$

$$\frac{7}{5} \div \frac{2}{3} = \frac{\frac{7}{5}}{\frac{2}{3}}$$ ← Rewriting the division as a fraction

$$= \frac{\frac{7}{5}}{\frac{2}{3}} \times \frac{\frac{3}{2}}{\frac{3}{2}}$$ ← "Multiplication by One"—this form of the number 1 is chosen because $\frac{3}{2}$ is the reciprocal of the divisor, $\frac{2}{3}$

$$= \frac{\frac{7}{5} \times \frac{3}{2}}{\frac{2}{3} \times \frac{3}{2}}$$

$$= \frac{\frac{7}{5} \times \frac{3}{2}}{1}$$ ← Definition of reciprocals

$$= \frac{7}{5} \times \frac{3}{2}$$

Dividing a fraction by another fraction: $\frac{7}{5} \div \frac{2}{3}$ (9)

It was shown in the last frame that

$$\frac{7}{5} \div \frac{2}{3} = \frac{7}{5} \times \frac{3}{2}$$

Completing the multiplication:

$$\frac{7}{5} \times \frac{3}{2} = \frac{7 \times 3}{5 \times 2} = \frac{21}{10}$$

Thus, $\frac{7}{5} \div \frac{2}{3} = \frac{21}{10}$. It has now been shown that *any* division can be done by multiplying the dividend by the reciprocal of the divisor. This includes division of a fraction by a whole number since every whole number can be written as a fraction.

Dividing a fraction by another fraction: (10)

Complete the following as shown:

A. $\frac{3}{7} \div \frac{5}{9} = \frac{3}{7} \times \frac{9}{5} = \frac{3 \times 9}{7 \times 5} = \frac{27}{35}$

B. $\frac{2}{5} \div \frac{2}{3} = \frac{2}{5} \times \frac{\bigcirc}{\square} = \frac{2 \times \bigcirc}{5 \times \square} = \frac{6}{10} \text{ or } \frac{3}{5}$

C. $\frac{7}{10} \div \frac{24}{100} = \frac{7}{10} \times \frac{100}{24} = \frac{7 \times 100}{10 \times 24} = \frac{\triangle}{\hexagon}$

Dividing a fraction by a whole number: (11)

Complete the following as shown:

A. $\frac{1}{5} \div 2 = \frac{1}{5} \times \frac{1}{2} = \frac{1 \times 1}{5 \times 2} = \frac{1}{10}$

B. $\frac{3}{4} \div 7 = \frac{3}{4} \times \frac{1}{7} = \frac{3 \times 1}{4 \times 7} = \frac{\triangle}{\hexagon}$

C. $\frac{2}{5} \div 22 = \frac{2}{5} \times \frac{\bigcirc}{\square} = \frac{2 \times \bigcirc}{5 \times \square} = \frac{2}{110} \text{ or } \frac{1}{55}$

Practice Exercises (12)

Divide each of the following, reducing all answers to lowest terms:

1. $\frac{4}{5} \div \frac{2}{3}$
2. $\frac{3}{10} \div \frac{4}{5}$
3. $\frac{7}{100} \div 10$
4. $17 \div \frac{1}{4}$
5. $7 \div \frac{3}{4}$
6. $\frac{25}{36} \div \frac{5}{6}$
7. $\frac{3}{7} \div \frac{3}{100}$
8. $\frac{5}{8} \div 3$
9. $\frac{27}{35} \div \frac{9}{7}$

In showing how division can be done by multiplying by the reciprocal of the divisor, the following expressions were used: (13)

$$\frac{\frac{7}{5}}{\frac{2}{3}} \text{ and } \frac{4}{\frac{1}{2}}$$

These numeral forms have a special name; they are called *complex fractions.*

A *complex fraction* is a fraction that has another fraction as its numerator and/or its denominator. Every division with fractions can be rewritten as a complex fraction and every complex fraction can be rewritten as a division. (14)

For example,

$$\frac{3}{4} \div \frac{5}{6} = \frac{\frac{3}{4}}{\frac{5}{6}} \qquad \frac{2}{\frac{1}{3}} = 2 \div \frac{1}{3}$$

$$\frac{\frac{3}{15}}{\frac{2}{7}} = \frac{3}{15} \div \frac{2}{7} \qquad \frac{1}{6} \div 5 = \frac{\frac{1}{6}}{5}$$

Complex fractions: (15)

Complete the following:

$$\frac{1}{2} \div \frac{2}{3} = \frac{\frac{\triangle}{\square}}{\frac{2}{3}} \qquad \frac{\frac{1}{7}}{\bigcirc} = \frac{1}{7} \div 3$$

Practice Exercises (16)

Rewrite each complex fraction as a division and each division as a complex fraction.

1. $\frac{3}{4} \div \frac{5}{6}$

2. $\frac{\frac{3}{15}}{\frac{2}{7}}$

3. $\frac{2}{3} \div 5$

4. $\frac{\frac{2}{3}}{7}$

5. $1 \div \frac{13}{25}$

6. $\frac{11}{\frac{8}{3}}$

POST-TEST

1. Write the reciprocal for each fraction.

   (a) $\frac{1}{4}$ (b) $\frac{6}{11}$

2. Divide: $1 \div \frac{12}{25}$

3. Divide: $\frac{7}{3} \div \frac{4}{5}$

4. Divide: $\frac{5}{11} \div 6$

5. Rewrite each complex fraction as a division and each division as a complex fraction.

   (a) $\frac{4}{15} \div \frac{3}{7}$ (b) $\dfrac{\frac{5}{6}}{\frac{1}{2}}$

LESSON 12
MIXED NUMBER OPERATIONS

Objectives:

1. Rewrite a mixed number as an improper fraction. (1–11)
2. Rewrite an improper fraction as a mixed number. (7–11)
3. Determine the sum of two or more mixed numbers. (13–17)
4. Determine the difference between two mixed numbers. (18–22)
5. Determine the product of two or more mixed numbers. (23–25)
6. Determine the quotient for two mixed numbers. (26–27)
7. Determine the value of a power of a mixed number. (28–30)
8. Write the reciprocal for a given mixed number. (31–33)

Vocabulary:

Mixed Number
Improper fraction

PRE-TEST

1. Rewrite each mixed number as a fraction and rewrite each fraction as a mixed number.

(a) $\frac{12}{5}$ (b) $3\frac{7}{9}$

(c) $1\frac{5}{7}$ (d) $\frac{50}{47}$

2. Perform the indicated operation on mixed numbers for each of the following, changing all improper fractions in answers to mixed numbers:

(a) $3\frac{1}{4} \times 2\frac{3}{5}$ (b) $9\frac{5}{12} - 6\frac{5}{6}$

(c) $2\frac{1}{4} + 7\frac{3}{10}$ (d) $5\frac{1}{3} \div 3\frac{2}{5}$

(e) $\left(1\frac{6}{7}\right)^2$

3. Write the reciprocal for the mixed number $8\frac{3}{5}$.

Mr. Cunningham is building a playhouse for his daughter and decides to build in a shelf. Measuring the distance between the windows, he finds that he can build a shelf $3\frac{3}{4}$ feet long. (1)

It is common to mix whole numbers and fractions in statements such as "a shelf $3\frac{3}{4}$ feet long." This is an example of the use of a mixed number.

A *mixed number* is the sum of a whole number and a fraction. Each mixed number can be changed to an equivalent fraction with a numerator larger than the denominator. The fractional form of a mixed number is often called an *improper fraction.* (2)

The mixed number $3\frac{3}{4}$ can be thought of as $3 + \frac{3}{4}$.

Rewriting $3\frac{3}{4}$ as an equivalent improper fraction: (3)

"Why"

$$3\frac{3}{4} = 3 + \frac{3}{4} \quad \leftarrow \text{Definition of a mixed number}$$

$$= \frac{3}{1} + \frac{3}{4}$$

$$= \left(\frac{3}{1} \times \frac{4}{4}\right) + \frac{3}{4} \quad \leftarrow \text{"Multiplication by 1"}$$

$$= \frac{12}{4} + \frac{3}{4}$$

$$= \frac{12 + 3}{4} \quad \leftarrow \text{Addition of fractions}$$

$$= \frac{15}{4}$$

Thus, the mixed number $3\frac{3}{4}$ is equivalent to the improper fraction $\frac{15}{4}$.

Rewriting $3\frac{3}{4}$ as an equivalent improper fraction: (4)

There is a short cut for the above procedure.

"How"

$$\overset{+}{3\frac{3}{4}} = \frac{15}{4}$$
$$\times$$

First: Multiply the denominator of the fraction times the whole number: $4 \times 3 = 12$.

Second: Add this product to the numerator of the fraction: 12 + 3 = 15.

Third: Write the equivalent improper fraction with this sum as its numerator: $\frac{15}{4}$.

Rewriting $15\frac{3}{8}$ as an equivalent improper fraction: 5

$$15\frac{3}{8} = \frac{(8 \times 15) + 3}{8}$$

$$= \frac{120 + 3}{8}$$

$$= \frac{123}{8}$$

Rewriting $3\frac{3}{7}$ as an equivalent improper fraction: 6

Complete the following:

$$3\frac{3}{7} = \frac{(7 \times 3) + \square}{\bigcirc}$$

$$= \frac{21 + \square}{\bigcirc}$$

$$= \frac{24}{7}$$

Rewriting $\frac{15}{4}$ as a mixed number: 7

"Why"

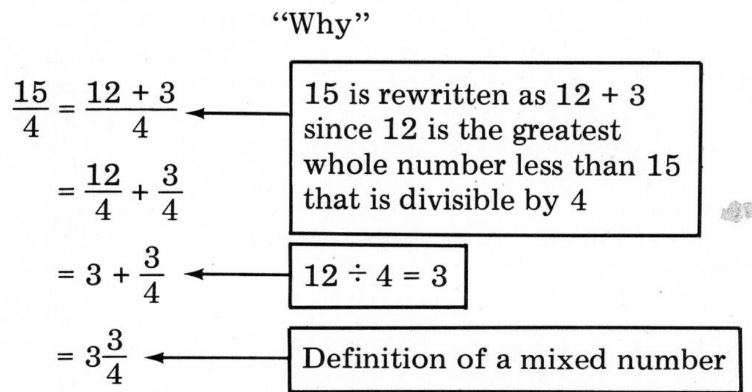

$\frac{15}{4} = \frac{12 + 3}{4}$
15 is rewritten as 12 + 3 since 12 is the greatest whole number less than 15 that is divisible by 4
$= \frac{12}{4} + \frac{3}{4}$
$= 3 + \frac{3}{4}$
12 ÷ 4 = 3
$= 3\frac{3}{4}$
Definition of a mixed number

Rewriting $\frac{15}{4}$ as a mixed number: ⑧

The short cut for the above procedure is as follows:

"How"

Remember that one meaning of $\frac{15}{4}$ is 15 ÷ 4.

$$\begin{array}{r} 3 \\ 4\overline{)15} \\ \underline{12} \\ 3 \end{array}$$

15 ÷ 4 = 3 R 3

In the quotient, the remainder, 3, is written as the numerator of a fraction with the divisor, 4, as the denominator.

$15 \div 4 = 3\frac{3}{4}$ ← Remainder (numerator) ← Divisor (denominator)

Rewriting $\frac{38}{9}$ as a mixed number: ⑨

$\frac{38}{9} = 38 \div 9$

$= 4\frac{2}{9}$ ← Remainder (numerator) ← Divisor (denominator)

$$\begin{array}{r} 4 \\ 9\overline{)38} \\ \underline{36} \\ 2 \end{array}$$

← Remainder

Rewriting $\frac{34}{5}$ as a mixed number: ⑩

Complete the following:

$\frac{34}{5} = 34 \div 5$

$= 6\frac{4}{5}$

$$\begin{array}{r} 6 \\ 5\overline{)34} \\ \underline{30} \\ 4 \end{array}$$

← Remainder

Practice Exercises

11

Rewrite each mixed number as a fraction and rewrite each fraction as a mixed number.

1. $8\frac{2}{3}$
2. $\frac{81}{5}$
3. $6\frac{2}{7}$
4. $\frac{49}{3}$
5. $18\frac{1}{2}$
6. $\frac{11}{4}$
7. $1\frac{1}{3}$
8. $\frac{28}{25}$

In the following frames the addition, subtraction, multiplication, and division of mixed numbers will be developed. The guideline to follow when performing any of these operations is to *first* rewrite the mixed numbers as equivalent improper fractions, then perform the indicated operation, and finally, rewrite the result as a mixed number. Alternate methods for performing additions and subtractions will be given. 12

Add: $4\frac{1}{5} + 2\frac{2}{3}$ 13

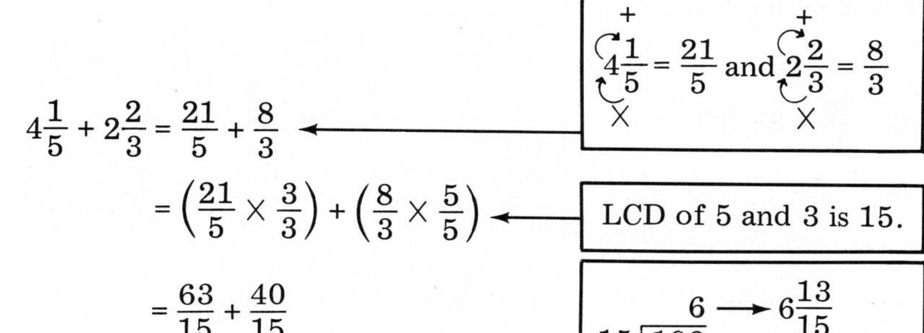

$$4\frac{1}{5} + 2\frac{2}{3} = \frac{21}{5} + \frac{8}{3}$$

$4\frac{1}{5} = \frac{21}{5}$ and $2\frac{2}{3} = \frac{8}{3}$

$$= \left(\frac{21}{5} \times \frac{3}{3}\right) + \left(\frac{8}{3} \times \frac{5}{5}\right)$$

LCD of 5 and 3 is 15.

$$= \frac{63}{15} + \frac{40}{15}$$

$$= 6\frac{13}{15}$$

$15\overline{)103}$ = 6, remainder 13: 90, 13 → $6\frac{13}{15}$

Add: $4\frac{1}{5} + 2\frac{2}{3}$ 14

Alternate Method:

$$\begin{array}{r} 4\frac{1}{5} \\ +2\frac{2}{3} \\ \hline \end{array} \rightarrow \begin{array}{r} 4 + \left(\frac{1}{5} \times \frac{3}{3}\right) \\ 2 + \left(\frac{2}{3} \times \frac{5}{5}\right) \\ \hline \end{array} \rightarrow \begin{array}{r} 4\frac{3}{15} \\ +2\frac{10}{15} \\ \hline 6\frac{13}{15} \end{array}$$

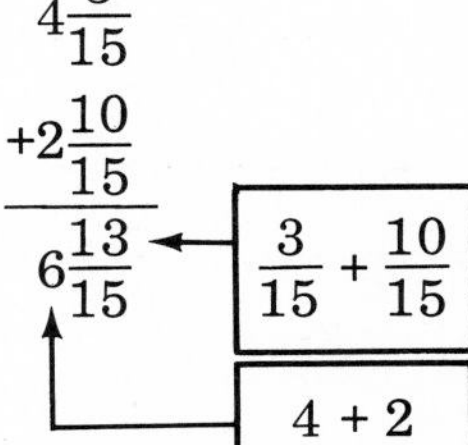

$\frac{3}{15} + \frac{10}{15}$

$4 + 2$

Add: $18\frac{4}{5} + 7\frac{1}{2}$ ⑮

Complete the following:

$$18\frac{4}{5} + 7\frac{1}{2} = \frac{94}{5} + \frac{15}{2} \leftarrow \boxed{\text{LCD is } 2 \times 5 = 10}$$

$$= \left(\frac{94}{5} \times \frac{\square}{\square}\right) + \left(\frac{15}{2} \times \frac{\bigcirc}{\bigcirc}\right)$$

$$= \frac{188}{10} + \frac{75}{10}$$

$$= \frac{\triangle}{10}$$

$$= 26\frac{3}{10}$$

$$10\overline{)263} = 26 \rightarrow 26\frac{3}{10}$$

$$\begin{array}{r} 26 \\ 10\overline{)263} \\ \underline{20} \\ 63 \\ \underline{60} \\ 3 \end{array}$$

Add: $18\frac{4}{5} + 7\frac{1}{2}$ ⑯

Complete the following:

Alternate Method:

$$\begin{array}{r} 18\frac{4}{5} \\ + \ 7\frac{1}{2} \\ \hline \end{array} \quad \rightarrow \quad \begin{array}{r} 18 + \left(\frac{4}{5} \times \frac{2}{2}\right) \\ 7 + \left(\frac{1}{2} \times \frac{5}{5}\right) \\ \hline \end{array} \quad \rightarrow \quad \begin{array}{r} 18\frac{\bigcirc}{10} \\ + \ 7\frac{\triangle}{10} \\ \hline 25\frac{13}{10} \end{array}$$

Rewriting the sum:

$$25\frac{13}{10} = 25 + \frac{13}{10}$$

$$= 25 + 1\frac{3}{10}$$

$$= 25 + 1 + \frac{3}{10}$$

$$= 26 + \frac{3}{10}$$

$$= 26\frac{3}{10}$$

Adding mixed numbers:

(17)

There is no "best" method; each method has its advantages and disadvantages. The alternate method is most useful when the numbers are large.

For example, adding $72\frac{3}{4}$ and $83\frac{4}{5}$:

$$
\begin{aligned}
72\frac{3}{4} + 83\frac{4}{5} &= \frac{291}{4} + \frac{419}{5} \\
&= \left(\frac{291}{4} \times \frac{5}{5}\right) + \left(\frac{419}{5} \times \frac{4}{4}\right) \\
&= \frac{1{,}455}{20} + \frac{1{,}676}{20} \\
&= \frac{3{,}131}{20} \\
&= 156\frac{11}{20}
\end{aligned}
$$

$$
\begin{array}{r l}
 & 156 \longrightarrow 156\frac{11}{20} \\
20 & \overline{)3{,}131} \\
 & \underline{20} \\
 & 113 \\
 & \underline{100} \\
 & 131 \\
 & \underline{120} \\
 & 11
\end{array}
$$

The equivalent fraction method is difficult because of the size of the numbers that result when the mixed numbers are changed to improper fractions.

Using the alternate method,

$$
\begin{array}{r}
72\frac{3}{4} \\
+83\frac{4}{5} \\
\hline
\end{array}
\quad \Rightarrow \quad
\begin{array}{r}
72 + \left(\frac{3}{4} \times \frac{5}{5}\right) \\
83 + \left(\frac{4}{5} \times \frac{4}{4}\right) \\
\hline
\end{array}
\quad \Rightarrow \quad
\begin{array}{r}
72\frac{15}{20} \\
+83\frac{16}{20} \\
\hline
155\frac{31}{20}
\end{array}
$$

$$
\begin{aligned}
155\frac{31}{20} &= 155 + \frac{31}{20} \\
&= 155 + 1\frac{11}{20} \\
&= 156\frac{11}{20}
\end{aligned}
$$

The alternate method has more reasonable numbers with which to work.

Subtract: $10\frac{4}{5} - 2\frac{3}{8}$ (18)

$$10\frac{4}{5} - 2\frac{3}{8} = \frac{54}{5} - \frac{19}{8} \longleftarrow \boxed{\text{LCD is } 2^3 \times 5 = 40}$$

$$= \left(\frac{54}{5} \times \frac{8}{8}\right) - \left(\frac{19}{8} \times \frac{5}{5}\right)$$

$$= \frac{432}{40} - \frac{95}{40}$$

$$= \frac{432 - 95}{40}$$

$$= \frac{337}{40}$$

$$= 8\frac{17}{40} \longleftarrow \boxed{\begin{array}{r} 8 \longrightarrow 8\frac{17}{40} \\ 40\overline{)337} \\ \underline{320} \\ 17 \end{array}}$$

Subtract: $10\frac{4}{5} - 2\frac{3}{8}$ (19)

Alternate method:

$$\begin{array}{r} 10\frac{4}{5} \\ \underline{-\ 2\frac{3}{8}} \end{array} \quad \Rightarrow \quad \begin{array}{r} 10 + \left(\frac{4}{5} \times \frac{8}{8}\right) \\ \underline{2 + \left(\frac{3}{8} \times \frac{5}{5}\right)} \end{array} \quad \Rightarrow \quad \begin{array}{r} 10\frac{32}{40} \\ \underline{-\ 2\frac{15}{40}} \\ 8\frac{17}{40} \end{array}$$

$$8\frac{17}{40} \longleftarrow \boxed{\frac{32}{40} - \frac{15}{40}}$$

$$8 \longleftarrow \boxed{10 - 2}$$

Subtract: $4\frac{1}{5} - 2\frac{2}{3}$ (20)

Complete the following:

$$4\frac{1}{5} - 2\frac{2}{3} = \frac{21}{5} - \frac{8}{3} \longleftarrow \boxed{\text{LCD is } 3 \times 5 = 15}$$

$$= \left(\frac{21}{5} \times \frac{\square}{\square}\right) - \left(\frac{8}{3} \times \frac{\bigcirc}{\bigcirc}\right)$$

$$= \frac{63}{15} - \frac{40}{15}$$

$$= \frac{63 - 40}{15}$$

$$= \frac{\triangle}{15}$$

$$= 1\frac{8}{15}$$

Subtract: $4\frac{1}{5} - 2\frac{2}{3}$ (21)

Alternate method:

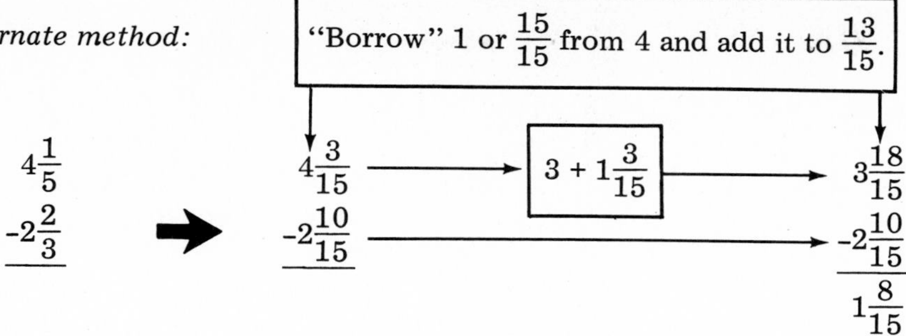

"Borrow" 1 or $\frac{15}{15}$ from 4 and add it to $\frac{13}{15}$.
$4\frac{1}{5}$
$-2\frac{2}{3}$
$4\frac{3}{15}$
$-2\frac{10}{15}$
$3 + 1\frac{3}{15}$
$3\frac{18}{15}$
$-2\frac{10}{15}$
$1\frac{8}{15}$

This last example shows the advantage of the equivalent fraction method if the numbers are small.

Subtract: $4\frac{9}{10} - 1\frac{5}{5}$ (22)

Complete the following:

$$4\frac{9}{10} - 1\frac{3}{5} = \frac{49}{\square} - \frac{8}{\bigcirc}$$

$$= \frac{49}{\square} - \left(\frac{8}{\bigcirc} \times \frac{2}{2}\right)$$

$$= \frac{49}{\square} - \frac{16}{\square}$$

$$= \frac{\triangle}{\square}$$

$$= 3\frac{3}{10}$$

To perform multiplication and division of mixed numbers, rewrite the mixed numbers as improper fractions and proceed in exactly the same manner as in multiplication and division of fractions. The results are always rewritten as mixed numbers, if necessary. (23)

Multiply: $3\frac{4}{5} \times 6\frac{2}{7}$ (24)

$$3\frac{4}{5} \times 6\frac{2}{7} = \frac{19}{5} \times \frac{44}{7}$$

$$= \frac{19 \times 44}{5 \times 7}$$

$$= \frac{836}{35}$$ ← Rewrite as a mixed number.

$$= 23\frac{31}{35}$$

Multiply: $5 \times 3\frac{2}{3}$ 25

Complete the following:

$$5 \times 3\frac{2}{3} = \frac{5}{1} \times \frac{\square}{3}$$

$$= \frac{\bigcirc}{3}$$ ← Rewrite as a mixed number.

$$= 18\frac{1}{3}$$

Divide: $4\frac{1}{2} \div 8\frac{1}{3}$ 26

$$4\frac{1}{2} \div 8\frac{1}{3} = \frac{9}{2} \div \frac{25}{3}$$

$$= \frac{9}{2} \times \frac{3}{25}$$

$$= \frac{27}{50}$$

Divide: $10\frac{1}{4} \div 4\frac{1}{4}$ 27

Complete the following:

$$10\frac{1}{4} \div 4\frac{1}{2} = \frac{\square}{4} \div \frac{\bigcirc}{2}$$

$$= \frac{\square}{4} \times \frac{2}{\bigcirc}$$

$$= \frac{82}{36}$$

$$= 2\frac{10}{36} \text{ or } 2\frac{5}{18}$$

To determine the power of a mixed number, first rewrite the mixed number as an improper fraction. (28)

For example, $\left(4\frac{1}{3}\right)^2$

$$\left(4\frac{1}{3}\right)^2 = \left(\frac{13}{3}\right)^2$$
$$= \left(\frac{13}{3} \times \frac{13}{3}\right)$$
$$= \frac{13 \times 13}{3 \times 3}$$
$$= \frac{169}{9}$$

← Rewrite as a mixed number.

$$= 18\frac{7}{9}$$

Determine the value of $\left(1\frac{7}{8}\right)^2$. (29)

Complete the following:

$$\left(1\frac{7}{8}\right)^2 = \left(\frac{\bigcirc}{8}\right)^2$$
$$= \frac{\bigcirc}{8} \times \frac{\bigcirc}{8}$$
$$= \frac{\bigcirc \times \bigcirc}{8 \times 8}$$
$$= \frac{225}{64}$$

← Rewrite as a mixed number.

$$= 3\frac{33}{64}$$

Practice Exercises (30)

Perform the indicated operations on mixed numbers.

1. $2\frac{3}{5} \times 6\frac{1}{3}$
2. $4\frac{5}{8} - 1\frac{3}{4}$
3. $10\frac{2}{7} \div 5\frac{1}{21}$
4. $9\frac{3}{10} + 4\frac{15}{16}$
5. $4\frac{1}{7} \div 2\frac{3}{14}$
6. $3\frac{1}{6} + 2\frac{2}{5}$
7. $3\frac{6}{7} \times 2\frac{4}{9}$
8. $3\frac{2}{3} - 2\frac{1}{5}$
9. $\left(1\frac{3}{4}\right)^2$
10. $\left(5\frac{2}{5}\right)^2$

To find the reciprocal of a mixed number, first change the number to fraction form. (31)

For example, what is the reciprocal of $3\frac{1}{5}$?

$$\text{Since}\quad 3\frac{1}{5} = \frac{16}{5},\ \text{the reciprocal is}\ \frac{5}{16}.$$

$$\text{Check: } 3\frac{1}{5} \times \frac{5}{16} = \frac{16}{5} \times \frac{5}{16} = 1.$$

What is the reciprocal of $2\frac{3}{16}$? 

Complete the following:

$$\text{Since}\quad 2\frac{3}{16} = \frac{\bigcirc}{\square},\ \text{the reciprocal is}\ \frac{\square}{\bigcirc}.$$

$$\text{Check: } 2\frac{3}{16} \times \frac{16}{35} = \frac{35}{16} \times \frac{16}{35} = 1.$$

Practice Exercises 

Write the reciprocal for each of the following:

1. $1\frac{2}{3}$
2. $7\frac{9}{10}$
3. $181\frac{1}{2}$
4. $16\frac{1}{4}$

POST-TEST

1. Rewrite each mixed number as a fraction and rewrite each fraction as a mixed number.
   (a) $\frac{25}{8}$
   (b) $6\frac{3}{7}$
   (c) $20\frac{1}{3}$
   (d) $\frac{100}{13}$
2. Perform the indicated on mixed numbers for each of the following:
   (a) $3\frac{3}{4} \times 5\frac{2}{3}$
   (b) $4\frac{3}{5} - 2\frac{1}{6}$
   (c) $7\frac{7}{12} + 9\frac{4}{15}$
   (d) $3\frac{1}{2} \div 2\frac{1}{2}$
   (e) $\left(3\frac{1}{3}\right)^3$
3. Write the reciprocal for the mixed number $11\frac{2}{3}$.

LESSON 13
FRACTIONAL PARTS AND PROBABILITY

Objectives:

1. Write a fractional part to describe an indicated portion of a given whole. (1–7)
2. Solve word problems that contain fractional parts. (8–13)
3. Determine the probability of an event or events given all possible outcomes. (14–21)

Vocabulary:

Fractional Parts
Probability

PRE-TEST

1. Given the following numbers:

   5, 6, 9, 11, 12, 14, 16, 17

   (a) Write the fractional part of the given numbers that are even numbers.
   (b) Write the fractional part of the given numbers that are prime numbers.
2. Solve the following word problem:
   A community is ethnically comprised of the following:
   $\frac{7}{12}$ Caucasian, $\frac{3}{12}$ Chicano, and the remainder Asian. If the community has a population of 240,000, how many persons of Asiatic descent does the community have?
3. Consider tossing a fair pair of dice.
   (a) What is the probability of tossing a sum of 6 on the upper faces?
   (b) What is the probability of tossing (on one toss) a sum of 7 *or* a sum of 11?

A fraction can be used to indicate what part of a whole or total another number is. For example, what part of a dozen eggs is 8 eggs? This is the same as asking, what *fractional part* of a dozen is 8? To answer this, think of 8 eggs out of 12 eggs per dozen as $\frac{8}{12}$ or $\frac{2}{3}$ of a dozen. (1)

Naming Fractional Parts: ②

To name a *fractional part*, write a fraction with the number of parts as the numerator and the total of number of parts in the equally divided whole as the denominator. If necessary, reduce this fraction to lowest terms.

Naming fractional parts: ③

Each of the figures below has been subdivided into a certain number of equal parts. The fraction below each of the figures names the fractional part of the whole that is shaded.

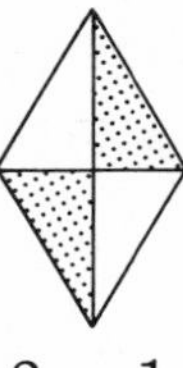

$\frac{2}{4}$ or $\frac{1}{2}$

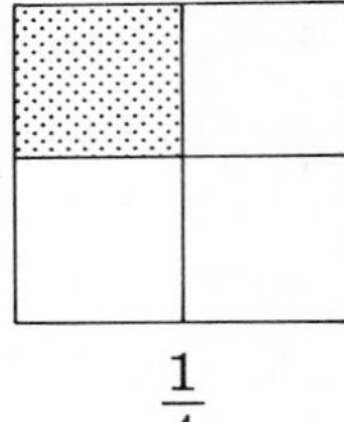

$\frac{1}{4}$

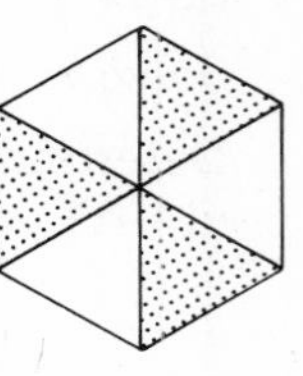

$\frac{3}{6}$ or $\frac{1}{2}$

Notice that each numerator indicates the number of shaded parts and each denominator indicates the number of equal subdivisions of the whole.

When a single die is thrown a number from 1 to 6 will appear on the top of the die. What fractional part of these numbers are prime numbers? ④

1, 2, 3, 4, 5, 6

The prime numbers between 1 and 6 are 2, 3, and 5. Thus, the prime numbers represent $\frac{3}{6}$ or $\frac{1}{2}$ of the numbers on the faces of a die.

Write a fraction to indicate what fractional part of the following letters are vowels: ⑤

A, B, C, D, E, F, G, H

The fractional part of the letters that are vowels is $\frac{\square}{\bigcirc}$ or $\frac{\triangle}{\text{⬡}}$, reduced to lowest terms.

(The vowels in the alphabet are A, E, I, O, and U.)

When two dice are thrown the sum of the dots showing on the uppermost faces can vary from a minimum of 2 to a maximum of 12. The table below shows all the possible sums and how they can occur. To simplify matters, the dice are labeled "red" and "green." ⑥

Dots Showing on the Upper Face of the Red Die

Dots showing on the Upper Face of the Green Die	1	2	3	4	5	6
1	2	3	4	5	6	7
2	3	4	5	6	7	8
3	4	5	6	7	8	9
4	5	6	7	8	9	10
5	6	7	8	9	10	11
6	7	8	9	10	11	12

The table represents all the possible outcomes when the two dice are thrown. What fractional part of all the outcomes results in a sum of 8? First, notice that there are 36 possible outcomes, 5 of which result in a sum of 8. Thus, $\frac{5}{36}$ of all the possible outcomes result in a sum of 8.

Complete the following using the table:

Write a fraction to indicate what fractional part of the outcomes result in a sum of 2.

Write a fraction to indicate what fractional part of the outcomes result in a sum of 7.

$\frac{\triangle}{\square}$ or $\frac{1}{6}$, reduced to lowest terms

---

Practice Exercises 7

Write the fractional part of the whole indicated by the shaded parts for each of the following:

1. 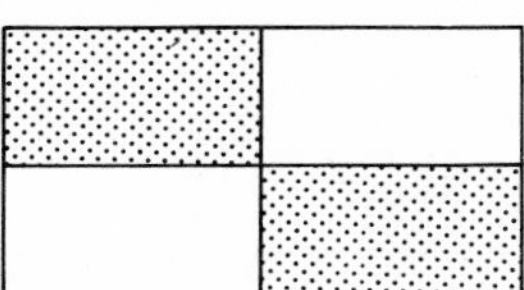

2. 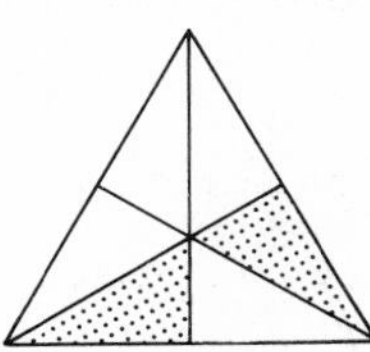

3. 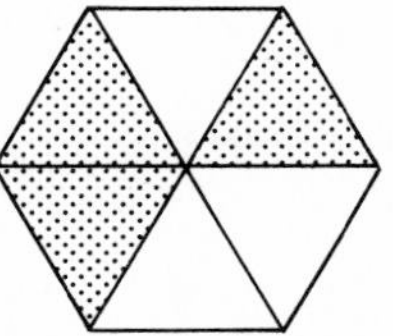

4. 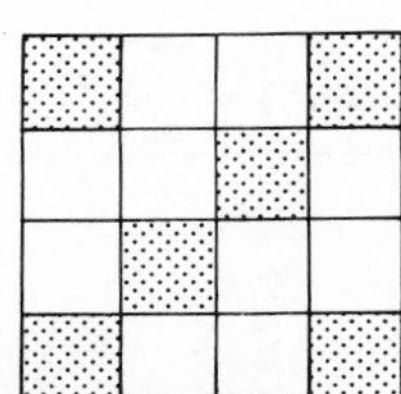

Relationships between fractional parts and the whole: (8)

If a gas tank is $\frac{2}{5}$ full, then it is $1 - \frac{2}{5}$ or $\frac{3}{5}$ empty, since the total capacity of the tank represents one whole tank.

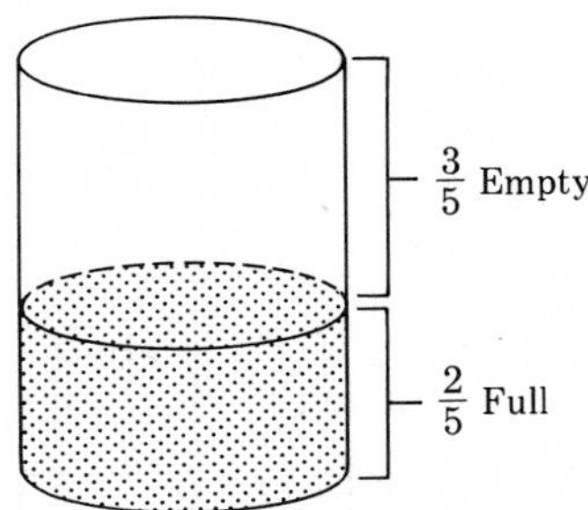

Any whole must be equal to the sum of its fractional parts.

$$\frac{2}{5} + \frac{3}{5} = 1$$

What fractional part of the following family budget is set aside for savings? 

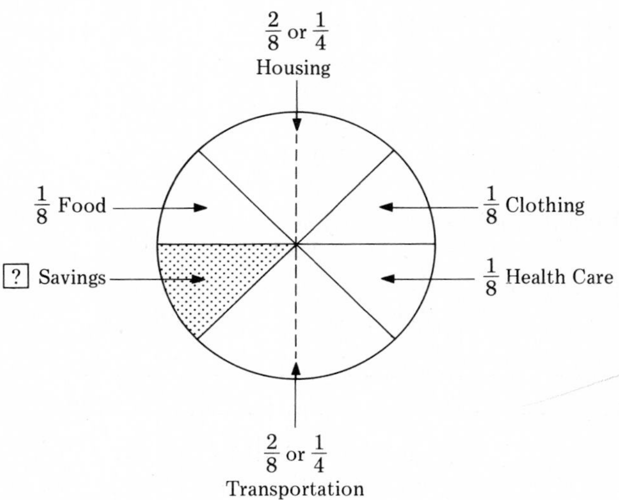

Recall from the last frame that the whole is the sum of its fractional parts.

$$\boxed{?} = 1 - \left(\frac{1}{8} + \frac{2}{8} + \frac{1}{8} + \frac{1}{8} + \frac{2}{8}\right)$$

$$\boxed{?} = 1 - \frac{7}{8}$$

$$\boxed{?} = \frac{1}{8}, \text{ fractional part for savings}$$

Word Problems with Fractional Parts: (10)

If a quality control technician examines 150 television sets and rejects 10 of them, what fractional part is acceptable?

How many sets are acceptable? 150 – 10 = 140

140 is what fractional part of 150?

$$\frac{140}{150} \text{ or } \frac{14}{15}$$

Thus, $\frac{14}{15}$ of the total number of television sets are acceptable.

Word Problems with Fractional Parts: (11)

If a gasoline tank holds 22 gallons and the gauge indicates that the tank is $\frac{1}{4}$ full, how many gallons are required to fill the tank?

The empty portion of the tank is $1 - \frac{1}{4}$ or $\frac{3}{4}$ of the whole tank.

Three-fourths of 22 gallons must be added to fill the tank.

$$\frac{3}{4} \text{ of 22 gallons} = \frac{3}{4} \times 22$$
$$= \frac{66}{4}$$
$$= \frac{33}{2} \text{ or } 16\frac{1}{2} \text{ gallons}$$

Recall that the word "of" indicates multiplication.

Word Problems with Fractional Parts: 

An automobile's exhaust has the following components, expressed as fractional parts. What fractional part of the exhaust is nitrous oxide?

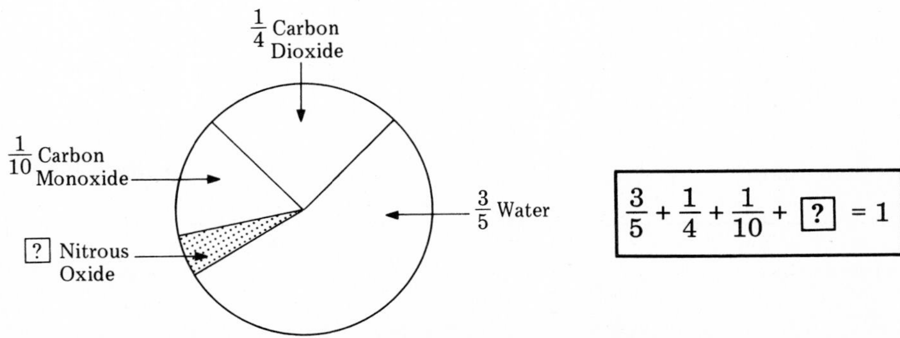

$$\boxed{?} = 1 - \left(\frac{3}{5} + \frac{1}{4} + \frac{1}{10}\right)$$

$$\boxed{?} = \frac{20}{20} - \left(\frac{12 + 5 + 2}{20}\right)$$

$$\boxed{?} = \frac{20}{20} - \frac{19}{20}$$

$$\boxed{?} = \frac{1}{20} \text{ nitrous oxide}$$

The known fractional parts are first added; this sum is then subtracted from the whole. Notice that these subdivisions each represent the reduced fractional parts of a whole divided into 20 equal parts; that is, the LCD is 20.

Practice Exercises 

Solve each word problem.

1. Leon's gasoline tank is $\frac{3}{8}$ full. How many gallons of gasoline will it take to fill the tank if it holds 20 gallons?
2. Grandfather Sutherland revised his will as follows:
$\frac{1}{5}$ of his estate is to go to each of his 3 grandchildren, $\frac{1}{4}$ to his daughter, and the remainder to the Cancer Society. If his estate has a 75,000-dollar value, how much will the Cancer Society receive?
3. The United States has a population of approximately 280,000,000 people, of which $\frac{1}{4}$ of the population is under 18 years, $\frac{3}{8}$ is 19 to 40 years of age, $\frac{1}{4}$ is 41 to 65 years of age, and the remainder of the population is over 65 years. Calculate the number of people in the United States who are over 65 years of age.

Probability: 

Whenever a coin is tossed, there are only two possible ways that it can land: "heads" up or "tails" up.

"Heads" Up	"Tails" Up

The chance that a tossed coin will land "heads" up is one out of the two possible ways it can land (the outcomes). Expressing this another way, it is said that the *probability* of getting a "head" is 1 out of 2 or $\frac{1}{2}$. The probability of getting a "tail" is also 1 out of 2 or $\frac{1}{2}$. If the coin were tossed many, many times, it would land "heads" up $\frac{1}{2}$ of the time and "tails" up the other $\frac{1}{2}$ of the time.

Probability is both an idea and a number. If someone asks what is the probability of a certain event, he is asking for a measure of the likelihood of the occurrence of that event. 

$$\text{Probability of an event} = \frac{\text{Number of favorable outcomes}}{\text{Total number of possible outcomes}}$$

Probability, the measure of likelihood, is given in numerical form. In fact, it represents the fractional part of the ways that an event can happen out of all possible similar events that could happen. The total outcomes are all the ways that the event can happen plus all the ways that the event can fail to happen.

All measures of probability are greater than or equal to zero and less than or equal to 1. 

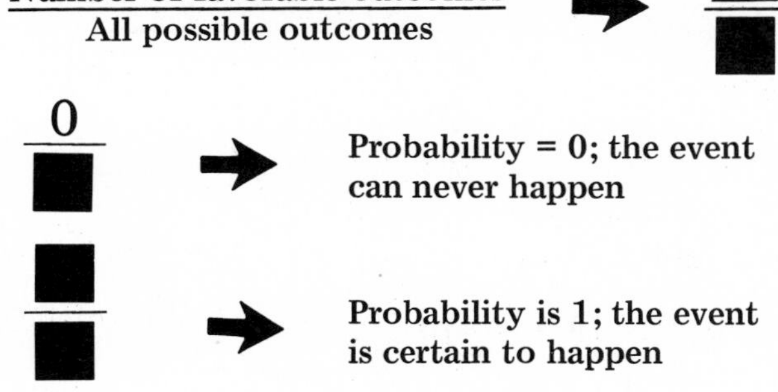
Number of favorable outcomes
All possible outcomes
0
Probability = 0; the event can never happen
Probability is 1; the event is certain to happen

Probability: 

Suppose a bag contains 4 red marbles and 6 white marbles. If a person draws only 1 marble, what is the probability that he will draw a red marble?

$$\text{Probability of drawing a red marble} = \frac{\text{4 red marbles}}{\text{4 red marbles + 6 white marbles}}$$

$$= \frac{4}{10} \text{ or } \frac{2}{5}$$

Total Number of favorable outcomes (numerator)

Total number of possible outcomes (denominator)

There are 4 ways to draw a red marble out of 10 possible draws (both red and white). The probability of drawing a red marble is 4 out of 10 or $\frac{2}{5}$.

Problems of Probability: (18)

Complete the following:

A bag contains 4 red marbles, 6 white marbles, and 2 blue marbles. What is the probability of selecting on the first draw:

A. A white marble?

Favorable outcomes

$$\text{Probability of a white marble} = \frac{6 \text{ white}}{4 \text{ red} + 6 \text{ white} + 2 \text{ blue}} = \frac{6}{12} = \frac{1}{2}$$

All possible Outcomes

B. A red marble?

$$\text{Probability of a red marble} = \frac{\bigcirc}{12} \text{ or } \frac{1}{3}$$

C. White or blue marble?

Favorable outcomes

$$\text{Probability of white or blue marble} = \frac{6 \text{ white} + 2 \text{ blue}}{4 \text{ red} + 6 \text{ white} + 2 \text{ blue}}$$

All possible outcomes

$$= \frac{8}{12} \text{ or } \frac{2}{3}$$

D. Red or white?

$$\text{Probability of red or white marble} = \frac{\bigcirc + \square}{12}$$

$$= \frac{10}{12} \text{ or } \frac{5}{6}$$

E. Red, white, or blue?

Favorable outcomes

$$\text{Probability of red, white, or blue marble} = \frac{4 \text{ red} + 6 \text{ white} + 2 \text{ blue}}{4 \text{ red} + 6 \text{ white} + 2 \text{ blue}}$$

All possible outcomes

$$= \frac{12}{12} \text{ or } 1$$

Probability Problems:

The chart below represents a standard deck of cards showing the four suits and the individual cards.

	Suit	Face Value												
	Clubs	Ace	2	3	4	5	6	7	8	9	10	Jack	Queen	King
Suits	Hearts	Ace	2	3	4	5	6	7	8	9	10	Jack	Queen	King
	Spades	Ace	2	3	4	5	6	7	8	9	10	Jack	Queen	King
	Diamonds	Ace	2	3	4	5	6	7	8	9	10	Jack	Queen	King
												Face Cards		

Complete the following:

If a single card is drawn from a shuffled deck, find

A. The probability that it is an 8 of clubs:

$$\frac{1}{52}$$

1 ← Numbers of 8's of clubs

52 ← Total number of cards in the deck

B. The probability that it is any diamond:

$$\frac{\triangle}{52} \text{ or } \frac{1}{4}$$

C. The probability that it is any face card:

$$\frac{\bigcirc}{\square} \text{ or } \frac{3}{13}$$

D. The probability that it is any red card (diamonds and hearts are "red" suits):

$$\frac{\hexagon}{\square} \text{ or } \frac{1}{2}$$

Probability Problems:

The table below is exactly the same as the table shown in Frame 6, showing all possible sums, and how they can occur, when two dice are rolled.

Points Showing on the Upper Face of the Red Die (columns); Points Showing on the Upper Face of the Green Die (rows)

	1	2	3	4	5	6
1	2	3	4	5	6	7
2	3	4	5	6	7	8
3	4	5	6	7	8	9
4	5	6	7	8	9	10
5	6	7	8	9	10	11
6	7	8	9	10	11	12

Complete the following:

If the two dice are rolled, find

A. The probability that the sum is 8:

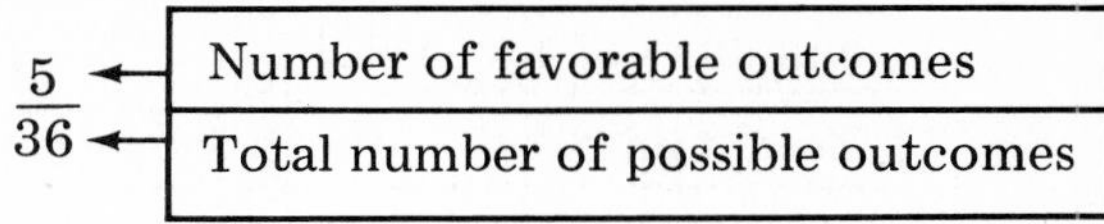

B. The probability that the sum is 9:

$\frac{\triangle}{\square}$ or $\frac{1}{9}$

C. The probability that the sum is 7:

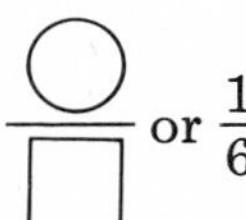 or $\frac{1}{6}$

Recall that probability is a measure of the likelihood of the occurrence of an outcome. Which is the more likely to occur: an 8, a 9, or a 7 appearing on a single roll of the dice?

Practice Exercises

1. Refer to the table in Frame 19 to find the following probabilities.
   Assume that the deck is well shuffled and that only one card is drawn. Also, recall that spades and clubs are black in color and that diamonds and hearts are red.
   (a) The probability of drawing any king
   (b) The probability of drawing any heart
   (c) The probability of drawing any black card
   (d) The probability of drawing any black *or* any red card
   (e) The probability of drawing any black ace
2. Refer to the table in Frame 20 to find the following probabilities.
   (a) The probability that the sum is 2
   (b) The probability that the sum is an even number
   (c) The probability that the sum is an odd number
   (d) The probability that the sum is a prime number greater than 5
   (e) The probability that the sum is either an odd number *or* an even number
3. Professor Welge, a psychologist, trained 50 mice to run through a maze. 10 were trained to make left turns, 15 were trained to make right turns, and 25 were trained to go straight ahead. If a mouse were selected at random, what is the probability it would do the following?
   (a) Turn right
   (b) Go straight ahead
   (c) Turn left
   (d) Turn right *or* go straight ahead
   (e) Turn left *or* turn right

4. While waiting at the bus station the Duncan brothers played cards and asked the following questions. Find the answers to their questions.
   (a) Danny asked, "What is the probability of drawing the queen of spades in one draw from a shuffled deck?"
   (b) Dennis questioned, "Which probability is greater: drawing a face card or drawing a 2, 3, 4, *or* 5?"
   (c) Donald wondered, "What is the probability of drawing either a heart *or* the queen of spades?"
   (d) Dunk, thinking about Danny's question, asked, "What is the probability that the queen of spades *will not* be drawn from the deck?"

POST-TEST

1. Given the following letters:

   A, B, C, D, E, F, G

   (a) Write the fractional part of the given letters that are vowels.
   (b) Write the fractional part of the given letters that are used to form the word "beef."
2. Solve the following word problem:

   A worker earns 300 dollars per week. If $\frac{1}{5}$ of the salary is withheld for income taxes, $\frac{2}{25}$ for retirement, and $\frac{1}{20}$ for savings bonds, what is the amount of the worker's take home pay for 1 week?
3. What is the probability of drawing either the king of hearts *or* a spade on a single draw from a standard deck of cards?

DRILL EXERCISES: UNIT 3

Perform the following additions and subtractions:

1. $\frac{2}{3} + \frac{5}{3}$

2. $\frac{5}{9} + \frac{2}{9}$

3. $\frac{1}{8} + \frac{4}{8}$

4. $\frac{6}{13} - \frac{5}{13}$

5. $\frac{11}{5} + \frac{3}{5}$

6. $\frac{5}{7} + \frac{13}{7}$

7. $\frac{4}{3} - \frac{1}{3}$

8. $\frac{11}{2} + \frac{4}{2}$

Perform the following additions and subtractions:

9. $\frac{4}{5} + \frac{2}{3}$

10. $\frac{7}{8} + \frac{3}{10}$

11. $\frac{3}{14} + \frac{1}{6}$

12. $2 + \frac{3}{10}$

13. $1 + \frac{4}{150}$

14. $\frac{5}{6} + \frac{1}{10}$

15. $\frac{5}{49} + \frac{15}{70}$

16. $\frac{3}{4} + \frac{5}{6} + \frac{2}{9}$

17. $\frac{4}{5} + \frac{3}{10} + \frac{2}{3}$

18. $\frac{3}{4} - \frac{5}{12}$

19. $\frac{7}{8} - \frac{3}{16}$

20. $\frac{5}{8} - \frac{1}{12}$

21. $\frac{7}{12} - \frac{3}{16}$

22. $\frac{13}{16} - \frac{7}{10}$

23. $3 - \frac{9}{20}$

24. $\frac{1}{2} - \frac{1}{6}$

25. $\frac{4}{5} - \frac{3}{4}$

26. $5 - \frac{3}{4}$

Perform the following divisions:

27. $\dfrac{\frac{1}{4}}{\frac{1}{3}}$

28. $\dfrac{\frac{2}{5}}{\frac{9}{16}}$

29. $\dfrac{\frac{7}{8}}{7}$

30. $\dfrac{6}{\frac{7}{8}}$

31. $\frac{1}{2} \div 10$

32. $\frac{1}{3} \div 36$

33. $8 \div \frac{1}{2}$

34. $15 \div \frac{1}{3}$

35. $\frac{1}{2} \div \frac{2}{3}$

36. $\frac{7}{8} \div \frac{7}{16}$

37. $\frac{3}{16} \div \frac{5}{12}$

38. $\frac{5}{12} \div \frac{3}{16}$

39. $\frac{13}{15} \div \frac{39}{5}$

40. $\frac{5}{11} \div \frac{2}{7}$

41. $\frac{7}{8} \div \frac{7}{8}$

42. $\frac{23}{4} \div \frac{23}{4}$

43. $\frac{3}{5} \div \frac{5}{3}$

44. $\frac{8}{7} \div \frac{7}{8}$

Rewrite each fraction as a mixed number:

45. $\frac{17}{3}$

46. $\frac{9}{4}$

47. $\frac{14}{9}$

48. $\frac{11}{6}$

49. $\frac{13}{10}$

50. $\frac{67}{12}$

51. $\frac{25}{16}$

52. $\frac{25}{8}$

Write each mixed number as a fraction:

53. $5\frac{3}{8}$

54. $5\frac{1}{7}$

55. $1\frac{1}{9}$

56. $9\frac{9}{10}$

57. $6\frac{2}{3}$

58. $3\frac{1}{14}$

59. $19\frac{3}{4}$

60. $18\frac{3}{5}$

Perform the indicated operations. Any improper fraction should be expressed as a mixed number.

61. $2\frac{1}{3} + 5\frac{3}{7}$

62. $4\frac{3}{20} + 2\frac{1}{8}$

63. $3\frac{4}{5} + 1\frac{1}{2}$

64. $4\frac{3}{7} + 1\frac{3}{4}$

65. $1\frac{1}{5} - \frac{4}{7}$

66. $1\frac{1}{9} - \frac{1}{27}$

67. $23 - 3\frac{1}{8}$

68. $15 - 12\frac{6}{7}$

69. $2\frac{2}{3} \times \frac{5}{8}$

70. $3\frac{2}{3} \times \frac{6}{5}$

71. $5\frac{3}{5} \times 4\frac{1}{7}$

72. $6\frac{2}{5} \times 3\frac{1}{8}$

73. $5\frac{2}{5} \div \frac{1}{8}$

74. $\frac{1}{2} \div 2\frac{1}{7}$

75. $4\frac{7}{8} \div 3\frac{1}{4}$

76. $4\frac{1}{6} \div 3\frac{1}{3}$

77. $\left(2\frac{1}{2} + 3\frac{1}{4}\right) \div \frac{7}{10}$

78. $\frac{7}{10} \div \left(\frac{3}{4} - \frac{1}{5}\right)$

79. Mr. Karl spent 800 dollars improving and repairing his house. He spent 300 dollars for paint, 200 dollars for wallpaper, 100 dollars for plumbing, and 100 dollars for wiring. The rest went for miscellaneous expenses. Write the fractional part of his expenses for each of the following: (a) paint, (b) wallpaper, (c) plumbing, (d) wiring, (e) miscellaneous expenses

80. Consider a standard deck of 52 cards. Write a fraction to represent that part of the deck which is: (a) clubs, (b) tens, (c) black, (d) face cards

81. In Pauls Valley, Oklahoma, the only television dealer in town sold 325 television sets of which 75 were color sets. What fractional part were black and white sets?

82. Mr. Duncan bought stock which closed at $38\frac{1}{2}$ dollars on Monday and it gained $1\frac{3}{4}$ dollars on the following day. What was the closed price on Tuesday?

83. A trucker agrees to haul $1\frac{3}{4}$ tons of copper wire, $4\frac{1}{3}$ tons of sheet steel, and $2\frac{1}{6}$ tons of lead pipe. What is the total weight of the cargo?

84. A copper tube $6\frac{1}{16}$ inches long must fit inside another piece of tubing which is $5\frac{7}{8}$ inches long. How much of the copper tubing must be cut away to be the same length as the other?

85. Find the thickness of the wall of the plastic pipe shown in the cross section below:

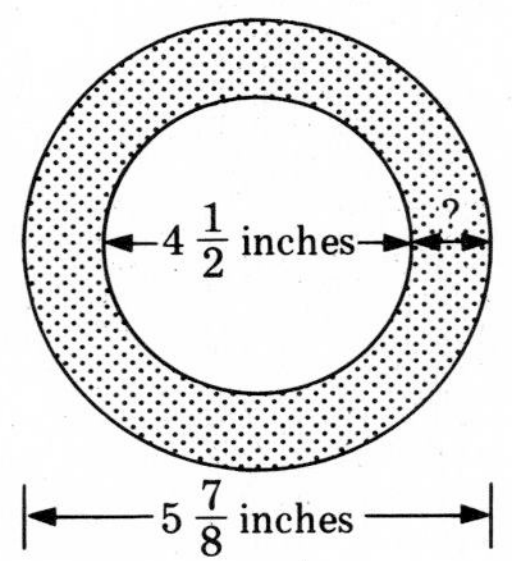

86. If a jet plane flies at 600 miles per hour for $1\frac{3}{4}$ hours, how far does it travel?

87. A house worth 42,000 dollars was assessed for $\frac{1}{4}$ of its value. What is the assessed value of the house?

88. How many pieces of wood, each $1\frac{3}{4}$ feet long, can be cut from a board 14 feet long?

89. The total weight of 3 sirloin steaks is $4\frac{1}{2}$ pounds. What is the weight of each of 6 equal servings?

90. What is the probability of rolling a pair of dice and obtaining a sum of 2 on the upper faces?

91. What is the probability of drawing a queen on one draw from a shuffled deck?

92. What is the probability of drawing a club on one draw from a shuffled deck?

93. What is the probability of rolling a sum of 8 *or* a sum of 7 on one toss of a pair of dice?

94. Ten cards, numbered from 1 to 10, are placed in a hat. What is the probability that on the first draw:
   (a) The card marked 5 is drawn?
   (b) A card marked with an odd number is drawn?
   (c) The number on the card will be greater than 4 but less than 9?

95. A bag contains 10 yellow chips, 8 black chips, and 5 red chips. What is the probability that the first draw is:

(a) A red chip?
(b) A yellow chip?
(c) A red *or* yellow chip?
(d) A green chip?
(e) A yellow, *or* black, *or* red chip?

SELF-TEST: UNIT 3

1. Add: $\frac{12}{23}+\frac{9}{23}$
   (a) $\frac{21}{46}$
   (b) $\frac{21}{23}$
   (c) $\frac{22}{23}$
   (d) $\frac{20}{23}$
   (e) None of the above

2. Subtract: $\frac{13}{42}-\frac{5}{42}$
   (a) $\frac{4}{21}$
   (b) $\frac{3}{16}$
   (c) $\frac{3}{7}$
   (d) $\frac{1}{6}$
   (e) None of the above

3. Divide: $12 \div \frac{3}{4}$.
   (a) 9
   (b) 16
   (c) $\frac{1}{9}$
   (d) $\frac{1}{16}$
   (e) None of the above

4. Find the sum: $\frac{1}{6}+\frac{2}{15}$
   (a) $\frac{1}{7}$
   (b) $\frac{1}{45}$
   (c) $\frac{13}{15}$
   (d) $\frac{3}{10}$
   (e) None of the above

5. Add: $\frac{1}{3}+\frac{2}{5}+\frac{3}{8}$
   (a) $\frac{97}{120}$
   (b) $\frac{219}{311}$

(c) $\frac{133}{120}$

(d) $\frac{7}{8}$

(e) None of the above

6. Subtract $\frac{3}{12}$ from $\frac{7}{8}$.

(a) $\frac{5}{8}$

(b) $\frac{3}{4}$

(c) $\frac{1}{2}$

(d) 1

(e) None of the above

7. Which of the following is equivalent to $\dfrac{\frac{5}{8}}{\frac{2}{3}}$?

(a) $\frac{2}{3} \div \frac{5}{8}$

(b) $\frac{5}{8} \div \frac{2}{3}$

(c) $\frac{2}{3} \div \frac{8}{5}$

(d) $\frac{5}{8} \div \frac{3}{2}$

(e) None of the above

8. The reciprocal of $\frac{7}{12}$ is:

(a) $\frac{12}{7}$

(b) $\dfrac{1}{\frac{12}{7}}$

(c) 12×7

(d) $7 \div 12$

(e) None of the above

9. Divide: $\frac{6}{15} \div \frac{7}{10}$

(a) $\frac{7}{25}$

(b) $\frac{1}{5}$

(c) $\frac{4}{7}$

(d) $\frac{7}{4}$

(e) None of the above

10. Write $\frac{36}{25}$ as a mixed number.

(a) $\frac{6}{5}$

(b) $1\frac{1}{5}$

(c) $1\frac{9}{25}$

(d) $1\frac{11}{25}$

(e) None of the above

11. Write $3\frac{8}{9}$ as a fraction.

(a) $\frac{20}{9}$

(b) $\frac{31}{9}$

(c) $\frac{35}{9}$

(d) $\frac{33}{9}$

(e) None of the above

12. Find the product: $1\frac{1}{2} \times 2\frac{3}{8}$

(a) $\frac{12}{19}$

(b) $9\frac{1}{2}$

(c) $2\frac{3}{8}$

(d) 3

(e) None of the above

13. Find the sum: $3\frac{2}{5} + 4\frac{7}{8}$

(a) $8\frac{2}{5}$

(b) $8\frac{11}{40}$

(c) $8\frac{23}{40}$

(d) $7\frac{9}{13}$

(e) None of the above

14. Divide: $3\frac{13}{14} \div 5\frac{5}{6}$

(a) $\frac{78}{175}$

(b) $22\frac{11}{12}$

(c) $\frac{106}{245}$

(d) $\frac{33}{49}$

(e) None of the above

15. Find the value of $\left(1\frac{1}{8}\right)^2$.

(a) $1\frac{1}{64}$

(b) $1\frac{9}{16}$

(c) $1\frac{17}{64}$
(d) $2\frac{1}{4}$
(e) None of the above

16. Find the reciprocal of $6\frac{3}{5}$.
(a) $\frac{5}{33}$
(b) $\frac{33}{5}$
(c) $6\frac{5}{3}$
(d) $\frac{5}{21}$
(e) None of the above

17. Given the list of numbers: 2, 5, 8, 11, 14, 15, 18. The fractional part of the list which is even numbers is:
(a) $\frac{1}{2}$
(b) $\frac{3}{7}$
(c) $\frac{5}{8}$
(d) $\frac{4}{7}$
(e) None of the above

18. There are 22,500 students at Golden West College. If $\frac{1}{10}$ of the students are under 18 years of age, $\frac{1}{5}$ are between 18 and 25 years, and $\frac{2}{5}$ are between 25 and 49, how many students at Golden West are over 49?
(a) 6,750
(b) 15,750
(c) 9,000
(d) 13,500
(e) None of the above

19. What is the probability of drawing the ace of spaces *or* any ten on a single draw from a standard deck of cards?
(a) $\frac{1}{13}$
(b) $\frac{1}{52}$
(c) $\frac{5}{52}$
(d) $\frac{4}{52}$
(e) None of the above

20. What is the probability of a coin's landing "heads" up *or* "tails" up on a single toss?
(a) $\frac{1}{4}$
(b) $\frac{1}{2}$
(c) 0
(d) 1
(e) None of the above

LESSON 10:

Pre-test

1. (a) $\frac{5}{6}$ (b) $\frac{11}{15}$
2. (a) $\frac{1}{5}$ (b) $\frac{1}{7}$
3. (a) $\frac{92}{105}$ (b) $\frac{37}{84}$
4. (a) $\frac{2}{5}$ (b) $\frac{11}{16}$

Practice Exercises

Frame 9

1. $\frac{5}{7}$ 2. $\frac{4}{3}$ or $1\frac{1}{3}$ 3. $\frac{7}{11}$ 4. 9

5. $\frac{6}{7}$ 6. $\frac{8}{13}$ 7. 1 8. $\frac{49}{53}$

Frame 18

1. $\frac{27}{40}$ 2. $\frac{41}{24}$ or $1\frac{17}{24}$ 3. $\frac{5}{21}$ 4. $\frac{20}{21}$

5. $\frac{17}{30}$ 6. $\frac{5}{4}$ or $1\frac{1}{4}$ 7. $\frac{7}{20}$ 8. $\frac{73}{60}$ or $1\frac{13}{60}$

Frame 21

1. $\frac{22}{45}$ 2. $\frac{9}{20}$ 3. $\frac{2}{105}$ 4. $\frac{47}{42}$ or $1\frac{5}{42}$

Post-test

1. (a) $\frac{12}{11}$ or $1\frac{1}{11}$ (b) $\frac{7}{8}$
2. (a) $\frac{5}{9}$ (b) $\frac{2}{7}$
3. (a) $\frac{61}{84}$ (b) $\frac{137}{108}$ or $1\frac{29}{108}$
4. (a) $\frac{19}{48}$ (b) $\frac{5}{9}$

LESSON 11:

Pre-test

1. (a) $\frac{97}{2}$ or $48\frac{1}{2}$ (b) $\frac{5}{1}$ or 5
2. $\frac{15}{2}$ or $7\frac{1}{2}$ 3. $\frac{12}{11}$ or $1\frac{1}{11}$

4. $\frac{3}{14}$

5. (a) $\dfrac{\frac{7}{9}}{\frac{3}{4}}$ (b) $\frac{1}{2} \div \frac{2}{3}$

Practice Exercises

Frame 3

1. $\frac{8}{5}$ or $1\frac{3}{5}$
2. $\frac{9}{1}$ or 9
3. $\frac{1}{7}$
4. $\frac{2}{7}$

Frame 12

1. $\frac{6}{5}$ or $1\frac{1}{5}$
2. $\frac{3}{8}$
3. $\frac{7}{1{,}000}$
4. 68
5. $\frac{28}{3}$ or $9\frac{1}{3}$
6. $\frac{5}{6}$
7. $\frac{100}{7}$ or $14\frac{2}{7}$
8. $\frac{5}{24}$
9. $\frac{3}{5}$

Frame 16

1. $\dfrac{\frac{3}{4}}{\frac{5}{6}}$
2. $\frac{3}{15} \div \frac{2}{7}$
3. $\dfrac{\frac{2}{3}}{5}$
4. $\frac{2}{3} \div 7$
5. $\dfrac{1}{\frac{13}{25}}$
6. $11 \div \frac{8}{3}$

Post-test

1. (a) 4 (b) $\frac{11}{6}$ or $1\frac{5}{6}$
2. $\frac{25}{12}$ or $2\frac{1}{12}$
3. $\frac{35}{12}$ or $2\frac{11}{12}$
4. $\frac{5}{66}$
5. (a) $\dfrac{\frac{4}{15}}{\frac{3}{7}}$ (b) $\frac{5}{6} \div \frac{1}{2}$

LESSON 12:

Pre-test

1. (a) $2\frac{2}{5}$ (b) $\frac{34}{9}$ (c) $\frac{12}{7}$ (d) $1\frac{3}{47}$
2. (a) $8\frac{9}{20}$ (b) $2\frac{7}{12}$ (c) $9\frac{11}{20}$ (d) $1\frac{29}{51}$ (e) $3\frac{22}{49}$
3. $\frac{5}{43}$

Practice Exercises

Frame 11

1. $\frac{26}{3}$
2. $16\frac{1}{5}$
3. $\frac{44}{7}$
4. $16\frac{1}{3}$

5. $\frac{37}{2}$ 6. $2\frac{3}{4}$ 7. $\frac{4}{3}$ 8. $1\frac{3}{25}$

Frame 30

1. $16\frac{7}{15}$ 2. $2\frac{7}{8}$ 3. $2\frac{2}{53}$ 4. $14\frac{19}{80}$

5. $1\frac{27}{31}$ 6. $5\frac{17}{30}$ 7. $9\frac{3}{7}$ 8. $1\frac{7}{15}$

9. $3\frac{1}{16}$ 10. $29\frac{4}{25}$

Frame 33

1. $\frac{3}{5}$ 2. $\frac{10}{79}$ 3. $\frac{2}{363}$ 4. $\frac{4}{65}$

Post-test

1. (a) $3\frac{1}{8}$ (b) $\frac{45}{7}$ (c) $\frac{61}{3}$ (d) $7\frac{9}{13}$

2. (a) $21\frac{1}{4}$ (b) $2\frac{13}{30}$ (c) $16\frac{17}{20}$ (d) $1\frac{2}{5}$ (e) $37\frac{1}{27}$

3. $\frac{3}{35}$

LESSON 13:

Pre-test

1. (a) $\frac{1}{2}$ (b) $\frac{3}{8}$ 2. 40,000 people

3. (a) $\frac{5}{36}$ (b) $\frac{8}{36}$ or $\frac{2}{9}$

Practice Exercises

Frame 7

1. $\frac{2}{4}$ or $\frac{1}{2}$ 2. $\frac{2}{6}$ or $\frac{1}{3}$ 3. $\frac{3}{6}$ or $\frac{1}{2}$ 4. $\frac{6}{16}$ or $\frac{3}{8}$

Frame 13

1. $12\frac{1}{2}$ gallons   2. 11,250 dollars   3. 35,000,000 people

Frame 21

1. (a) $\frac{1}{13}$ (b) $\frac{1}{4}$ (c) $\frac{1}{2}$ (d) 1   (e) $\frac{1}{26}$

2. (a) $\frac{1}{36}$ (b) $\frac{1}{2}$ (c) $\frac{1}{2}$ (d) $\frac{2}{9}$ (e) 1

3. (a) $\frac{3}{10}$ (b) $\frac{1}{2}$ (c) $\frac{1}{5}$ (d) $\frac{4}{5}$ (e) $\frac{1}{2}$

4. (a) $\frac{1}{52}$ (b) Probability of drawing a 2, 3, 4, or 5 is greater.

(c) $\frac{14}{52}$ or $\frac{7}{26}$ (d) $\frac{51}{52}$

Post-test

1. (a) $\frac{2}{7}$ (b) $\frac{3}{7}$
2. 201 dollars
3. $\frac{14}{52}$ or $\frac{7}{26}$

Answers To Drill Exercises: Unit 3

1. $\frac{7}{3}$ or $2\frac{1}{3}$	2. $\frac{7}{9}$	3. $\frac{5}{8}$	4. $\frac{1}{13}$
5. $\frac{14}{5}$ or $2\frac{4}{5}$	6. $\frac{18}{7}$ or $2\frac{4}{7}$	7. 1	8. $\frac{15}{2}$ or $7\frac{1}{2}$
9. $\frac{22}{15}$ or $1\frac{7}{15}$	10. $\frac{47}{40}$ or $1\frac{7}{40}$	11. $\frac{8}{21}$	12. $\frac{23}{10}$ or $2\frac{3}{10}$
13. $\frac{77}{75}$ or $1\frac{2}{75}$	14. $\frac{14}{15}$	15. $\frac{31}{98}$	16. $\frac{65}{36}$ or $1\frac{29}{36}$
17. $\frac{53}{30}$ or $1\frac{23}{30}$	18. $\frac{1}{3}$	19. $\frac{11}{16}$	20. $\frac{13}{24}$
21. $\frac{19}{48}$	22. $\frac{9}{80}$	23. $\frac{51}{20}$ or $2\frac{11}{20}$	24. $\frac{1}{3}$
25. $\frac{1}{20}$	26. $\frac{17}{4}$ or $4\frac{1}{4}$	27. $\frac{3}{4}$	28. $\frac{32}{45}$
29. $\frac{1}{8}$	30. $\frac{3}{28}$	31. $\frac{1}{20}$	32. $\frac{1}{108}$
33. 16	34. 45	35. $\frac{3}{4}$	36. 2
37. $\frac{9}{20}$	38. $\frac{20}{9}$ or $2\frac{2}{9}$	39. $\frac{1}{9}$	40. $\frac{35}{22}$ or $1\frac{13}{22}$
41. 1	42. 1	43. $\frac{9}{25}$	44. $\frac{64}{49}$ or $1\frac{15}{49}$
45. $5\frac{2}{3}$	46. $2\frac{1}{4}$	47. $1\frac{5}{9}$	48. $1\frac{5}{6}$
49. $1\frac{3}{10}$	50. $5\frac{7}{12}$	51. $1\frac{9}{16}$	52. $3\frac{1}{8}$
53. $\frac{43}{8}$	54. $\frac{36}{7}$	55. $\frac{10}{9}$	56. $\frac{99}{10}$
57. $\frac{20}{3}$	58. $\frac{43}{14}$	59. $\frac{79}{4}$	60. $\frac{93}{5}$
61. $7\frac{16}{21}$	62. $6\frac{11}{40}$	63. $5\frac{3}{10}$	64. $6\frac{5}{28}$
65. $\frac{22}{35}$	66. $1\frac{2}{27}$	67. $19\frac{7}{8}$	68. $2\frac{1}{7}$
69. $1\frac{2}{3}$	70. $4\frac{2}{5}$	71. $23\frac{1}{5}$	72. 20
73. $43\frac{1}{5}$	74. $\frac{7}{30}$	75. $1\frac{1}{2}$	76. $1\frac{1}{4}$
77. $8\frac{3}{14}$	78. $1\frac{3}{11}$		

79. (a) $\frac{3}{8}$ (b) $\frac{1}{4}$ (c) $\frac{1}{8}$ (d) $\frac{1}{8}$ (e) $\frac{1}{8}$

80. (a) $\frac{1}{4}$ (b) $\frac{1}{13}$ (c) $\frac{1}{2}$ (d) $\frac{3}{13}$

81. $\frac{10}{13}$ black and white sets

82. $40\frac{1}{4}$ dollars

83. $8\frac{1}{4}$ tons

84. $\frac{3}{16}$ inches

85. $\frac{11}{16}$ inches

86. 1,050 miles

87. 10,500 dollars

88. 8 pieces

89. $\frac{3}{4}$ pounds

90. $\frac{1}{36}$

91. $\frac{1}{13}$

92. $\frac{1}{4}$

93. $\frac{11}{36}$

94. (a) $\frac{1}{10}$ (b) $\frac{1}{2}$ (c) $\frac{2}{5}$

95. (a) $\frac{5}{23}$ (b) $\frac{10}{23}$ (c) $\frac{15}{23}$ (d) 0 (e) 1

Answers To Self-Test: Unit 3

1. (b)
2. (a)
3. (b)
4. (d)
5. (c)
6. (a)
7. (b)
8. (a)
9. (c)
10. (d)
11. (c)
12. (e)
13. (b)
14. (d)
15. (c)
16. (a)
17. (d)
18. (a)
19. (c)
20. (d)

UNIT FOUR

An Introduction to Algebra

Lesson 14	MATHEMATICAL PHRASES AND SENTENCES
Lesson 15	SOLVING EQUATIONS
Lesson 16	PROPORTIONS
Lesson 17	FORMULAS

LESSON 14
MATHEMATICAL PHRASES AND SENTENCES

Objectives:

1. Translate an English sentence into an equation. (1–8)
2. Determine if an equation is true or false. (9–11)
3. Write an open sentence for a given word problem, specifying the variable. (12–22)

Vocabulary:

Mathematical Phrases	Variable
Equivalent Phrases	Open Sentence
Equation	Solution
Algebra	

PRE-TEST

1. Translate each English sentence into an equation.
   (a) The square of 9 minus 50 is 31.
   (b) The number 51 is 3 more than the product of 6 and 8.
2. Write true or false for each equation.
   (a) $6^3 - 10^2 = 116$
   (b) $\sqrt{(3^2 + 4^2)} = 7$
3. Restate the following word problem as an open sentence letting N represent the number of ballots cast at the polls:
   The 2,200 votes cast in a municipal election consisted of the ballots cast at the polls and 315 absentee ballots. How many votes were cast at the polls?

A *mathematical phrase* is an expression that names a specific number. ①

Mathematical Phrase		*Number*
$3 + 2$	names	5
5×7	names	35
$6(4 + 7)$	names	66
$5^3 - 10^2$	names	25

Not every numeric expression is a mathematical phrase. An attempt to divide by zero, which is impossible, is an example of a numeric expression that is *not* a mathematical phrase. ②

Since $(5 + 3) \div 0$ is impossible,
$(5 + 3) \div 0$ does not name a number.
Therefore, $(5 + 3) \div 0$ is not a mathematical phrase.

More than one mathematical phrase can name the same number: ③

$3 + 2 \longrightarrow 5$ | $7 \times 5 \longrightarrow 35$

$20 \div 4 \longrightarrow 5$ | $6^2 - 1 \longrightarrow 35$

$\sqrt{(2 + 3)^2} \longrightarrow 5$ | $\sqrt[3]{64} \times 8 + 3 \longrightarrow 35$

Mathematical phrases that name the same number are called *equivalent phrases.* ④

Since $3 + 2$ names 5 and $20 \div 4$ names 5,
$3 + 2$ and $20 \div 4$ are equivalent phrases.
Since $6^2 - 1$ names 35 and 7×5 names 35,
$6^2 - 1$ and 7×5 are equivalent phrases.

The equivalence of two phrases can be shown by the use of the = sign:

$3 + 2 = 20 \div 4$ $\qquad$ $6^2 - 1 = 7 \times 5$

An *equation* is a number sentence that shows the equivalence of two phrases. ⑤

Equation

$3 \times 5 = 11 + 4$

Left Phrase (3 × 5) — Right Phrase (11 + 4)

Both phrases name the same number, 15.

Algebra can be called the shorthand of mathematics. Notice how the following word phrases are translated into mathematical phrases using symbols. Symbols have the advantage of being more compact and easier to write. ⑥

Word Phrase		*Mathematical Phrase*
"The sum of five and nine"	⟷	$5 + 9$
"The square of seven"	⟷	7^2
"The product of six and three"	⟷	6×3

Sentences can also be translated into *equations:* 7

"Three plus two is equal to five" ⟷ $3 + 2 = 5$

"Five squared less one is twenty-four" ⟷ $5^2 - 1 = 24$

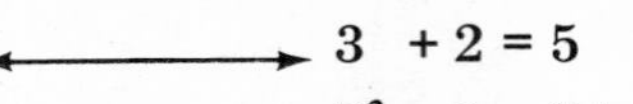

Notice that the word "is" translates as "=" (equals). An equation is the translation of a sentence having a form of "to be"; for example: is; are; was; were.

Practice Exercises 8

Translate each sentence into an equation.

1. The sum of twelve and seventeen is twenty-nine.
2. Eleven is the difference between thirty-four and twenty-three.
3. Three-fifths of thirty is eighteen.
4. The cube of three is five squared increased by two.

Equations are either true or false. If an equation is *true*, both the left and right phrases must name the same number. If both phrases do *not* name the same number, the equation is *false*. 9

True Equations	*False Equations*
$7 + 3 = 10$	$3 \times 5 = 17$ because $3 \times 5 = 15$
$6^2 - 35 = 1$	$4^3 - \sqrt[3]{27} = 51$ because $4^3 - \sqrt[3]{27} = 64 - 3$
	$= 61$

Determine if the following equations are true or false by completing the missing information: 10

A. $\sqrt{(3^2 + 4^2)} = 7$

Since $3^2 + 4^2 = \square$

and $\sqrt{\square} = \bigcirc$,

$\bigcirc \neq 7$

The equation is false since $5 \neq 7$.

B. $4^3 - 6 \times 7 = 22$

Since $4^3 = 64$

and $6 \times 7 = \triangle$,

$64 - \triangle = \hexagon$

The equation is true since $22 = 22$.

The symbol "≠" means "is not equal to" or "is not equivalent to."

Practice Exercises 11

Write true or false for each equation.

1. $(5 \times 2) \times 3 = 5 \times (2 \times 3)$
2. $4 \times 0 = 4 + 0$
3. $6 + 2 = 2(3 + 1)$
4. $4\left(\frac{3}{8} + \frac{1}{16}\right) = \left(4 \times \frac{3}{8}\right) + \left(4 \times \frac{1}{16}\right)$

A *variable* is any symbol used to represent an unspecified number. In algebra, letters are commonly used as variables. The choice of what letter is used as a variable is left up to the individual. (12)

$N + 3$	means	**Some number added to 3**
$3N$	means	**Some number multiplied by 3**

Recall from the definition of multiplication that $3N$ also means $N + N + N$. In the expression, $N + 3$, the variable N is understood to mean $1N$.

An *open sentence* is an equation that contains a variable. It is called an *open* sentence because the variable is *open* for replacement by a number. (13)

$N + 3 = 9$	means	**Some number added to 3 is 9**
$3N = 6$	means	**Some number multiplied by 3 is 6**

An open sentence is an equation that is neither true nor false.

A variable in an open sentence must be replaced by a number before it can be determined whether the resulting equation is true or false. (14)

$$N + 3 = 9$$

If N is replaced by 6, the equation $6 + 3 = 9$ is true.

If N is replaced by 2, the equation $2 + 3 = 9$ is false.

$$3N = 6$$

If N is replaced by 2, the equation $3 \times 2 = 6$ is true.

If N is replaced by 6, the equation $3 \times 6 = 6$ is false.

A *solution* for an open sentence is a number that makes the resulting equation true when that number is used to replace the variable. (15)

6	is the solution for	$N + 3 = 9$	since	$6 + 3 = 9$.
2	is the solution for	$3N = 6$	since	$3 \times 2 = 6$.

Any open sentence may be used to state, clearly and concisely, a variety of word problems. (16)

For example, consider the following word problems:

A. The sum of a number and 3 is 9. Find the number.

B. The committee studying trends in urban development has 9 members. If 3 of the members are men, how many members are women?

C. Richard Michael wishes to build a porch onto his house. The city building codes require a 3-foot clearance between the porch and his neighbor's property line. What is the maximum width that the porch could be if his house is 9 feet from the property line?

Each of these questions could be answered by finding the solution for the open sentence:

$$N + 3 = 9$$

In using the open sentence $N + 3 = 9$ to answer each of the questions above, the only difference is in what N represents in each problem. 17

In Example A, N stands for a specific addend.

In Example B, N stands for the specific number of women committee members.

In Example C, N stands for the specific width of the porch.

When an open sentence is used to solve a word problem, the variable will always represent the unknown quantity. It is necessary to state exactly what variable is being used and what it represents.

Writing open sentences for word problems: 

The sum of two consecutive whole numbers is 49. Find the smaller of the two numbers.

First specify the variable.

Let N = the smaller of the two consecutive numbers.

Then,

$N + 1$ = the next consecutive whole number

And the sum of these numbers is

$N + (N + 1)$

Since the sum is given as 49, the open sentence is

$N + (N + 1) = 49$

Writing open sentences for word problems: 

Five more than three times some number is seventeen. Find the number.

First: Let N = the number.

Then: $3N$ is 3 times the number and
$3N + 5$ is 5 more than 3 times the number.

Thus, the open sentence is $3N + 5 = 17$

Writing open sentences for word problems:

One-half the difference between some number and seven is twenty-four. Find the number.

First: Let N = the number.

Then: $N - 7$ is the difference and

$\frac{1}{2}(N - 7)$ is one-half of the difference.

Thus, the open sentence is $\frac{1}{2}(N - 7) = 24$

Writing open sentences for word problems:

Mary is twice as old as her sister. If the sum of their ages is twenty-four, how old is her sister?

Complete the following:

First: Let A = Mary's sister's age.

Then: ______ = Mary's age and

______ = the sum of their ages.

Thus, the open sentence is $A + 2A = 24$.

Practice Exercises

Write an open sentence for each of the following word problems using the specified variable.

1. A number increased by 5 is 31. Let N = the number.
2. The sum of two consecutive whole numbers is 17. What is the smaller number? Let N = the smaller number.
3. Glenn's bowling score of 145 is 10 more than one-half of James' score. What is James' score? Let J = James' score.
4. The Golden West basketball team has won 3 times as many games as it has lost this season. If the total number of games played is 84, how many games have they lost? Let L = the number of games lost.

POST-TEST

1. Translate each English sentence into an equation.
   (a) The cube of 5 increased by 25 is 150.
   (b) The number 436 is one-half of the difference between 922 and 50.
2. Determine whether the following equations are true or false.
   (a) $\sqrt{(6^2 + 8^2)} = 14$
   (b) $\sqrt{64} - 3 = 1$
3. Restate the following word problem as an open sentence, letting X represent the number of operations performed in June.

   Twice as many operations were performed at the Community Hospital in July as had been performed in June. The total number of operations for the 2-month period was 624. How many operations were performed in June?

LESSON 15
SOLVING EQUATIONS

Objectives:

1. Solve an equation that has a single variable. (1–13)
2. Solve an equation in which the variable appears more than once. (14–20)
3. Write and solve an equation for a word problem. (21–24)

PRE-TEST

1. Solve the following equations:
   (a) $5(3X - 1) = 15$
   (b) $4W - W = 36$
2. Toby is 3 times as old as his sister. The sum of their ages is 24. How old is Toby?

Writing an equation is the first step in determining the answer to a word problem. It is still necessary to solve the equation. The purpose of this lesson is to develop a strategy for solving any simple equation. (1)

For example, consider the equation

$$2N + 3 = 15$$

What does "solve the equation, $2N + 3 = 15$" mean? "Solve the equation" means "find the solution." Recall that the solution was defined as any number that makes a true equation when it is used to replace the variable. The equation will be solved when the following question can be answered: N is what number? (2)

$$N = ?$$

In order to answer this question about the number N it is useful to examine the equation $2N + 3 = 15$ in order to determine the sequence of operations that has been performed with the number N. Looking at the equation, we see that (3)

$2N + 3 = 15$ means **3 more than 2 times the number N is 15**

Since N stands for a number, this is the only possible interpretation of the equation according to the rules for order of operation.

In order to be able to answer the question $N = ?$, it is necessary to "undo" this sequence of operations. If we use the fact that inverse operations "undo" each other, what happens to the equivalent phrases of the equation when the sequence of operation is reversed? 4

$$2N + 3 = 15$$

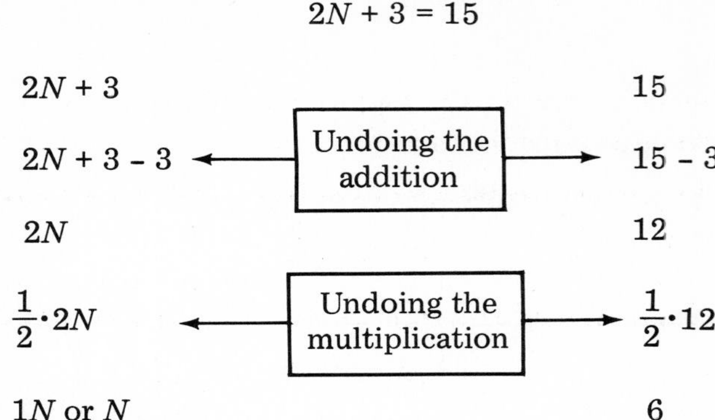

Since the addition was done last, it is "undone" first by subtracting 3 from both phrases. Then, the multiplication by 2 is "undone" by dividing both phrases by 2, which is most easily accomplished by multiplying by $\frac{1}{2}$, the reciprocal of 2.

Since the same operations were performed on both equivalent phrases of the equation $2N + 3 = 15$, is it reasonable to assume that $N = 6$? If this statement can be made, then the equation has been solved; the solution is 6. 5

$$2(6) + 3 \stackrel{?}{=} 15$$
$$12 + 3 \stackrel{?}{=} 15$$
$$15 = 15$$

When N is replaced by 6, the resulting equation is true. The solution is 6.

Solve the equation: $\frac{1}{2}X - 4 = 5$ 6

$\frac{1}{2}X - 4 = 5$ means [4 subtracted from the product $\frac{1}{2}$ times X is 5]

$\frac{1}{2}X - 4 + 4 = 5 + 4$ ← [Adding 4 to "undo" the subtraction]

$\frac{1}{2}X = 9$ ← [Rewriting]

$2 \cdot \frac{1}{2}X = 2 \cdot 9$ ← [Multiplying by the reciprocal of $\frac{1}{2}$ to "undo" the multiplication]

$X = 18$

Checking the solution by replacing X by 18 in the original equation:

$$\frac{1}{2}(18) - 4 \stackrel{?}{=} 5$$
$$9 - 4 \stackrel{?}{=} 5$$
$$5 = 5$$

Notice that it is more convenient to work on both sides of the equation at the same time than with the two phrases independently.

Solve the equation: $3(Y - 2) = 15$ ⑦

$3(Y - 2) = 15$ means 3 times the difference, $Y - 2$ is 15

$\frac{1}{3} \cdot 3(Y - 2) = \frac{1}{3} \cdot 15$ ← Multiplying by the reciprocal of 3 to "undo" the multiplication by 3

$Y - 2 = 5$

$Y - 2 + 2 = 5 + 2$ ← Adding 2 to "undo" the subtraction

$Y = 7$

The solution can be checked by replacing Y by 7 in the original equation.

Solve the equation: $\frac{Z}{5} = 17$ ⑧

$\frac{Z}{5} = 17$ means Z divided by 5 is 17

$5 \cdot \frac{Z}{5} = 5 \cdot 17$ ← Multiplying by 5, the reciprocal of $\frac{1}{5}$, since $\frac{Z}{5} = \frac{1}{5}Z$

$Z = 85$

Solve the equation $X + 4\frac{3}{4} = 7$ by completing the following: ⑨

$$X + 4\frac{3}{4} = 17$$
$$X + 4\frac{3}{4} - \square = 17 - \square$$
$$X = \bigcirc$$

Find the solution for the equation $7X + 14 = 140$ by completing the following: ⑩

$$7X + 14 = 140$$

$$7X + 14 - \square = 140 - \square$$

$$7X = 126$$

$$\frac{1}{\bigcirc} \cdot 7X = \frac{1}{\bigcirc} \cdot 126$$

$$X = \triangle$$

Solve the equation $\frac{3}{4}N = 21$ by completing the following: ⑪

$$\frac{3}{4}N = 21$$

$$\frac{\square}{\triangle} \cdot \frac{3}{4}N = \frac{\square}{\triangle} \cdot 21$$

$$N = \bigcirc$$

Solve the equation $\frac{Y}{10} = 41$ by completing the following: ⑫

$$\frac{Y}{10} = 41$$

$$\square \cdot \frac{Y}{10} = \square \cdot 41$$

$$Y = \bigcirc$$

Practice Exercises ⑬

Solve each of the following equations and check each solution.

1. $X - 5\frac{1}{3} = 11$
2. $Y + \frac{7}{8} = 14$
3. $5N + 4 = 12$
4. $\frac{2}{5}X - \frac{3}{5} = 1$
5. $2(Z - 3) = 1\frac{1}{2}$
6. $\frac{X}{4} = 6$

Before equations such as $3W + 2W = 60$ and $5X - X = 24$ can be solved, it will be necessary to deal with the indicated operations $3W + 2W$ and $5X - X$. (14)

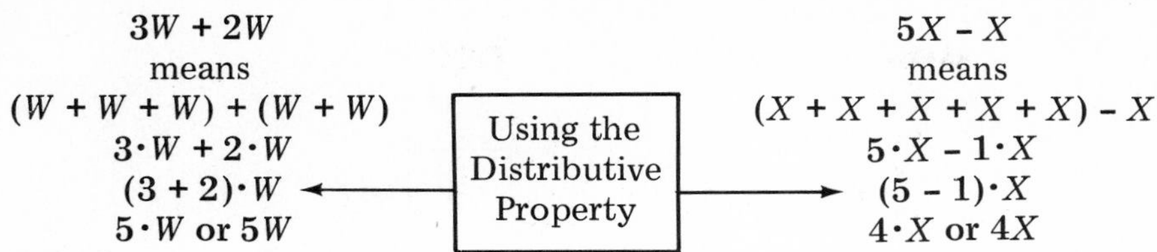

3W + 2W
means
(W + W + W) + (W + W)
3·W + 2·W
(3 + 2)·W
5·W or 5W
Using the Distributive Property
5X - X
means
(X + X + X + X + X) - X
5·X - 1·X
(5 - 1)·X
4·X or 4X

Solving $3W + 2W = 60$, an equation in which the variable occurs more than once. (15)

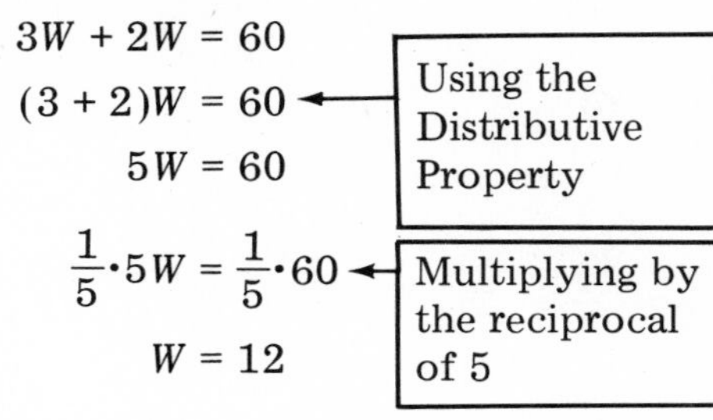

3W + 2W = 60
(3 + 2)W = 60
5W = 60
1/5·5W = 1/5·60
W = 12
Using the Distributive Property
Multiplying by the reciprocal of 5

Check:

$$3(12) + 2(12) \stackrel{?}{=} 60$$
$$36 + 24 \stackrel{?}{=} 60$$
$$60 = 60$$

Solve the equation $5X - X = 24$: (16)

$$5X - X = 24$$
$$(5 - 1)X = 24 \quad \leftarrow \text{Distributive Property}$$
$$4X = 24$$
$$\frac{1}{4}\cdot 4X = \frac{1}{4}\cdot 24 \quad \leftarrow \text{Multiplying by the reciprocal of 4}$$
$$X = 6$$

Check:

$$5(6) - 6 \stackrel{?}{=} 24$$
$$30 - 6 \stackrel{?}{=} 24$$
$$24 = 24$$

Solve the equation $N + (N + 1) = 13$: (17)

$$N + (N + 1) = 13$$
$$(N + N) + 1 = 13 \quad \leftarrow \text{Regrouping}$$
$$2N + 1 = 13 \quad \leftarrow \text{Since } N + N = 1N + 1N = (1 + 1)N$$
$$2N + 1 - 1 = 13 - 1$$
$$2N = 12$$
$$\frac{1}{2}\cdot 2N = \frac{1}{2}\cdot 12$$
$$N = 6$$

Solve the equation $7A + A = 56$ by completing the following: (18)

$$7A + A = 56$$

$$\square \; A = 56$$

$$\frac{1}{\square} \cdot \square \; A = \frac{1}{\square} \cdot 56$$

$$A = \bigcirc$$

Solve the equation $7N - 3N = 26$ by completing the following: (19)

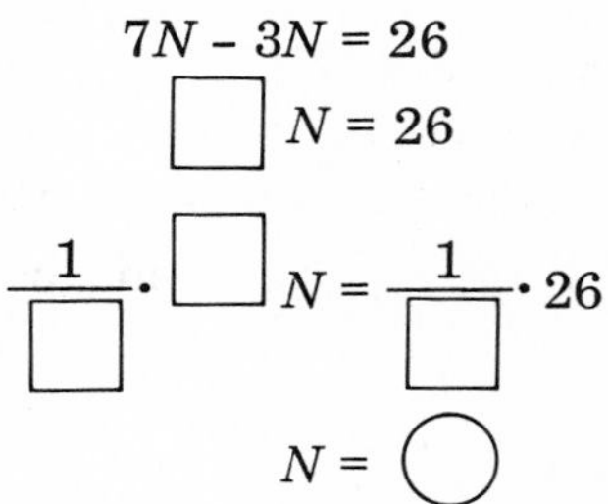

Practice Exercises 

Solve each of the following equations:

1. $7X - X = 30$
2. $4Y + 5Y = 23$
3. $N + \frac{3}{4}N + 5 = 12$
4. $\frac{17}{4}R - 3R = 25$

A rope 12 feet long is to be cut into two pieces, one piece 3 feet longer than the other. What is the length of each piece of rope? (21)

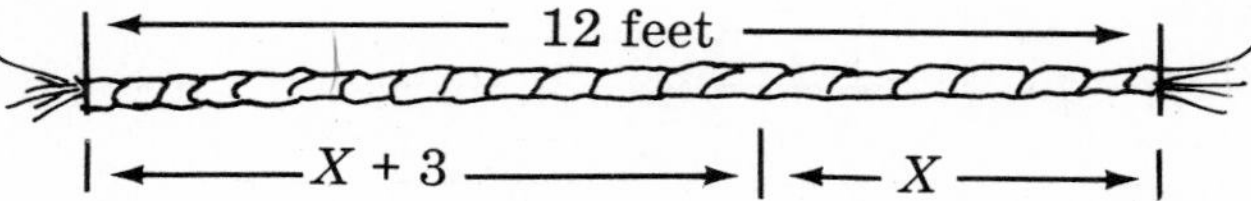

Writing the equation: Let X = the length of the shorter piece of rope in feet.
Then, $X + 3$ = the length of the longer piece in feet.

12 feet = $X + (X + 3)$ feet

Solving the equation:

$$12 = X + (X + 3)$$
$$12 = (X + X) + 3$$
$$12 = 2X + 3$$
$$12 - 3 = 2X + 3 - 3$$
$$9 = 2X$$
$$\frac{1}{2} \cdot 9 = \frac{1}{2} \cdot 2X$$
$$\frac{9}{2} = X$$

Thus, the shorter piece of rope is $\frac{9}{2}$ or $4\frac{1}{2}$ feet long.

The length of the longer piece is $4\frac{1}{2} + 3 = 7\frac{1}{2}$ feet long.

Checking the solution: $12 \stackrel{?}{=} 4\frac{1}{2} + \left(4\frac{1}{2} + 3\right)$

$$12 \stackrel{?}{=} \frac{9}{2} + \left(\frac{9}{2} + \frac{6}{2}\right)$$

$$12 \stackrel{?}{=} \frac{24}{2}$$

$$12 = 12$$

Richie knows that he has 3 times as many coins as Harvey. Harvey knows that Richie has 24 more coins than he does. How many coins does each boy have?  22

Writing the equation: Let C = the number of coins that Harvey has.

Then, $3C$ = the number of coins that Richie has.

$$24 \text{ coins} = 3C - C$$

Solving:

$$24 = 3C - C$$

$$24 = 2C$$

$$\frac{1}{2} \cdot 24 = \frac{1}{2} \cdot 2C$$

$$12 = C$$

Thus, Harvey has 12 coins and Richie has $3(12) = 36$ coins.

Checking: $36 - 12 = 24$; the solution is correct.

Seven inches is cut from a length of copper tubing. When the remaining piece of tubing is cut in half, each of the 2 equal pieces is 14 inches long. How long was the original piece of tubing before any cuts were made?  23

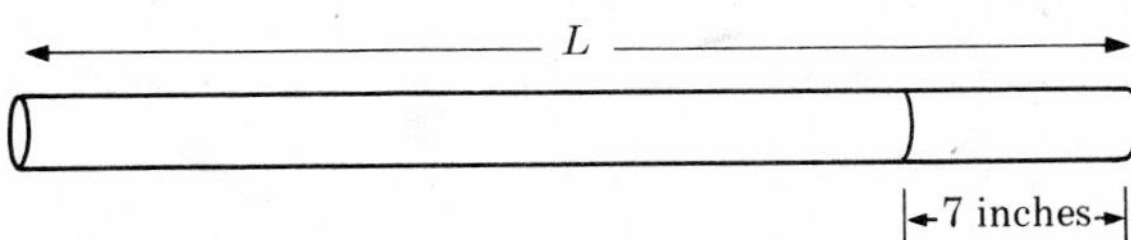

Writing the equation: Let L = the length of the original piece of copper tubing in inches.

And $L - 7$ = the length of the tubing in inches after the 7-inch piece was cut off.

Then, $\frac{1}{2}(L - 7)$ = the length of each half after the final cut.

$$\frac{1}{2}(L - 7) \text{ inches} = 14 \text{ inches}$$

Solving:

$$\frac{1}{2}(L - 7) = 14$$

$$2 \cdot \frac{1}{2}(L - 7) = 2 \cdot 14$$

$$L - 7 = 28$$

$$L - 7 + 7 = 28 + 7$$

$$L = 35$$

Checking:

$$\frac{1}{2}(35 - 7) \stackrel{?}{=} \frac{1}{2}(28)$$

$$\frac{1}{2}(28) \stackrel{?}{=} 14$$

$$14 = 14$$

The correct solution is 35 inches.

Practice Exercises

Solve and check each of the following word problems.

1. Find two consecutive whole numbers whose sum is 17.
2. A 61-foot board is cut into five equal pieces with 1 foot left over. Find the length of each of the five pieces of board.
3. The school basketball team has won twice as many games at home as it has won on the road. If the team has won a total of 15 games, how many were won at home?

POST-TEST

1. Solve the following equations:
   (a) $3(M + 2) = 27$
   (b) $5X + (X + 1) = 7$
2. A 15-foot board is to be cut into two pieces so that the shorter piece is 2 feet shorter than the longer piece. Find the lengths of both pieces of board.

LESSON 16
PROPORTIONS

Objective:

1. Write a proportion for a given word problem and solve. (1–11)

Vocabulary:

Proportion	Extremes
Means	Cross Product

PRE-TEST

1. Write a proportion for each of the following and solve:
   (a) The ratio of 2 to 3 is the same as the ratio of 10 to what number?
   (b) If an 8-ounce jar of roasted almonds costs 96 cents, what is the cost of 1 ounce?
   (c) Find the height of a tree that casts a 46-foot shadow if a man who is 6 feet tall standing next to the tree casts a 4-foot shadow.

While shopping, Mrs. Cook finds that one brand of soup is priced at 48 cents for 3 cans. Another equally good brand is priced at 85 cents for 5 cans. Which is the better buy? (1)

To answer this question, it is useful to know the unit price of both brands of soup. One way of finding the unit prices is to set up and solve a proportion.

Definition of a proportion: (2)

A *proportion* is a statement that expresses the equivalence of two ratios.

For example, consider the ratios, $\frac{2}{3}$ and $\frac{8}{12}$.

Since

$$\frac{8}{12} = \frac{\overset{1}{\cancel{2}} \times \overset{1}{\cancel{2}} \times 2}{\underset{1}{\cancel{2}} \times \underset{1}{\cancel{2}} \times 3}$$

$$= \frac{2}{3}$$

The two ratios are equivalent and the following proportion may be written:

$$\frac{2}{3} = \frac{8}{12}$$

Or, alternately,

$$2:3 = 8:12$$

This proportion is read 2 is to 3 as 8 is to 12.

The four parts of a proportion: ③

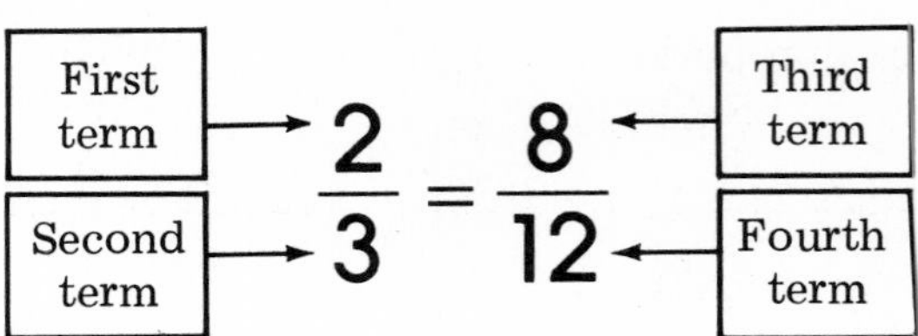

The first and fourth terms are called the *extremes*; 2 and 12 are the extremes. The second and third terms are called the *means;* 3 and 8 are the means.

The cross product of a proportion: ④

$$\frac{2}{3} \times \frac{8}{12}$$

The *cross products* of the proportion are the following:

The product of the means: $3 \times 8 = 24$
The product of the extremes: $2 \times 12 = 24$

In any true proportion the product of the means is equal to the product of the extremes. That is, the cross products are equal. This fact will be used whenever proportions are used to solve word problems.

Returning to the problem in Frame 1, it is now possible to write the two proportions needed to determine the unit prices. The unit prices are related to the following proportions: ⑤

48 cents is to 3 cans as *X* cents is to 1 can
85 cents is to 5 cans as *Y* cents is to 1 can

The following tables may be used to say the same thing. Similar tables will be used throughout this lesson.

Cost	48	X
Cans	3	1

Cost	85	Y
Cans	5	1

The tables are used to set up the following proportions:

$$\frac{48}{3} = \frac{X}{1} \qquad \frac{85}{5} = \frac{Y}{1}$$

or or

$$16 = X \qquad 17 = Y$$

At 16 cents per can, the brand selling at 3 cans for 48 cents is the better buy since the brand priced at 5 cans for 85 cents is selling for 17 cents per can.

Using a proportion to solve a word problem: (6)

If an automobile travels 286 miles on 22 gallons of gasoline, how far will it travel on 10 gallons?

It is possible to set up more than one proportion to solve this problem as the following tables show:

	First Trip	Second Trip
Miles	286	X
Gallons	22	10

	Miles	Gallons
First Trip	286	22
Second Trip	X	10

Setting up the proportions from the tables, letting X = number of miles:

$$\frac{286}{22} = \frac{X}{10} \qquad \frac{286}{X} = \frac{22}{10}$$

Forming the cross products:

$$22X = 286 \cdot 10 \qquad 22X = 286 \cdot 10$$

Both equations are identical. These examples show that there is more than one way to correctly set up a proportion for a problem. In fact, there are four proportions all having the same cross products. Any of the four could be used to solve this word problem.

Solving:

$$22X = 286 \cdot 10$$

$$\frac{1}{22} \cdot 22X = \frac{1}{22} \cdot 286 \cdot 10$$ ← Multiplying by the reciprocal of 22

$$1X = \frac{286 \cdot 10}{22}$$

$$X = \frac{\overset{1}{\cancel{22}} \cdot 13 \cdot 10}{\underset{1}{\cancel{22}}}$$ ← Factoring and cancelling

$$X = 13 \cdot 10 \text{ or } 130$$

Therefore, if it takes 22 gallons of gas to travel 286 miles, then a trip of 130 miles can be made on 10 gallons of gas.

Using a proportion to solve a word problem: (7)

A buttermilk bread recipe calls for 2 teaspoons of salt and 5 cups of flour. If 12 cups of flour were used, how much salt would be required?

Let X = the number of teaspoons of salt.

	Basic Recipe	Increased Recipe
Teaspoons of Salt	2	X
Cups of Flour	5	12

$$\frac{2}{5} = \frac{X}{12}$$

$$5X = 2 \cdot 12$$

$$\frac{1}{5} \cdot 5X = \frac{1}{5} \cdot 2 \cdot 12$$

$$1X = \frac{2 \cdot 12}{5}$$

$$X = \frac{24}{5} \text{ or } 4\frac{4}{5} \text{ teaspoons of salt}$$

Using a proportion to solve a word problem: (8)

If 3 loaves of bread sell for 96 cents, how much does 1 loaf cost?

Complete the following:

Let C be the cost of one loaf of bread.

Number of Loaves	3	1
Cost in Cents	96	C

$$\frac{3}{96} = \frac{1}{C}$$

$$\square\, C = 1 \cdot \bigcirc$$

$$\frac{1}{\square} \cdot \square\, C = \frac{1}{\square} \cdot 1 \cdot \bigcirc$$

$$1C = \frac{\bigcirc}{\square}$$

$$C = \hexagon \text{ cents}$$

Using a proportion to solve a word problem: (9)

A seminar has a ratio of 3 men to every 2 women. How many men are there if there is a total of 35 people in the seminar?

This problem is slightly different in that the ratio of men to women is given, but what is unknown is the ratio of men to the total number of people in the seminar.

Think: 3 men + 2 women = 5 people.

Let M = the total number of men in the seminar.

Number of Men	3	M
Number of People	5	35

$$\frac{3}{5} = \frac{M}{35}$$

$$5M = 3 \cdot 35$$

$$\frac{1}{5} \cdot 5M = \frac{1}{5} \cdot 3 \cdot 35$$

$$1M = \frac{3 \cdot 35}{5}$$

$$M = \frac{3 \cdot \overset{1}{\cancel{5}} \cdot 7}{\underset{1}{\cancel{5}}}$$

M = 21 men in the seminar

Using a proportion to solve a word problem: (10)

There are 5 voters registered as Democrats for every 4 voters registered as Republicans in the local precinct. If there is a total of 729 registered Democrats and Republicans, how many are Democrats?

Complete the following:

Think: 5 Democrats + 4 Republicans = $\triangle$ Democrats and Republicans.

Let D = the total number of Democrats in the precinct.

Democrats	5	D
Democrats and Republicans	$\triangle$	729

$$\frac{5}{\triangle} = \frac{D}{729}$$

$$\triangle\, D = 5 \cdot 729$$

$$\frac{1}{\triangle} \cdot \triangle\, D = \frac{1}{\triangle} \cdot 5 \cdot 729$$

$D = \square$ Democrats in the precinct

Practice Exercises

Write a proportion for each of the following and solve.

1. If a student finds that he can read 12 pages of a certain novel in 20 minutes, how long will it take him to read the 360 total pages in the book at the same rate?
2. A 50-pound sack of fertilizer will cover 2,000 square feet of lawn. How many pounds of fertilizer will it take to cover a 7,000-square foot lawn?
3. If a car can travel 18 miles on 1 gallon of gas on the open highway, how many gallons of gas will it use on a trip of 450 miles?
4. The ratio of veterans to non-veterans on the campus is 4 to 11. If there are 7,500 students on campus, how many are veterans?

POST-TEST

1. Write a proportion and solve each of the following:
   (a) The ratio of 4 to 7 is the same as the ratio of what number to 102?
   (b) If a box of detergent weighing 24 ounces costs 72 cents, what is the price per ounce?
   (c) If a car travels an average of 200 miles in 3 hours, how far can it travel in 8 hours at the same average rate?

LESSON 17
FORMULAS

Objective:

1. Solve a given formula for the value of an unknown quantity. (1-13)

Vocabulary:

Formula

PRE-TEST

1. Use the Distance Formula, $D = RT$, to find each of the following:
   (a) The distance when the rate is 348 miles per hour and the time is $3\frac{1}{2}$ hours.
   (b) The time when the distance is 5,000 miles and the rate is 600 miles per hour.
2. The number of Widgets that a store owner may expect to sell is related to the price he charges for them. He uses the following formula, where N = number of widgets sold at price P:

$$N = \frac{4{,}000}{P^2}$$

   (a) How many widgets may he expect to sell if he charges 2 dollars for each widget?
   (b) How many at 5 dollars per widget?

If a plane flies for $2\frac{1}{2}$ hours at 550 miles per hour, how far has it flown? (1)

Is it possible to solve this word problem with just the information that is given in the problem? It could be solved using a proportion, but it is much easier to find the answer if the Distance Formula is known.

Distance, rate (average speed), and time are related as follows: (2)

$$\text{Distance} = \text{Rate} \times \text{Time}$$

That is, the total distance traveled is determined by multiplying the average speed or rate by the total time spent in traveling.

A simpler way to convey this relationship is to use a formula. A *formula* is an equation that expresses the relationship between quantities in a specific situation. For example, ③

Distance Formula

$$D = RT$$

Formulas look like any other equations, except that they usually have more than one variable. This formula, and many others, will be used in this lesson. There will be no need to memorize the formulas since the purpose of this lesson is to learn to evaluate a given formula.

Evaluating a formula: ④

Before the formula can be used, it is necessary to correctly identify the given information. Referring to the problem in Frame 1, we see that

$$R = 550 \text{ miles per hour}$$

$$T = 2\frac{1}{2} \text{ hours}$$

Once the given information has been identified, these values can be substituted for the variables in the formula.

$$D = RT$$

$$D = 550 \cdot 2\frac{1}{2}$$

Solving the equation:

$$D = 550 \cdot \frac{5}{2}$$

$$= \frac{550 \cdot 5}{2}$$

$$= 1{,}375$$

The total distance flown in $2\frac{1}{2}$ hours at 550 miles per hour is 1,375 miles.

While on a hunting trip, David walked due east for 3 miles from the road to a river. He turned due north and followed the river for 4 miles. He would like to return to his car by the shortest possible route. What is the shortest distance between where David is and where his car is parked on the road? ⑤

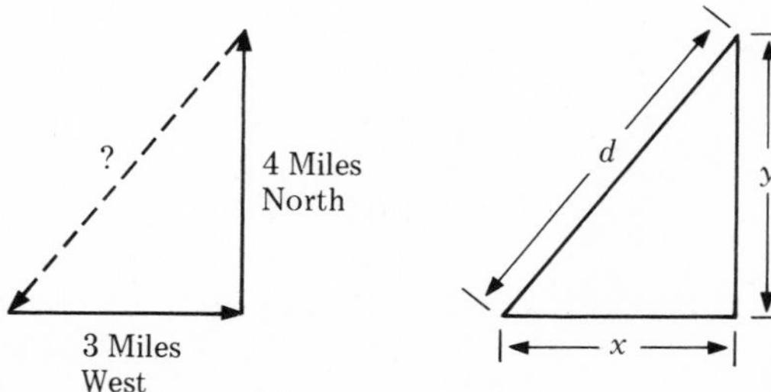

The shortest distance, d, between David and the car can be determined by using the following formula:

$$d = \sqrt{(x^2 + y^2)}$$

The illustrations help to identify the given information in this problem and how the information is to be used:

$$x = 3$$

$$y = 4$$

Substituting: $d = \sqrt{(3^2 + 4^2)}$

Solving: $d = \sqrt{(9 + 16)}$

$= \sqrt{25}$

$= 5$ miles

The shortest distance between David and his car is 5 miles.

The formula used to determine the safe distance, S, in feet between 2 cars traveling at R miles per hour is (6)

$$S = \frac{1}{25}R^2 + R$$

Find the safe distance between 2 cars traveling at the rate of 30 miles per hour.

Substituting R = 30: $S = \frac{1}{25} \cdot 30^2 + 30$

Solving: $S = \frac{900}{25} + 30$

$= 36 + 30$

$= 66$ feet

The drivers should maintain a distance of 66 feet between cars.

The height, H, of a ball, thrown upward with an initial velocity of 96 feet per second, at time, T, is given by the formula: (7)

$$H = 96T - 16T^2$$

Find the height of the ball when $T = 3$ seconds.

Substituting: $H = 96 \cdot 3 - 16 \cdot 3^2$

Solving: $H = 96 \cdot 3 - 16 \cdot 9$

$= 288 - 144$

$= 144$ feet

The formula, $C = \frac{1}{40} \cdot R^3$, gives the fuel cost per hour, C, of operating a ship traveling at R miles per hour. Find the hourly fuel cost of a ship traveling at 10 miles per hour. (8)

Substituting: $C = \frac{1}{40} \cdot 10^3$

Solving:

$$C = \frac{1}{40} \cdot 1{,}000$$

$$= \frac{1{,}000}{40}$$

$$= 25 \text{ dollars}$$

The time, T, in seconds for an object to strike the ground when dropped from height, H, in feet is given by the formula: (9)

$$T = \sqrt{\frac{H}{16}}$$

How long will it take for a stone dropped from a height of 80 feet to strike the ground?

Complete the following:

Substituting: $T = \sqrt{\frac{\square}{16}}$

Solving: $T = \sqrt{\bigcirc}$

The time, T, is between $\triangle$ seconds and $\hexagon$ seconds.

If a 234-mile drive is made at an average speed of 45 miles per hour, how long will it take? (10)

Since this problem deals with a distance, a rate, and a time, it is necessary to use the Distance Formula: $D = RT$.

Given information:

$D = 234$ miles

$R = 45$ miles per hour

Substituting: $234 = 45T$

Solving for T:

$$\frac{1}{45} \cdot 234 = \frac{1}{45} \cdot 45T$$

$$\frac{234}{45} = 1T$$

$$5\frac{1}{5} = T$$

The time it takes to drive 234 miles at 45 miles per hour is $5\frac{1}{5}$ hours.

The pressure being exerted on a scuba diver's body determines how deep he can safely descend. (The pressure being exerted on any body at sea level is $14\frac{7}{10}$ pounds per square inch.) A formula relating the pressure, P, in pounds per square inch, and the depth, D, in feet, is (11)

$$P = \frac{1}{2}D + 15$$

If a diver finds that the maximum pressure that he can withstand with no ill effects is 125 pounds per square inch, what is his maximum diving depth?

Given information: $P = 125$ pounds per square inch

Substituting: $125 = \frac{1}{2}D + 15$

Solving for D:

$$125 - 15 = \frac{1}{2}D + 15 - 15$$

$$110 = \frac{1}{2}D$$

$$2 \cdot 110 = 2 \cdot \frac{1}{2}D$$

$$220 = D$$

His maximum diving depth is 220 feet.

The formula for converting temperatures in degrees Celsius to degrees Fahrenheit is (12)

$$F = \frac{9}{5}C + 32$$

If water boils at 212 degrees Fahrenheit (at sea level), find the boiling point of water in degrees Celsius.

Complete the following:

Given: $F = 212$ degrees

Substituting: $\square = \frac{9}{5}C + 32$

Solving:

$$\square - 32 = \frac{9}{5}C + 32 - 32$$

$$\bigcirc = \frac{9}{5}C$$

$$\frac{5}{9} \cdot \bigcirc = \frac{5}{9} \cdot \frac{9}{5}C$$

$$⬡ = C$$

Water boils at ⬡ degrees Celsius (at sea level).

Practice Exercises

1. The price, P, in dollars, and the quantity, Q, demanded for a certain commodity are related by the formula:

$$4Q + P = 10$$

(a) Find Q when P is 2 dollars.
(b) Find P when Q is 1.

2. Referring to the Distance Formula (given in Frame 3):

(a) Find the distance when the rate is 50 miles per hour and the time is $3\frac{1}{2}$ hours.

(b) Find the rate when the distance is 105 miles and the time is $2\frac{1}{3}$ hours.

3. Refer to the formula relating the height of a ball and the time (given in Frame 7) to find the height of the ball at time $T = 1\frac{1}{2}$ seconds.

POST-TEST

1. Use the Celsius–Fahrenheit Conversion Formula to find each of the following:

$$F = \frac{9}{5}C + 32$$

(a) The degrees Fahrenheit, F, if the temperature is 25 degrees Celsius, C.
(b) The normal body temperature, $98\frac{3}{5}$ degrees Fahrenheit, in degrees Celsius.

2. Use the Safe Distance Formula, $D = \frac{1}{25}R^2 + R$, where D is the safe distance between 2 cars, in feet, traveling at R miles per hour, to find the following:
(a) The safe distance when $R = 50$ miles per hour.
(b) The safe distance when $R = 70$ miles per hour.

DRILL EXERCISES: UNIT 4

Translate each English sentence into an equation.

1. The square of 5 decreased by 4 is 21.
2. The number 95 is 39 more than the product of 7 and 8.
3. The square of 5 is the square root of 625.
4. The number 45 increased by the product of 4 and 3 is 57.
5. 3 less than the square root of 49 is 24 divided by 6.
6. 8 more than the cube root of 64 is the square root of 144.

Write true or false for each of the following equations:

7. $6^2 + 5^2 = 61$
8. $7^2 - 2^2 = 44$
9. $\sqrt{(4^2 - 3^2)} = 1$
10. $\sqrt{(4^2 - 3^2)} = \sqrt{7}$
11. $\sqrt[3]{8} + \sqrt{4} = 4$
12. $7 \times 8 + \sqrt{4} = 64 - 5$

Rewrite the following word problems as open sentences using the specified variables. Do not solve.

13. A number, N, increased by 48 is 72.
14. The sum of two consecutive whole numbers is 47. Let N represent the smaller number.
15. Four more than three times some number, N, is sixteen.
16. Think of a number N. If you add the number to itself and then add 5, you get 13. What is the number?
17. In the month of July, Mr. Jones earned 500 dollars more than twice as much as he earned in the month of June. If he earned 1,000 dollars in July, how much did he earn in June? Let X be the amount earned in June.

Find the solution to each of the following equations:

18. $X - 3 = 8$
19. $M - 4 = 0$
20. $5 + N = 11$
21. $X + 3{,}465 = 5{,}674$
22. $N - 3{,}675 = 260$
23. $6X = 42$
24. $73M = 146$
25. $\frac{W}{4} = 28$
26. $\frac{X}{25} = 9$
27. $\frac{1}{10} = \frac{N}{3{,}000}$
28. $4X - 1 = 17$
29. $\frac{7}{9}N = \frac{6}{5}$
30. $\frac{9}{13} + X = \frac{14}{13}$
31. $R + 7\frac{2}{3} = 17\frac{1}{2}$
32. $X - 4\frac{1}{3} = 7\frac{3}{4}$
33. $\frac{1}{3} + Y = \frac{4}{7}$
34. $5X - \frac{1}{2} = \frac{3}{4}$
35. $X + X + \frac{3}{16} = \frac{7}{8}$
36. $4X - 2X = 6\frac{1}{3}$
37. $3T - 1\frac{2}{5} = 3\frac{1}{4}$

Write and solve an equation for each of the following:

38. Separate 112 into two parts so that one part is 17 more than the other.

39. Jim's father is twice as old as Jim is now. In five years, the sum of their ages will be 70 years. What is Jim's age now?

40. A newspaper article states that $\frac{3}{5}$ of a particular crop was damaged by the heavy rains. If 2,319 acres were damaged, approximately how many acres did the crop cover?

41. The sum of a number and 5 times the number is 138. Find the number.

42. Find a number such that when 3 times the number is increased by 12, the result is 54.

43. A house and lot together cost 35,000 dollars. The cost of the house was seven times the cost of the lot. How much did the lot cost?

44. Leta spent three times as much money on criminology books as she did on math books. Together all books cost 48 dollars. How much did she spend on the criminology books?

45. A 20-foot board is to be cut into two pieces so that the shorter piece is 2 feet shorter than the longer piece. Find the lengths of both pieces of board.

46. Six inches are to be cut from a piece of plastic pipe. The remaining piece is divided into thirds and each of the three equal parts is 16 inches long. How long was the original piece of pipe before any cuts were made?

47. If Gramsomamso has three times as many yards of green cloth as she has blue cloth, and together she has 16 yards of cloth, how many yards of blue cloth does she have?

Write a proportion and solve the following word problems:

48. If 3 gallons of paint cover a surface of 834 square feet, how many gallons at the same rate will be needed to paint a surfact containing 2,919 square feet?

49. If an automobile used 8 gallons of gasoline in traveling 116 miles, how many gallons will it use at the same speed in going 522 miles?

50. A steel beam 7 feet in length weighs 525 pounds. Find the weight of a similar steel beam $2\frac{1}{3}$ feet in length.

51. A cargo plane travels 1,204 miles in $3\frac{1}{2}$ hours. How far will the plane travel in $4\frac{1}{4}$ hours at the same speed?

52. If 2 pints of punch serve 3 people, how many pints are needed to serve 24 people?

53. How many radios can be assembled in $5\frac{1}{4}$ hours if 240 radios can be assembled in 7 hours?

54. If 1 inch represents 150 miles on a map, how many inches will represent a distance of 375 miles?

55. If a real estate tax is 14 dollars for each 1,000-dollar valuation, what tax must be paid on a home valued at 25,000 dollars?

56. If a 2-pound weight stretches a spring $\frac{1}{2}$ inch, how far will a 5-pound weight stretch the spring?

57. A recipe for 4 servings calls for 3 tablespoons of sugar. How much sugar is needed for 5 servings?

58. A seminar has a ratio of 2 men to 3 women. How many men are in the seminar if there are 35 members?

59. If 3 slices of bread contain 40 grams of carbohydrates, how many grams of carbohydrates will be contained in 7 slices of the same bread?

60. If 3 cans of soup sell for 48 cents, how much will 5 cans cost at the same unit price?

Use the given formulas to solve each of the following:

61. The formula $A + B + C = 180$ states that the sum of the angles of a triangle is 180 degrees. If angle A is 30 degrees, and angle B is 120 degrees, how many degrees are in angle C?

62. The horsepower (H) of an electric motor is equal to voltage (V) times amperes (I) divided by 70.

$$H = \frac{VI}{70}$$

(a) Find the horsepower (H) when $V = 35$ and $I = 20$.

(b) Find the voltage (V) when $H = \frac{25}{7}$ and $I = 20$.

(c) Find the amperes (I) when $H = 2\frac{6}{7}$ and $V = 20$.

63. The amount of energy (E) in mass (M) is given by the equation:

$$E = 200{,}000M$$

(a) Find the energy (E) when $M = 33$.
(b) Find the mass (M) when $E = 2{,}000{,}000$.

64. Refer to Frame 5, Lesson 17, and solve the following problems:
(a) A man walks 6 miles due east and then 8 miles due north. What is the shortest distance to his starting point?
(b) A hunter walks 5 miles due west and then 5 miles due south. What is the shortest distance to his starting point?

65. A formula relating pressure (P), in pounds per square inch, and depth of water (D), in feet, is given by:

$$P = \frac{1}{2}D + 15$$

(a) Find the pressure (P) at a depth (D) of 35 feet.
(b) Find the depth (D) when pressure (P) is 100 pounds per square inch.
(c) Find the depth (D) when pressure (P) is 200 pounds per square inch.

66. Use the formula $F = \frac{9}{5}C + 32$ to find the following. (Refer to Frame 12, Lesson 17, for the meaning of the formula.)
(a) F when $C = 200$
(b) C when $F = 50$

SELF-TEST: UNIT 4

Choose the correct equation for each of the following English sentences.

1. 4 more than the square of 7 is 53.
   (a) $(7 + 4)^2 = 53$
   (b) $7 + 4 = 53$
   (c) $7^2 + 4 = 53$
   (d) $49 = 53 - 4$
   (e) None of the above

2. The number 23 increased by the product of 3 and 4 is 35.
   (a) $23 + 3 \times 4 = 35$
   (b) $23 \times 3 + 4 = 35$
   (c) $23 = 35 - 12$
   (d) $23 + 3 + 4 = 35$
   (e) None of the above

3. 2 less than the cube root of 8 is 0.
   (a) $\sqrt[3]{(8 - 2)} = 0$
   (b) $(8 - 2)^3 = 0$
   (c) $\sqrt[3]{8} - 2 = 0$
   (d) $8^3 - 2 = 0$
   (e) None of the above

The following equations are either true or false:

4. $\sqrt{36} + 4 = 22$ is
   (a) True
   (b) False

5. $3^2 + 6^2 = 90 \div 2$ is
   (a) True
   (b) False

6. $3 \times 8 + 7 = 33 - 2$ is
   (a) True
   (b) False

Translate each of the following into open sentences:

7. The sum of my present age, Y years, and my age 7 years from now is 50.
   (a) $Y + (Y - 7) = 50$
   (b) $Y - Y + 7 = 50$
   (c) $Y + (Y + 7) = 50$
   (d) $Y - Y - 7 = 50$
   (e) None of the above

8. One-half the difference between some number, N, and 8 is 63.
   (a) $\frac{1}{2}N - 8 = 63$
   (b) $\frac{1}{2}(N - 8) = 62$
   (c) $N - 8 = \frac{1}{2}(63)$

(d) $N - \frac{1}{2}(8) = 63$

(e) None of the above

Find the solution for each of the following equations:

9. $W + 15 = 28$
   (a) 43
   (b) 15
   (c) 13
   (d) 28
   (e) None of the above

10. $\frac{X}{33} = 9$
   (a) 297
   (b) 33
   (c) 102
   (d) 96
   (e) None of the above

11. $6X - 11 = 31$
   (a) 11
   (b) 42
   (c) $3\frac{2}{3}$
   (d) 7
   (e) None of the above

12. $5X - 3X = 18$
   (a) 21
   (b) 36
   (c) 9
   (d) 2
   (e) None of the above

13. $\frac{5}{M} = \frac{8}{12}$
   (a) $7\frac{1}{2}$
   (b) 60
   (c) $3\frac{1}{3}$
   (d) 15
   (e) None of the above

Solve the following word problems:

14. Find the smallest number of a set of three consecutive whole numbers such that the sum of the three numbers is 72. (Let X be the smallest number.)
   (a) 21
   (b) 25
   (c) 24
   (d) 22
   (e) None of the above

15. In a local bond election, 1,266 more *yes* votes were cast than *no* votes. If N *no* votes were cast and the total number of votes cast was 6,214, find the number of *no* votes cast.
(a) 4,948
(b) 633
(c) 2,474
(d) 3,740
(e) None of the above

Solve each of the following problems by using a proportion:

16. If 3 pounds of ground beef cost $1\frac{1}{2}$ dollars, then how many pounds of ground beef will $2\frac{1}{2}$ dollars purchase?
(a) 4
(b) 6
(c) 7
(d) 5
(e) None of the above

17. On a map 2 inches represent 5 miles. If two points on the same map are $3\frac{1}{2}$ inches apart, then how many miles apart are the points?
(a) $16\frac{2}{3}$
(b) $1\frac{1}{3}$
(c) $8\frac{1}{3}$
(d) $6\frac{2}{3}$
(e) None of the above

Use the given formulas to solve each of the following.

18. If $P = 3L + 2W$, find W when $P = 45$ and $L = 5$.
(a) 45
(b) 15
(c) 30
(d) 20
(e) None of the above

19. If $V = XM^2$, find V when $X = 2$ and $M = 5$.
(a) 100
(b) 50
(c) 20
(d) $\frac{1}{3}$
(e) None of the above

20. If $H = 16T^2 - 7T$, find H when $T = 3$.
(a) 411
(b) 42
(c) 9
(d) 123
(e) None of the above

LESSON 14:

Pre-test

1. (a) $9^2 - 50 = 31$ (b) $51 = 6 \times 8 + 3$
2. (a) True (b) False
3. $N + 315 = 2{,}200$

Practice Exercises

Frame 8

1. $12 + 17 = 29$
2. $11 = 34 - 23$
3. $\frac{3}{5}(30) = 18$
4. $3^3 = 5^2 + 2$

Frame 11

1. True
2. False
3. True
4. True

Frame 22

1. $N + 5 = 31$
2. $N + (N + 1) = 17$
3. $\frac{1}{2}J + 10 = 145$
4. $3L + L = 84$

Post-test

1. (a) $5^3 + 25 = 150$ (b) $436 = \frac{1}{2}(922 - 50)$
2. (a) False (b) False
3. $2X + X = 624.$

LESSON 15:

Pre-test

1. (a) $X = \frac{4}{3}$ or $1\frac{1}{3}$ (b) $W = 12$
2. 18 years

Practice Exercises

Frame 13

1. $X = 16\frac{1}{3}$
2. $Y = 13\frac{1}{8}$
3. $N = \frac{8}{5}$ or $1\frac{3}{5}$
4. $X = 4$
5. $Z = 3\frac{3}{4}$
6. $X = 24$

Frame 20

1. $X = 5$ 2. $Y = \frac{23}{9}$ or $2\frac{5}{9}$ 3. $N = 4$ 4. $R = 20$

Frame 24

1. 8 and 9 2. 12 feet 3. 10 games

Post-test

1. (a) $M = 7$ (b) $X = 1$
2. $8\frac{1}{2}$ feet and $6\frac{1}{2}$ feet

LESSON 16:

Pre-test

1. (a) 15 (b) 12 cents (c) 69 feet

Practice Exercises

Frame 11

1. 600 minutes or 10 hours
2. 175 pounds
3. 25 gallons
4. 2,000 veterans

Post-test

1. (a) $58\frac{2}{7}$ (b) 3 cents (c) $533\frac{1}{3}$ miles

LESSON 17:

Pre-test

1. (a) 1,218 miles (b) $8\frac{1}{3}$ hours
2. (a) 1,000 widgets (b) 160 widgets

Practice Exercises

Frame 13

1. (a) $Q = 2$ (b) $P = 6$ dollars
2. (a) $D = 175$ miles (b) $R = 45$ miles per hour
3. (a) $H = 108$ feet

Post-test

1. (a) $F = 77$ degrees (b) $C = 37$ degrees
2. (a) $D = 150$ feet (b) $D = 266$ feet

Answers To Drill Exercises: Unit 4

1. $5^2 - 4 = 21$
2. $95 = 39 + 7 \times 8$
3. $5^2 = \sqrt{625}$
4. $45 + 4 \times 3 = 57$
5. $\sqrt{49} - 3 = 24 \div 6$
6. $\sqrt[3]{64} + 8 = \sqrt{144}$
7. True
8. False
9. False
10. True
11. True
12. False
13. $N + 48 = 72$
14. $N + (N + 1) = 47$
15. $3N + 4 = 16$
16. $(N + N) + 5 = 13$
17. $2X + 500 = 1{,}000$
18. $X = 11$
19. $M = 4$
20. $N = 6$
21. $X = 2{,}209$
22. $N = 3{,}935$
23. $X = 7$
24. $M = 2$
25. $W = 112$
26. $X = 225$
27. $N = 300$
28. $X = 4\frac{1}{2}$
29. $N = 1\frac{19}{35}$
30. $X = \frac{5}{13}$
31. $R = 9\frac{5}{6}$
32. $X = 12\frac{1}{12}$
33. $Y = \frac{5}{21}$
34. $X = \frac{1}{4}$
35. $X = \frac{11}{32}$
36. $X = 3\frac{1}{6}$
37. $T = 1\frac{11}{20}$
38. $47\frac{1}{2}$; $64\frac{1}{2}$
39. 20 years
40. 3,865 acres
41. 23
42. 14
43. 4,375 dollars
44. 36 dollars
45. 11 feet, 9 feet
46. 54 inches
47. 4 yards
48. $10\frac{1}{2}$ gallons
49. 36 gallons
50. 175 pounds
51. 1,462 miles
52. 16 pints
53. 180 radios
54. $2\frac{1}{2}$ inches
55. 350 dollars
56. $1\frac{1}{4}$ inches
57. $3\frac{3}{4}$ tablespoons
58. 14 men
59. $93\frac{1}{3}$ grams
60. 80 cents
61. $C = 30$ degrees
62. (a) $H = 10$ (b) $V = 12\frac{1}{2}$ (c) $I = 10$
63. (a) $E = 6{,}600{,}000$ (b) $M = 10$
64. (a) 10 miles (b) $\sqrt{50}$ miles, a little more than 7 miles
65. (a) $P = 32\frac{1}{2}$ pounds per square inch (b) $D = 170$ feet (c) $D = 370$ feet
66. (a) $F = 392$ degrees (b) $C = 10$ degrees

Answers To Self-Test: Unit 4

1. (c)
2. (a)
3. (c)
4. (b)
5. (a)
6. (a)
7. (c)
8. (e)
9. (c)
10. (a)
11. (d)
12. (c)
13. (a)
14. (e)
15. (c)
16. (d)
17. (e)
18. (b)
19. (b)
20. (d)

UNIT FIVE

Decimals and Decimal Operations

Lesson 18	DECIMAL NUMERATION
Lesson 19	DECIMAL OPERATIONS
Lesson 20	DECIMAL DIVISION
Lesson 21	ADDITIONAL DECIMAL TOPICS

LESSON 18
DECIMAL NUMERATION

Objectives:

1. Rewrite a decimal as a fraction or a mixed number. (1–11)
2. Rewrite a decimal in expanded notation, using the exponential forms of the powers of ten. (12–14)
3. Determine the name of the place-value position of specified digits in a given decimal. (15–17)
4. Rewrite a decimal in word form. (18–22)

Vocabulary:

Decimal Numeral (Decimal)

Decimal Point

PRE-TEST

1. Rewrite 3.02 as a mixed number.
2. Rewrite 23.8075 in expanded notation, using the exponential forms of the powers of ten.
3. Name the place-value position of each of the underlined digits.
   A. $48.0\underline{4}8$
   B. $503.503\underline{7}$
4. Rewrite 407.0308 in word form.

The addition and subtraction of fractions and mixed numbers require that close attention be paid to the denominators. ①

$$\frac{13}{25} + \frac{19}{47} \qquad\qquad 12\frac{71}{123} - 3\frac{3}{404}$$

Some fractions have denominators that make computation extremely difficult.

Certain fractions have denominators that are easier to deal with, for example, fractions that have denominators that are powers of ten. ②

$$\frac{3}{10} - \frac{17}{100} \qquad\qquad 1\frac{21}{100} + 6\frac{97}{1{,}000}$$

To change any of these fractions to other equivalent fractions with the LCD as their denominators, it is only necessary to multiply the numerator and the denominator by the same power of 10.

Recall from the meaning of a fraction that one interpretation of $\frac{3}{10}$ and $\frac{17}{100}$, for example, is (3)

$$\frac{3}{10} \quad \text{means} \quad 3 \times \frac{1}{10} \qquad\qquad \frac{17}{100} \quad \text{means} \quad 17 \times \frac{1}{100}$$

The Powers of Ten Pattern can be extended to include fractions such as $\frac{1}{10}$ and $\frac{1}{100}$: (4)

$$\begin{array}{lll} \vdots & & \\ 10^2 = 100 & & \text{one hundred} \\ 10^1 = 10 & & \text{ten} \\ 10^0 = 1 & & \text{one} \\ \frac{1}{10^1} = \frac{1}{10} & & \text{one-tenth} \\ \frac{1}{10^2} = \frac{1}{100} & & \text{one-hundredth} \\ \vdots & & \end{array}$$

A *decimal numeral* is a numeral that represents the sum of a whole number and a fraction that has a denominator that is a power of ten. A *decimal point* is used to separate the whole number portion of the decimal numeral from the fractional portion. Decimal numerals are commonly called *decimals.* (5)

For example, the decimal numeral, 12.13:

whole number		fractional number
12	.	13

The decimal numeral 12.13 is read "twelve *and* thirteen-hundredths." The decimal point is always read as "and."

The meaning of a decimal numeral: (6)

$$12.13 = 12\frac{13}{100} \quad \text{means} \quad 12 + .13 = 12 + \frac{13}{100}$$

$$0.3 = \frac{3}{10} \quad \text{means} \quad 0 + .3 = 0 + \frac{3}{10}$$

Every whole number can be written as a decimal numeral by placing a decimal point to the right of the digit in the one's position. 7

For example,

13	can be written	**13. or 13.0**
115	can be written	**115. or 115.0**

A zero may be written to the right of the decimal point (.) so that it will not be mistaken for a period.

The Powers of Ten Pattern using Decimal Notation: 8

⋮	⋮	
$10^3 =$	1,000.0	one thousand
$10^2 =$	100.0	one hundred
$10^1 =$	10.0	ten
$10^0 =$	1.0	one
$\frac{1}{10^1} =$	0.1	one-tenth
$\frac{1}{10^2} =$	0.01	one-hundredth
$\frac{1}{10^3} =$	0.001	one-thousandth
⋮	⋮	

Notice that the number of digits to the *right* of the decimal point is counted by the exponent that appears in the denominator of the equivalent fraction.

When you rewrite a decimal as a fraction (or mixed number), the fractional part of the decimal numeral becomes the numerator of the equivalent fraction. The denominator is 10 raised to a power such that the exponent counts the number of digits to the right of the decimal point in the decimal numeral. 9

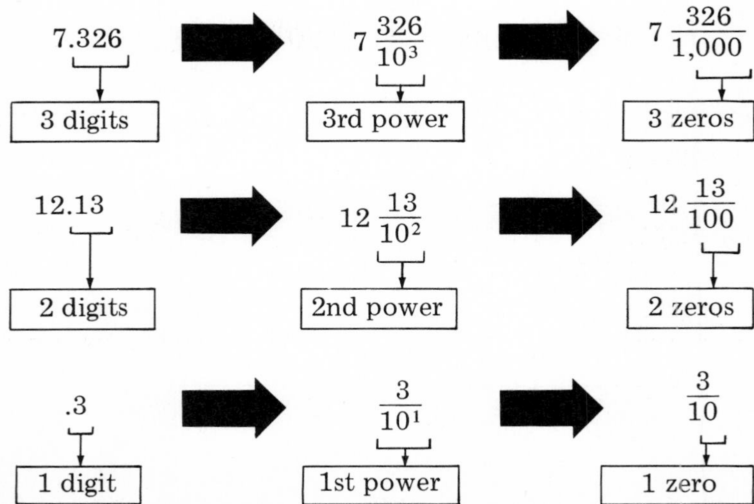

Rewriting decimals as fractions or mixed numbers: 

Complete the following:

A. $3.01 = 3\frac{1}{10^2} = 3\frac{1}{\square}$

B. $0.073 = \frac{73}{10^3} = \frac{73}{\bigcirc}$

C. $181.4 = 181\frac{\triangle}{10^1} = 181\frac{\triangle}{\hexagon}$ or $181\frac{2}{5}$

Practice Exercises (11)

Rewrite each decimal as a fraction or mixed number. Reduce all fractions to lowest terms.

1. 0.5	2. 0.0001	3. 7.12
4. 103.03	5. 1.41	6. 0.25

Rewriting a decimal numeral in expanded notation: 7.326 

$$
\begin{aligned}
7.326 &= 7 + 0.326 \\
&= 7 + 0.3 + 0.02 + 0.006 \\
&= 7 + \frac{3}{10^1} + \frac{2}{10^2} + \frac{6}{10^3} \\
&= (7 \times 10^0) + \left(3 \times \frac{1}{10^1}\right) + \left(2 \times \frac{1}{10^2}\right) + \left(6 \times \frac{1}{10^3}\right)
\end{aligned}
$$

Rewrite 32.85 in expanded notation: (13)

Complete the following:

$$
\begin{aligned}
32.85 &= 32 + 0.85 \\
&= 30 + 2 + 0.8 + 0.05 \\
&= 30 + 2 + \frac{8}{10^1} + \frac{5}{10^2} \\
&= \left(\square \times 10^1\right) + \left(\bigcirc \times 10^0\right) + \left(\triangle \times \frac{1}{10^1}\right) + \left(\hexagon \times \frac{1}{10^2}\right)
\end{aligned}
$$

Practice Exercises 

Rewrite each of the following in expanded notation, using the exponential forms of the powers of ten:

1. 53.832 2. 9.1548
3. 0.3027 4. 0.0904

The place-value names for decimal numerals: 

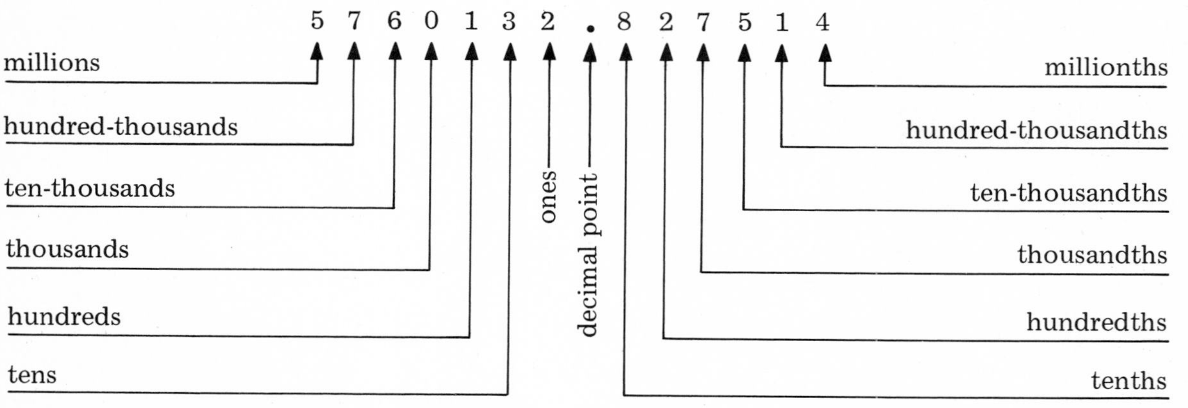

Note that the place-value names, tens and tenths, hundreds and hundredths, and so on, are symmetrical about the *one's position* rather than about the decimal point.

Write the correct place-value names for each of the digits in 367.214: 

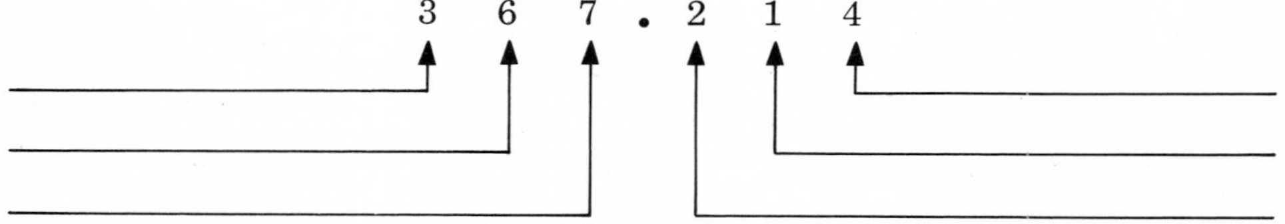

Practice Exercises 

Name the place value of each underlined digit in the following decimal numerals:

1. $42.3\underline{1}72$ __________ 2. $26\underline{3}.2171$ __________
3. $4321.627\underline{1}$ __________ 4. $831.27\underline{2}1$ __________
5. $\underline{4}267.2472$ __________ 6. $0.0031\underline{5}$ __________

Rewriting a decimal numeral in word form: 7.326 (18)

$$\text{Since } \mathbf{7.326 = 7\frac{326}{1{,}000}}$$

The decimal numeral is read and written exactly as the equivalent fraction is:

Seven *and* three hundred twenty-six thousandths

Remember that the decimal point is read and written as "and."

Rewriting a decimal in word form: (19)

It is not necessary to write the equivalent fractional form of a decimal before writing it in word form. The fractional portion of the decimal can be named by determining the place value of the last digit to the right of the decimal point.

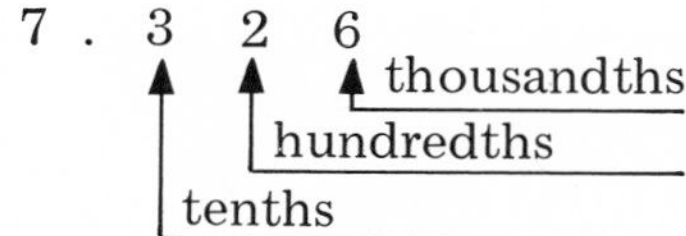

Since 6, the last digit to the right of the decimal point, is in the *thousandths* position, the fractional portion of the decimal is written three hundred twenty-six *thousandths.*

Rewriting a decimal in word form: (20)

12.13 is written **twelve *and* thirteen hundredths** since 3 is in the hundredths position.

1 2 . 1 3
- 3 → hundredths
- 1 → tenths

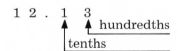

181.4 is written **one hundred eighty-one *and* four tenths** since 4 is in the tenths position.

1 8 1 . 4
- 4 → tenths

Rewriting a decimal in word form: (21)

Complete the following by writing the correct place-value name:

17 . 0 1 (0 → tenths, 1 → hundredths) is written seventeen and one ____________________

7 . 5 (5 → tenths) is written seven and five ____________________

Practice Exercises

Rewrite the following decimal numerals in word form:

1. 0.71
2. 3.0036
3. 100.001
4. 267.7

POST-TEST

1. Rewrite 0.0025 as a fraction, reduced to lowest terms.
2. Rewrite 17.021 in expanded notation, using the exponential forms of the powers of ten.
3. Name the place-value position of each of the underlined digits.
   (a) $83.\underline{7}05$
   (b) $170.09\underline{2}4$
4. Rewrite 906.00906 in word form.

LESSON 19
DECIMAL OPERATIONS

Objectives:

1. Determine the sum of two or more decimals. (1–6)
2. Determine the difference between two decimals. (7–11)
3. Determine the product of two or more decimals. (12–19)
4. Determine the value of a power of a decimal. (15)
5. Multiply a decimal by a power of ten using the decimal-shift method. (22–26)

Vocabulary:

Placeholder
Decimal-shift Method

PRE-TEST

1. Find each sum.
   (a)
   $$\begin{array}{r} 45.32 \\ 91.07 \\ +40.93 \\ \hline \end{array}$$
   (b) 3.5 + 7.08 + 11.109

2. Find each difference.
   (a)
   $$\begin{array}{r} 11.308 \\ -\ 7.979 \\ \hline \end{array}$$
   (b) 0.056 - 0.0328

3. Find each product.
   (a)
   $$\begin{array}{r} 9.87 \\ \times\ .56 \\ \hline \end{array}$$
   (b) 8.3 × 0.0039

4. Determine the value of $(0.1)^2$
5. Multiply the following by using the decimal-shift method: 4.3081 × 100,000

Adding decimals: 5.38 + 1.21 ①

"How" → "Why"

$$\begin{array}{r} 5.38 \\ +1.21 \\ \hline 6.59 \end{array}$$

$$\begin{array}{r} (5 \times 1) + \left(3 \times \frac{1}{10}\right) + \left(8 \times \frac{1}{100}\right) \\ +(1 \times 1) + \left(2 \times \frac{1}{10}\right) + \left(1 \times \frac{1}{100}\right) \\ \hline (6 \times 1) + \left(5 \times \frac{1}{10}\right) + \left(9 \times \frac{1}{100}\right) \end{array}$$

To add decimals, add the digits of like-place value. Note the alignment of the decimal points in the addends and the sum.

Adding decimals when carrying is necessary: 7.84 + 3.17 ②

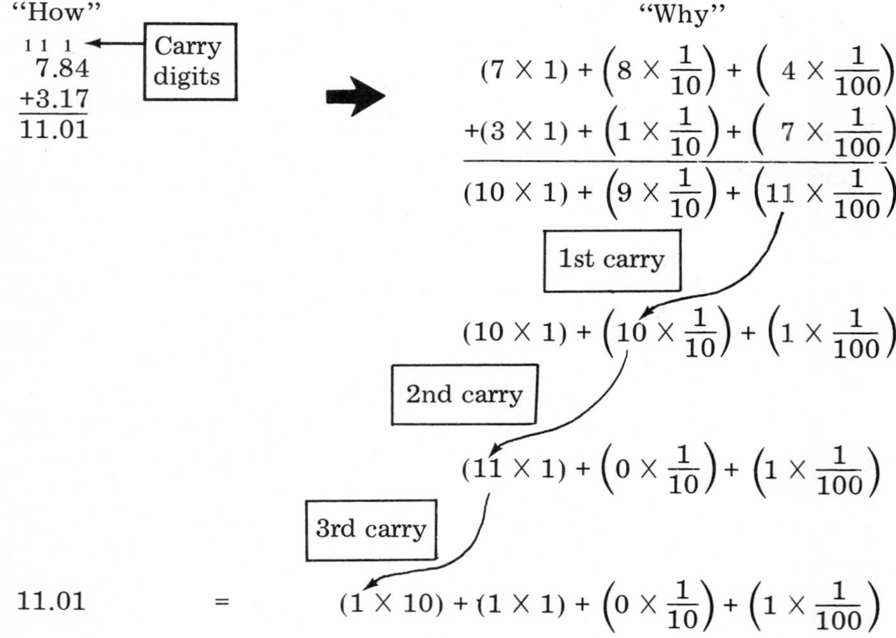

$$11.01 = (1 \times 10) + (1 \times 1) + \left(0 \times \frac{1}{10}\right) + \left(1 \times \frac{1}{100}\right)$$

Notice that the digit 1 appears in the ten's position in the sum even though neither addend has a digit in this place-value position.

Adding decimals: 3.1 + 12.537 ③

"How"

$$\begin{array}{r} 3.1 \\ +12.537 \\ \hline \end{array} \quad \rightarrow \quad \begin{array}{r} 3.100 \\ +12.537 \\ \hline 15.637 \end{array}$$

"Why"

$$\begin{array}{r} 3.1 \\ +12.537 \\ \hline \end{array} \rightarrow \begin{array}{r} 3\frac{1}{10} \\ +12\frac{537}{1,000} \\ \hline \end{array} \rightarrow \begin{array}{r} 3\frac{100}{1,000} \\ +12\frac{537}{1,000} \\ \hline 15\frac{637}{1,000} \end{array}$$

To add decimals whose fractional portions have differing numbers of digits, simply write as many additional zeros as necessary after the last digit to the right of the decimal point.

Adding decimals: 4.5 + 22.17 + 1.084 (4)

$$\begin{array}{r} 4.5 \\ 22.17 \\ +\ 1.084 \\ \hline \end{array} \rightarrow \begin{array}{r} \overset{1}{4}.500 \\ 22.170 \\ +\ 1.084 \\ \hline 27.754 \end{array}$$

First write the problem, aligning the decimal points; then write in the necessary zeros in the fractional portions of the decimals.

Adding decimals: 23.604 + 527 + 181.61 (5)

Complete the following by writing in the necessary zeros and adding:

$$\begin{array}{r} 23.604 \\ 527. \\ +181.61 \\ \hline \end{array}$$

Practice Exercises (6)

Calculate the sum for each of the following:

1. $\begin{array}{r} 25.607 \\ 3.6 \\ +20.004 \\ \hline \end{array}$

2. $\begin{array}{r} 0.0071 \\ 0.1763 \\ +0.92 \\ \hline \end{array}$

3. 356 + 0.26 + 3.71

4. 27.301 + 9.92 + 0.0007

Subtracting decimals: 7.46 – 1.24 (7)

"How"

$$\begin{array}{r} 7.46 \\ -1.24 \\ \hline 6.22 \end{array}$$

"Why"

$$\begin{array}{rl} & (7 \times 1) + \left(4 \times \frac{1}{10}\right) + \left(6 \times \frac{1}{100}\right) \\ \boxed{\text{Subtract}} & (1 \times 1) + \left(2 \times \frac{1}{10}\right) + \left(4 \times \frac{1}{100}\right) \\ \hline & (6 \times 1) + \left(2 \times \frac{1}{10}\right) + \left(2 \times \frac{1}{100}\right) \end{array}$$

The digits of like place value are subtracted.

Subtracting decimals when borrowing is necessary: 1.34 - 0.27 

"How" "Why"

$$\begin{array}{r} 1\,.\,\overset{2}{\not{3}}\,\overset{1}{4} \\ -\,0\,.\,2\,7 \\ \hline \end{array} \rightarrow \boxed{\text{Subtract}} \begin{array}{r} 1 + 0.3 + 0.04 \\ 0.2 + 0.07 \\ \hline \end{array} \rightarrow \boxed{\text{Subtract}} \begin{array}{r} 1 + 0.2 + 0.14 \\ 0.2 + 0.07 \\ \hline 1 + 0.0 + 0.07 \end{array}$$

Since 0.07 cannot be subtracted from 0.04, it is necessary to "borrow" one-tenth from 0.3 and add it to 0.04.

Subtracting decimals: 2.2 - 0.14 

"How"

$$\begin{array}{r} 2.2 \\ -0.14 \\ \hline \end{array} \rightarrow \begin{array}{r} 2\,.\,2\,0 \\ -\,0\,.\,1\,4 \\ \hline 2\,.\,0\,6 \end{array}$$

"Why"

$$\begin{array}{r} 2.2 \\ -0.14 \\ \hline \end{array} \rightarrow \begin{array}{r} 2\frac{2}{10} \\ -\frac{14}{100} \\ \hline \end{array} \rightarrow \begin{array}{r} 2\frac{20}{100} \\ -\frac{14}{100} \\ \hline 2\frac{6}{100} \end{array}$$

If the fractional portions of the decimals being subtracted have differing numbers of digits to the right of the decimal point, the necessary zeros can be written to the right of the last digit.

Subtracting decimals: 6 - 0.524 

Complete the following by writing in the necessary zeros and subtracting:

$$\begin{array}{r} 6. \\ -0.524 \\ \hline \end{array}$$

Practice Exercises 

Find each difference.

1. $\begin{array}{r} 27.602 \\ -13.871 \\ \hline \end{array}$

2. $\begin{array}{r} 12.572 \\ -\ \ 6.784 \\ \hline \end{array}$

3. 578.21 - 31.7573

4. 0.7 - 0.125

Multiplying decimals: 3.2 × 6.4 (12)

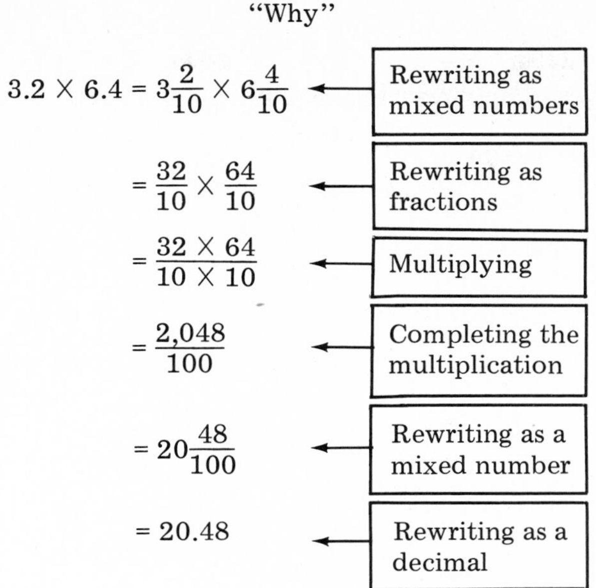

"Why"
$3.2 \times 6.4 = 3\frac{2}{10} \times 6\frac{4}{10}$
Rewriting as mixed numbers
$= \frac{32}{10} \times \frac{64}{10}$
Rewriting as fractions
$= \frac{32 \times 64}{10 \times 10}$
Multiplying
$= \frac{2{,}048}{100}$
Completing the multiplication
$= 20\frac{48}{100}$
Rewriting as a mixed number
$= 20.48$
Rewriting as a decimal

Multiplying decimals: 3.2 × 6.4 (13)

"How"

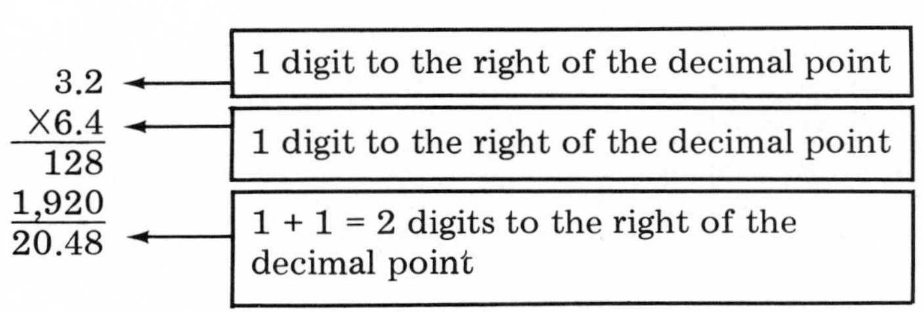

To multiply 3.2 × 6.4, calculate the product as if 32 were being multiplied by 64. Then, place the decimal point in the product by counting the number of digits to the right of the decimal point in each factor. Add these numbers and place the decimal point in the product by counting over this number of places from the right.

Multiplying decimals: 32.42 × 6.7 (14)

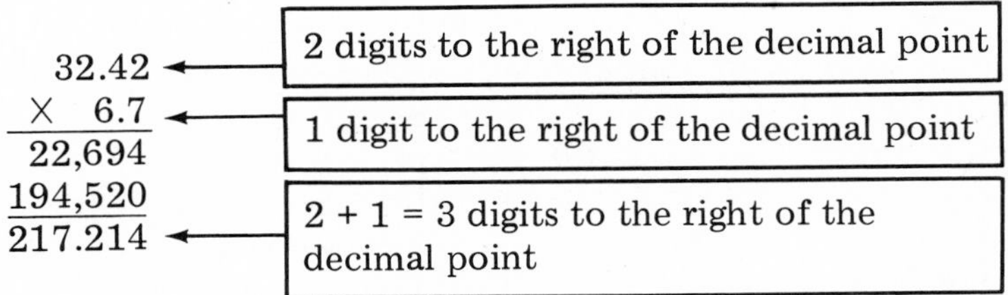

The placement of the decimal point may be checked by thinking hundredths times tenths is thousandths: $\frac{1}{100} \times \frac{1}{10} = \frac{1}{1{,}000}$.

Raising decimals to powers: 

$(0.2)^2$ means $(0.2)(0.2)$

$(0.03)^2$ means $(0.03)(0.03)$

To raise a decimal to a power, simply rewrite as a multiplication and determine the product.

$\mathbf{(0.2)^2 = (0.2)(0.2) = 0.04}$ $\mathbf{(0.03)^2 = (0.03)(0.03) = 0.0009}$

Multiplying decimals: 2.1 × 0.57 

Complete the following by correctly placing the decimal point in the product:

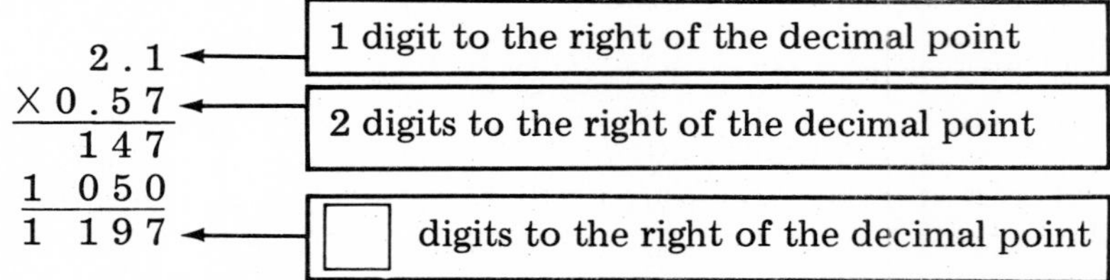

Check by thinking tenths times hundredths is thousandths.

Multiplying decimals: 0.07 × 0.6 

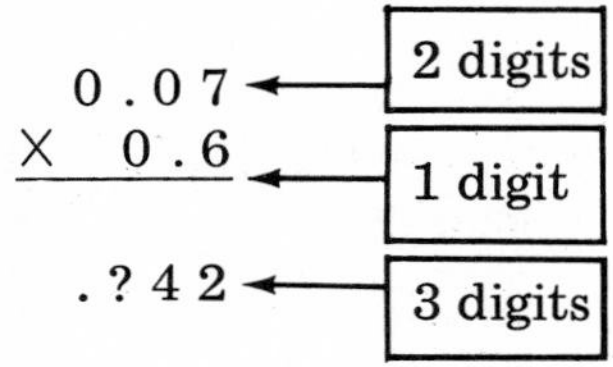

$$\begin{array}{r} 0.07 \\ \times\ \ 0.6 \\ \hline 0.042 \end{array}$$

A zero must be written before the 4 in order to correctly position the decimal point in the product. As many zeros as are necessary must be written to the left of the digits in the product if the correct number of decimal places in the product exceeds the number of digits.

Multiplying decimals: 0.105 × 0.08 

Complete the following by correctly placing the decimal point in the product, writing in the necessary number of zeros:

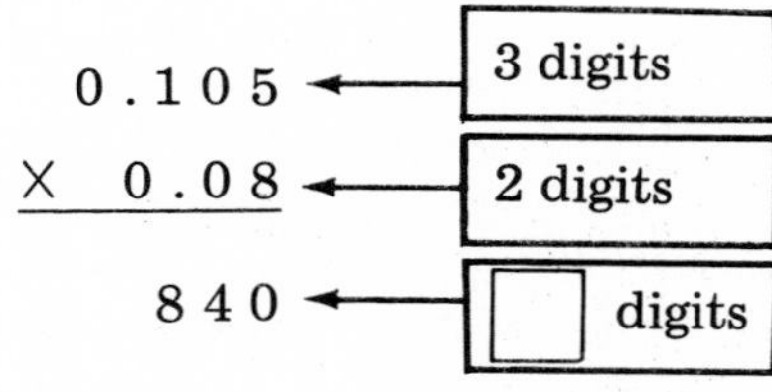

Practice Exercises

19

Perform each indicated multiplication.

1. $\begin{array}{r} 2.6 \\ \underline{\times 0.3} \end{array}$

2. $\begin{array}{r} 781.1 \\ \underline{\times\ 2.43} \end{array}$

3. $\begin{array}{r} 31.02 \\ \underline{\times\ 0.04} \end{array}$

4. $\begin{array}{r} 0.63 \\ \underline{\times 0.07} \end{array}$

5. 0.003×0.21

6. $(0.24)^2$

Zeros in decimal products: 0.00840

It was determined in Frame 18 that the product of 0.105 and 0.08 is 0.00840. The two zeros between the decimal point and the digit 8 are absolutely necessary as *placeholders*. That is, the zeros "hold" the unfilled place-value positions, tenths and hundredths.

$$0\,.\,\underbrace{\underline{0}\,\underline{0}}_{\text{placeholders}}\,8\,4\,0$$

Zeros in decimal products: 0.00840

Consider this product once more. While the zeros between the decimal point and the digit 8 are essential, what about the zero to the right of the digit 4? Is it also necessary?

$$0\,.\,0\,0\,8\,4\,\underline{0}$$

It can be shown by using cancellation that 0.00840 = 0.0084:

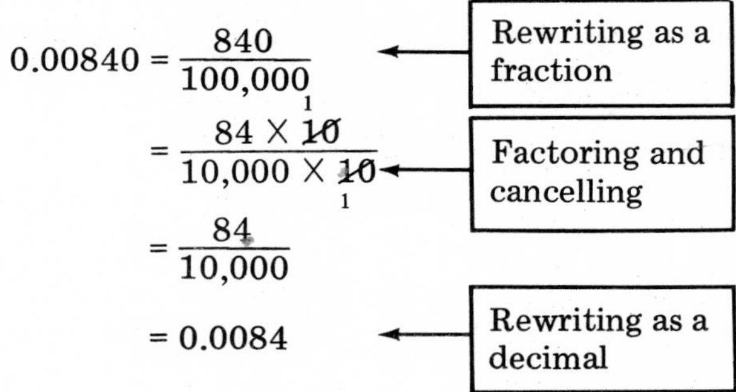

0.00840 = 840/100,000
= (84 × 10)/(10,000 × 10)
= 84/10,000
= 0.0084
Rewriting as a fraction
Factoring and cancelling
Rewriting as a decimal

In the product, 0.00840, the last zero may be omitted since 0.00840 = 0.0084. In fact, in any decimal numeral all zeros to the right of the last non-zero digit may be omitted without changing the value of the numeral.

Multiplying a decimal by a power of ten: 4.631 × 10 (22)

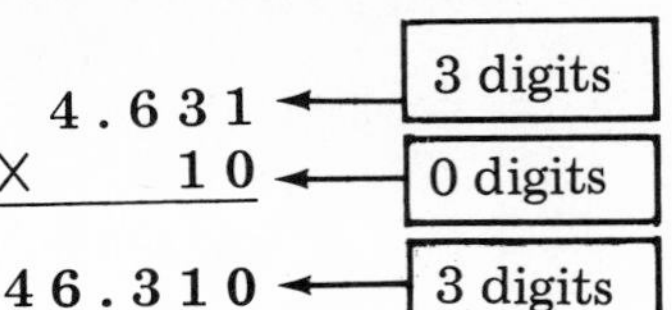

It was shown in the last frame that 46.310 may be written as 46.31. Compare this product with the factor, 4.631:

4.631

46.31

Notice that multiplying by 10 shifted the decimal point one place to the right.

Multiplying 4.631 by a power of ten: (23)

Consider the following patterns:

4.631 × 10	4.631 × 10^1	46.31
1 zero →	1st power →	1 place
4.631 × 100	4.631 × 10^2	463.1
2 zeros →	2nd power →	2 places
4.631 × 1,000	4.631 × 10^3	4631.0
3 zeros →	3rd power →	3 places
4.631 × 10,000	4.631 × 10^4	46310.0
4 zeros →	4th power →	4 places

When a decimal is multiplied by a power of ten, the decimal point is shifted to the right the number of places counted by the exponent of the power of ten. The number of places that the decimal point is shifted is also counted by the number of zeros in the power of ten. Notice in the last example that it was necessary to write an additional zero to the right of the digit 1 in order to correctly position the decimal point.

Multiplying 34 by a power of ten:  (24)

Since any whole number can be written as a decimal numeral,

34 × 10 ⟶ 34.0 × 10 ⟶ 340

34 × 100 ⟶ 34.0 × 100 ⟶ 3400

34 × 1,000 ⟶ 34.0 × 1,000 ⟶ 34000

Any whole number can be multiplied by a power of ten by using the *decimal-shift method*.

Multiplying by a power of ten: (25)

Complete the following by correctly placing the decimal point in the product, writing in any necessary zeros as needed:

A. 83.1279 × 100 = 8 3 1 2 7 9

B. 0.0045 × 1,000 = 0 0 4 5

C. 51 × 10,000 = 5 1

Practice Exercises (26)

Multiply the following by using the decimal-shift method:

1. 23.741 × 100 = ____________
2. 0.00671 × 1,000 = ____________
3. 8.2671 × 10 = ____________
4. 36.87 × 100,000 = ____________
5. 0.02 × 1,000 = ____________
6. 7.36 × 10,000 = ____________
7. 81 × 100 = ____________

POST-TEST

1. Find each sum.

   (a)
   $$\begin{array}{r} 103.62 \\ 85.97 \\ +\ 70.06 \\ \hline \end{array}$$

   (b) 0.03 + 1.8 + 0.379

2. Find each difference.

   (a)
   $$\begin{array}{r} 33.333 \\ -17.578 \\ \hline \end{array}$$

   (b) 2.07 − 1.758

3. Find each product.

   (a)
   $$\begin{array}{r} 57.08 \\ \times\ \ 1.9 \\ \hline \end{array}$$

   (b) 0.018 × 0.0027

4. Determine the value of $(0.3)^3$.
5. Multiply the following by using the decimal-shift method:
   0.0003491 × 100,000

LESSON 20
DECIMAL DIVISION

Objectives:

1. Determine the quotient for two decimals. (1–13)
2. Round a decimal to a specified place-value position. (14–28)
3. Divide a decimal by a power of ten using the decimal-shift method. (29–34)
4. Rewrite a fraction as a decimal, rounding to a specified place-value position if the decimal is non-terminating. (35–41)
5. Rewrite a repeating decimal as a fraction. (42–52)

Vocabulary:

Rounding	Terminating Decimal
Test Digit	Repeating Decimal

PRE-TEST

1. Divide until a zero remainder is reached: $3.52 \div 1.6$
2. Round 3,742.0726 to the nearest thousandth.
3. Divide by using the decimal-shift method: $1781.5 \div 1{,}000$
4. Rewrite $\frac{4}{7}$ as a decimal, correct to the nearest thousandth.
5. Rewrite $0.\overline{3}$ as a fraction reduced to lowest terms.

During one 4-month period from the beginning of January to the end of April, the total measured rainfall in one northern California community was 20.2 inches. What was the average monthly rainfall during this period of time? (1)

In order to determine the average monthly rainfall, it is necessary to divide the 20.2 inches of rain by the number of months, 4.

Dividing a decimal by a whole number: (2)

$$20.2 \div 4 = ?$$

Since 20.2 inches is between 20 inches and 24 inches,

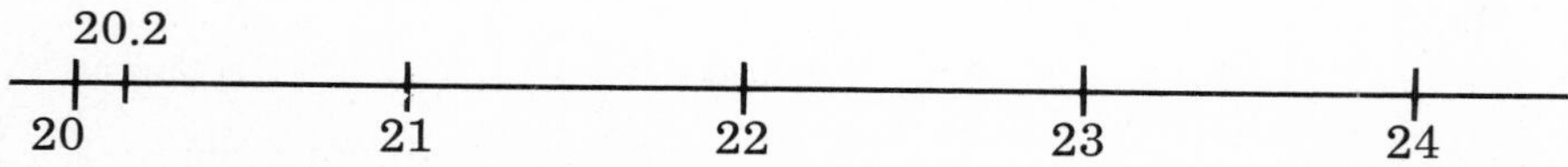

And since 20 ÷ 4 = 5 and 24 ÷ 4 = 6, a reasonable estimate of the quotient is that it is a number between 5 and 6.

Dividing a decimal by a whole number: 20.2 ÷ 4 (3)

It is possible to find the quotient by rewriting 20.2 as a mixed number and then dividing by 4.

$$20.2 \div 4 = 20\frac{2}{10} \div 4$$

$$= \frac{202}{10} \div 4$$ ← Rewriting as a fraction

$$= \frac{202}{10} \times \frac{1}{4}$$ ← Multiplying by the reciprocal of 4

$$= \frac{101}{20}$$ ← Completing the multiplication

$$= 5\frac{1}{20}$$ ← Rewriting as a mixed number

The mixed number $5\frac{1}{20}$ may be written as a decimal by using "Multiplication by One":

$$5\frac{1}{20} = \frac{101}{20} \times \frac{5}{5} = \frac{505}{100} = 5\frac{5}{100} \text{ or } 5.05$$

Thus, 20.2 ÷ 4 = 5.05, a number between 5 and 6, as estimated.

Dividing a decimal by a whole number: 20.2 ÷ 4 (4)

The division of a decimal by a whole number is done exactly as a division of a whole number by another whole number once the decimal point is positioned in the quotient.

$$4\overline{)20.2} \quad \Rightarrow \quad 4\overline{)20\overset{.}{\uparrow}2}$$

The decimal point is placed in the quotient immediately above the decimal point in the dividend. Then, proceeding with the division, we get

```
     5 . 0
 4 ) 2 0 . 2
     2 0
     ---
         2
         0
         -
         2  ← Non-zero remainder
```

If the division is stopped at this point, it would be necessary to write the quotient as $5.0\frac{1}{2}$, a rather unusual, but acceptable, numeral form. But since it was shown in a previous lesson that additional zeros can be written to the right of the digit 2, it is possible to continue the division:

```
      5.05
  4 ⟌20.20
     20
      2
      0
      20
      20
       0  ← Zero remainder
```

The quotient, 5.05, is identical to the quotient obtained when $20\frac{2}{10}$ was divided by 4.

Dividing a decimal by a whole number: 2.64 ÷ 8 ⑤

```
8⟌2.64   ➡   8⟌2↑64   ➡      .33
                           8⟌2.64
                             2 4
                               24
                               24
                                0
```

In this example, no additional zeros are needed to carry out the division to a zero remainder. To check a decimal division that has a zero remainder, multiply the divisor by the quotient: 8 × .33 = 2.64. Since this product is equal to the dividend, the quotient is correct.

Dividing a decimal by a whole number: 11.73 ÷ 20 ⑥

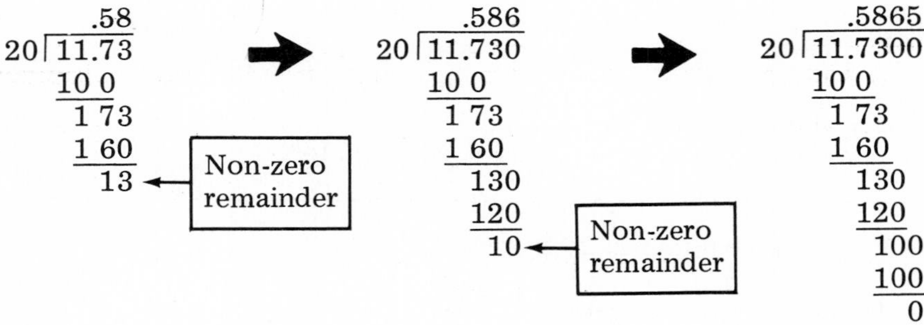

```
     .58                   .586                   .5865
20⟌11.73        ➡     20⟌11.730        ➡      20⟌11.7300
   10 0                   10 0                   10 0
    1 73                   1 73                   1 73
    1 60                   1 60                   1 60
      13 ← Non-zero          130                    130
           remainder         120                    120
                              10 ← Non-zero          100
                                   remainder         100
                                                       0
```

It is necessary to write in two additional zeros in order to carry out this division to a zero remainder.

Dividing a decimal by a whole number: 0.0036 ÷ 6 ⑦

"How"

```
      .   6               .0006
  6⟌0↑0036     ➡      6⟌0↑0036
         36                  36
          0                   0
```

"Why"

$$0.0036 \div 6 = \frac{36}{10{,}000} \div 6 = \frac{36}{10{,}000} \times \frac{1}{6} = \frac{6}{10{,}000} = 0.0006$$

One or more zeros may have to be written between the decimal point and the first non-zero digit in the quotient in order to locate the digit in the correct place-value position.

Dividing a decimal by a whole number: $0.153 \div 17$ ⑧

$17\overline{)0.153}$ (dot above the 1)	➡	Think: 1 can't be divided by 17. Write a zero in the quotient immediately above the 1.	→ $\begin{array}{r} .0 \\ 17\overline{)0.153} \end{array}$
$\begin{array}{r} .0 \\ 17\overline{)0.153} \end{array}$	➡	Think: 15 can't be divided by 17. Write a second zero in the quotient immediately above the 5.	→ $\begin{array}{r} .00 \\ 17\overline{)0.153} \end{array}$
$\begin{array}{r} .00 \\ 17\overline{)0.153} \end{array}$	➡	Think: 153 can be divided by 17; $153 \div 17 = 9$. Write the digit 9 in the quotient immediately above the 3.	→ $\begin{array}{r} .009 \\ 17\overline{)0.153} \\ \underline{153} \\ 0 \end{array}$

Checking the quotient: $0.009 \times 17 = 0.153$. Thus, $0.153 \div 17 = 0.009$

Practice Exercises ⑨

Divide each of the following, carrying out the division to a zero remainder:

1. $4\overline{)125.6}$
2. $12\overline{)1.404}$
3. $200\overline{)0.28}$
4. $35\overline{)59.535}$

Dividing a decimal by another decimal: $0.64 \div 0.8$ ⑩

"How"

$.8\overline{).64}$ (decimal points moved one place right) ➡ $\begin{array}{r} .8 \\ 8.\overline{)6.4} \\ \underline{6\ 4} \\ 0 \end{array}$

"Why"

$0.64 \div 0.8 = \dfrac{0.64}{0.8}$	←	Meaning of a fraction
$= \dfrac{0.64}{0.8} \times \dfrac{10}{10}$	←	"Multiplication by One"
$= \dfrac{0.64 \times 10}{0.8 \times 10}$	←	Multiplying
$= \dfrac{6.4}{8}$	←	Decimal-shift multiplication
$= 6.4 \div 8$	←	Rewriting the fraction as a division

Since 0.64 ÷ 0.8 = 6.4 ÷ 8, the division can be done in exactly the same way that a decimal is divided by a whole number once the decimal point is correctly located in the quotient. Shift the decimal point in the divisor to the right until it represents a whole number. Shift the decimal point in the dividend the same number of places to the right. The decimal point in the quotient is directly above the decimal point in the dividend.

Dividing a whole number by a decimal: 132 ÷ 0.03 (11)

Since 132 = 132.0, the division is done exactly as it was in the example in the last frame.

$0.03\overline{)132}$ ➡ $.03.\overline{)132.00.0}$ ➡ $3\overline{)13{,}200.0}$

```
   4,400.0 or 4,400
3 |13,200.0
   12
    12
    12
     0
```

Checking the quotient: 4,400 × 0.03 = 132.00; the quotient is correct.

Dividing a whole number by a decimal: 7 ÷ 0.14 

Complete the following by correctly placing the decimal point in the dividend and the quotient:

$0.14\overline{)7}$ ➡

```
           5 0 0
0 . 1 4 |7 0 0 0
           7 0
```

Checking the quotient: ☐ × 0.14 $\stackrel{?}{=}$ 7

Practice Exercises (13)

Divide each of the following, carrying out the division to a zero remainder:

1. $0.055\overline{)132}$
2. $0.08\overline{)9.824}$
3. $3.6\overline{)54}$
4. $4.8\overline{)27.792}$

Divide: 5.7 ÷ 0.9 (14)

$0.9\overline{)5.7}$ ➡

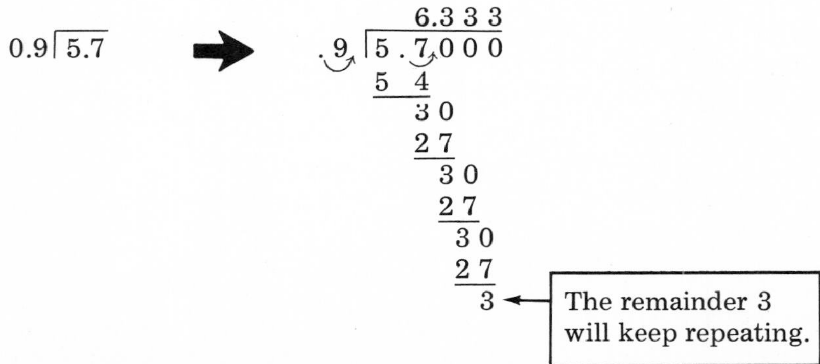

No matter how many additional zeros are written in the dividend, this division, and most decimal divisions, will never divide to a zero remainder. For this reason, most decimal divisions specify the desired accuracy of the quotient by some statement such as, "Correct to the nearest hundredth," for example. This specifies that the quotient is to be *rounded* to the hundredth's position.

Rounding a decimal: 6.333 to the nearest hundredth (15)

The first thing that must be done when rounding a number is to identify the specified place-value position. The next digit to the right of the specified place value is called the *test digit.*

6.333

hundredths ↑ ↑ test digit

Rounding a decimal: 6.333 to the nearest hundredth (16)

When a decimal is rounded, an exact number is being replaced by an approximation. In order to choose the best approximation for 6.333, correct to the nearest hundredth, it it necessary to decide whether 6.333 is closer in value to 6.33 or 6.34.

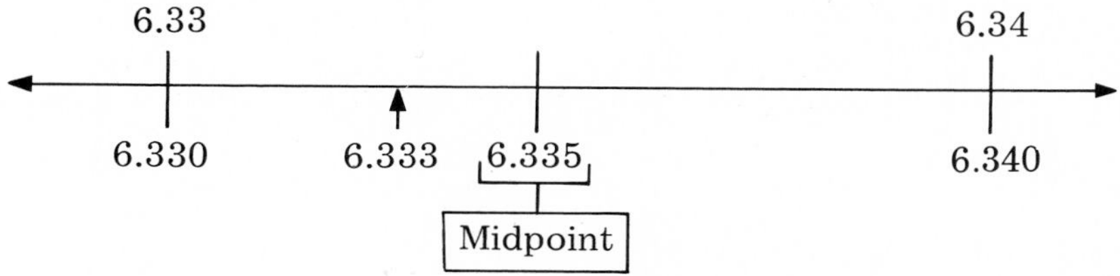

The midpoint between 6.33 and 6.34 is critical in determining the best approximation for 6.333. Since 6.333 is less than the midpoint value, it is best approximated by 6.33.

Rounding a decimal: 6.333 to the nearest hundredth (17)

$$6.333 \approx 6.33$$

The symbol $\approx$ means "is approximately."

To round any number: (18)

If the test digit is 0, 1, 2, 3, or 4 (the number is less than the midpoint value).
Round down by deleting all digits to the right of the specified place-value position.

If the test digit is 5, 6, 7, 8, or 9 (the number is greater than or equal to the midpoint value).
Round up by deleting all digits to the right of the specified place-value position *and* replace the digit in that position with the next larger number.

Round 0.973 to the nearest hundredth: (19)

0 . 9 7 3   7 is in the hundredths position; 3 is the test digit

Since the test digit is 3, **0.973 ≈ 0.97,** to the nearest hundredth.

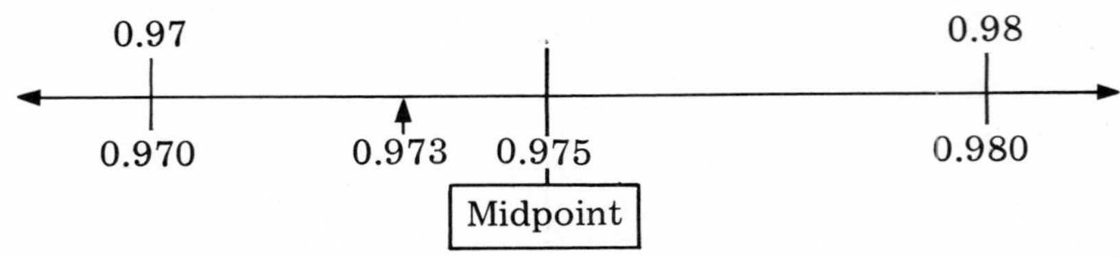

Round 0.973 to the nearest tenth: (20)

0 . 9 7 3   9 is the tenths position; 7 is the test digit

Since 7 is the test digit, 0.973 is to be rounded up. This means that 9 is to be replaced by the next larger number, 10. Write 0 in the tenths position and "carry" the 1 to the next place-value position:

0.973 ≈ 1.0

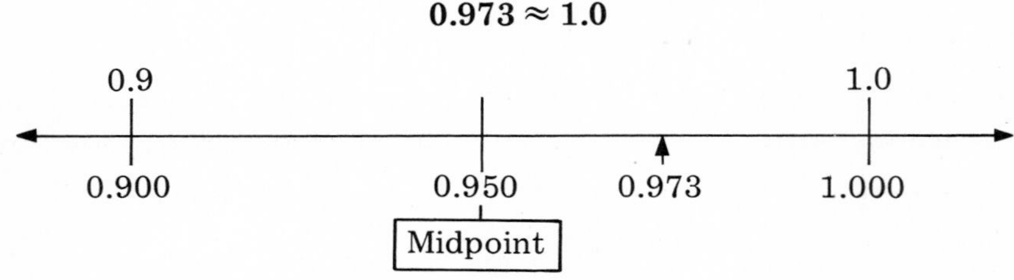

Round 1,263 to the nearest hundred: (21)

1 2 6 3   2 is in the hundred's position; 6 is the test digit

Since the test digit is six, 1,263 ≈ 1,300.

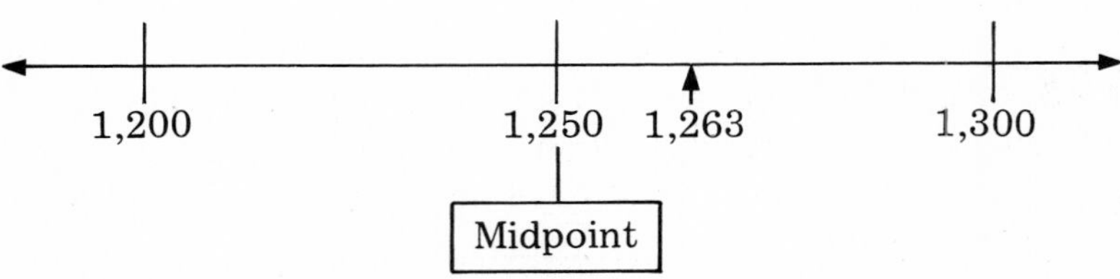

Round 3.3961 to the nearest hundredth: (22)

Complete the following:

3 . 3 9 6 1 — ○ is in the hundredths position; △ is the test digit

Thus, 3.3961 ≈ □, correct to the nearest hundredth.

Round 171.83 to the nearest one:

Complete the following:

1 7 1 . 8 3 — ○ is in the ones position; △ is the test digit

Thus, 171.83 ≈ □, correct to the nearest one.

Practice Exercises

Round each number to the specified place-value position.

1. 38.094 to the nearest hundredth
2. 0.0375 to the nearest thousandth
3. 42.856 to the nearest tenth
4. 5.0585 to the nearest one
5. 423.8897 to the nearest thousandth
6. 0.009398 to the nearest ten-thousandth

Rounding a quotient:

When a decimal division problem specifies the accuracy of the quotient, carry the division out until the quotient has a digit in the specified place-value position *and* a test digit; then round. For example, if the quotient is to be correct to the nearest hundredth, the division is continued until the quotient has a digit in the thousandths position, the test digit.

Find the quotient, correct to the nearest tenth: 1.253 ÷ 0.4.

```
                    3 . 1 3
0.4)1.253  →   .4.)1 .2.5 3   →   3.13 ≈ 4.1
                   1  2
                      5            Rounding the
                      4              quotient
                      1 3
                      1 2
                        1
```

Find the quotient, correct to the nearest hundredth: $3.96 \div 1.7$ 

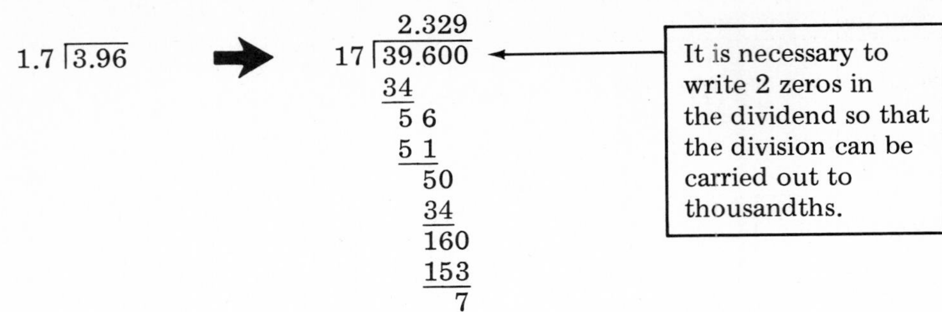

1.7 ⟌3.96
2.329
17 ⟌39.600
34
5 6
5 1
50
34
160
153
7
It is necessary to write 2 zeros in the dividend so that the division can be carried out to thousandths.

Thus, $3.96 \div 1.7 \approx 2.33$ since the digit in the thousandths position is 9.

Practice Exercises 

Find each quotient, correct to the nearest thousandth.

1. 0.6 ⟌5
2. 0.26 ⟌3.478
3. 2.06 ⟌0.0721
4. 0.7 ⟌0.02

Dividing by a power of ten: 

Consider the following divisions:

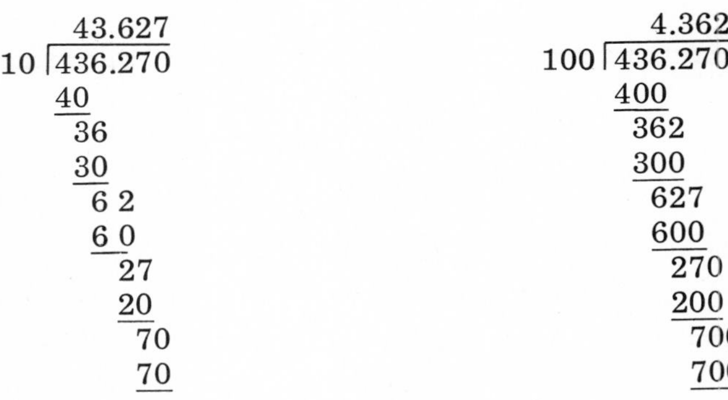

Dividing by a power of ten: 

Compare the quotients and the dividends:

436.27 ÷ 10 = 43.627 **436.27 ÷ 100 = 4.3627**

Notice that when 436.27 is divided by 10, the decimal point in the dividend is shifted one place to the left:

4 3 6 . 2 7

And when 436.27 is divided by 100, the decimal point is shifted two places to the left:

4 3 6 . 2 7

Dividing 436.27 by a power of ten: (31)

Consider the following patterns:

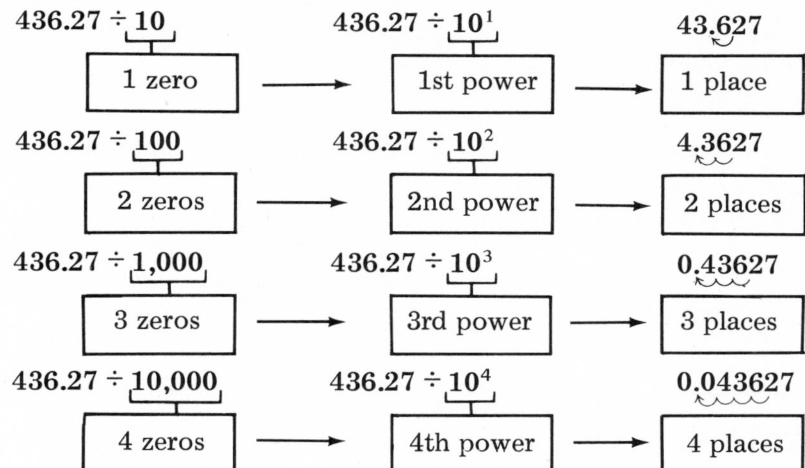

To divide by a power of ten, shift the decimal point to the left the number of places counted by the exponent of the power of ten. (The number of places that the decimal point is shifted is also counted by the number of zeros in the divisor.) Notice in the last example that it is necessary to write a zero before the digit 4 in order to correctly position the decimal point.

Dividing 1,452 by a power of ten: 

Since any whole number can be written as a decimal,

1,452 ÷ 10 →	1,452.0 ÷ 10 →	1 4 5 . 2
1,452 ÷ 100 →	1,452.0 ÷ 100 →	1 4 . 5 2
1,452 ÷ 1,000 →	1,452.0 ÷ 1,000 →	1 . 4 5 2

Any whole number can be divided by a power of ten by using the decimal-shift method.

Dividing by a power of ten: 

Complete the following by correctly placing the decimal point in the quotient, writing in any additional zeros as needed:

A. 831.2 ÷ 100 = 8 3 1 2

B. 0.45 ÷ 10 = 4 5

C. 51 ÷ 1,000 = 5 1

Practice Exercises

Divide by using the decimal-shift method.

1. $15.39 \div 1{,}000$
2. $0.83 \div 100$
3. $179.37 \div 10$
4. $831 \div 10{,}000$

Rewriting a fraction as a decimal:

One of the meanings of the fraction $\frac{2}{5}$ is $2 \div 5$. To rewrite $\frac{2}{5}$ as an equivalent decimal, simply carry out the indicated division.

$$\underset{\text{Fraction}}{\frac{2}{5}} \;\rightarrow\; \begin{array}{r} 0.4 \\ 5\overline{)2.0} \\ \underline{2\,0} \\ 0 \end{array} \;\rightarrow\; \underset{\text{Decimal}}{0.4}$$

Rewriting a fraction as a decimal:

To rewrite any fraction as an equivalent decimal, carry out the indicated division.

As many zeros as are necessary can be written to the right of the decimal point.

Rewrite $\frac{1}{9}$ as a decimal:

$$\frac{1}{9} \;\rightarrow\; \begin{array}{r} 0.1111\ldots \\ 9\overline{)1.0000\ldots} \\ \underline{9} \\ 10 \\ \underline{9} \\ 10 \\ \underline{9} \\ 10 \\ \underline{9} \\ 1 \end{array}$$

The remainder 1 keeps repeating.

Again, no matter how far this division is carried out, it will never reach a zero remainder.

Repeating decimals:

The decimal equivalents of fractions such as $\frac{1}{9}$ are called *repeating decimals* because the digit 1 keeps repeating in the quotient. If when the numerator is divided by the denominator, the division never results in a zero remainder, the decimal is a *repeating decimal.*

$$\frac{1}{9} \rightarrow 0.1111\ldots \quad \text{or} \quad 0.\overline{1}$$

The bar written above the 1 indicates that the digit repeats endlessly in the quotient, just as the notation ". . ." does.

Terminating decimals:

The decimal equivalents of fractions such as $\frac{2}{5}$ are called *terminating decimals* since the division can be carried out to a zero remainder. Every terminating decimal is also a repeating decimal if zero is considered a repeating digit.

$$\frac{2}{5} \rightarrow 0.4 \rightarrow 0.4000\ldots \quad \text{or} \quad 0.4\overline{0}$$

Repeating zeros are not usually written.

Rewrite the following as decimals, correct to the nearest thousandth:

Complete the following:

A. $\frac{5}{8} \rightarrow 8\overline{)5.0000}$ B. $\frac{2}{3} \rightarrow 3\overline{)2.0000}$

Are the decimal equivalents of $\frac{5}{8}$ and $\frac{2}{3}$ terminating?

Practice Exercises

Convert each fraction to a decimal, correct to the nearest thousandth.

1. $\frac{3}{8}$
2. $\frac{5}{6}$
3. $\frac{3}{16}$
4. $\frac{7}{17}$
5. $\frac{5}{3}$
6. $\frac{3}{4}$

Rewriting decimals as fractions: 

Rewriting a terminating decimal as a fraction follows directly from the definition of a decimal numeral, as shown in Lesson 18.

For example,

$$0.4 = \frac{4}{10} \text{ or } \frac{2}{5}$$

$$3.2 = 3\frac{2}{10} \text{ or } 3\frac{1}{5}$$

Rewriting a repeating decimal as a fraction requires a little more work.

To rewrite a repeating decimal as a fraction, the following equivalences will be used: 

$$\frac{1}{9} = 0.\overline{1} \quad \text{or} \quad 0.1111\ldots$$

$$\frac{1}{99} = 0.\overline{01} \quad \text{or} \quad 0.010101\ldots$$

$$\frac{1}{999} = 0.\overline{001} \quad \text{or} \quad 0.001001001\ldots$$

Each of these equivalences can be varified by dividing each numerator by its denominator.

Rewrite $0.\overline{3} = 0.3333\ldots$ as a fraction: 

$$0.3333\ldots = 3 \times (0.1111\ldots) \leftarrow \begin{array}{r} .1111\ldots \\ \times 3 \\ \hline 0.3333\ldots \end{array}$$

$$= 3 \times \frac{1}{9} \leftarrow \text{Since } \frac{1}{9} = 0.1111\ldots$$

$$= \frac{3}{9} \text{ or } \frac{1}{3}$$

Thus, $0.\overline{3} = \frac{1}{3}$.

Rewrite $0.\overline{7} = 0.777\ldots$ as a fraction: 45

$$0.777\ldots = 7 \times (0.111\ldots)$$

$$= 7 \times \frac{1}{9}$$

$$= \frac{7}{9}$$

Rewrite $0.\overline{03} = 0.030303\ldots$ as a fraction: (46)

$$0.030303\ldots = 3 \times (0.010101\ldots)$$

$$\begin{array}{r} 0.001001\ldots \\ \times 3 \\ \hline 0.030303\ldots \end{array}$$

$$= 3 \times \frac{1}{99}$$

Since $\frac{1}{99} = 0.010101\ldots$

$$= \frac{3}{99} \text{ or } \frac{1}{33}$$

Rewrite $0.\overline{162} = 0.162162\ldots$ as a fraction: (47)

$$0.162162\ldots = 162 \times (0.001001\ldots)$$

$$\begin{array}{r} 0.001001\ldots \\ \times 162 \\ \hline 0.162162\ldots \end{array}$$

$$= 162 \times \frac{1}{999}$$

Since $\frac{1}{999} = 0.001001\ldots$

$$= \frac{162}{999} \text{ or } \frac{6}{37}$$

Rewriting a repeating decimal as a fraction: (48)

Consider the following patterns:

$$\mathbf{0.\overline{1} = 0.111\ldots = \frac{1}{9}} \qquad \mathbf{0.\overline{11} = 11 \times (0.0101\ldots) = \frac{11}{99}}$$

$$\mathbf{0.\overline{2} = 2 \times (0.111\ldots) = \frac{2}{9}} \qquad \mathbf{0.\overline{12} = 12 \times (0.0101\ldots) = \frac{12}{99}}$$

$$\mathbf{0.\overline{3} = 3 \times (0.111\ldots) = \frac{3}{9}} \qquad \mathbf{0.\overline{13} = 13 \times (0.0101\ldots) = \frac{13}{99}}$$

$$\mathbf{0.\overline{4} = 4 \times (0.111\ldots) = \frac{4}{9}} \qquad \mathbf{0.\overline{14} = 14 \times (0.0101\ldots) = \frac{14}{99}}$$

$$\mathbf{0.\overline{161} = 161 \times (0.001001\ldots) = \frac{161}{999}}$$

$$\mathbf{0.\overline{162} = 162 \times (0.001001\ldots) = \frac{162}{999}}$$

$$\mathbf{0.\overline{163} = 163 \times (0.001001\ldots) = \frac{163}{999}}$$

$$\mathbf{0.\overline{164} = 164 \times (0.001001\ldots) = \frac{164}{999}}$$

Converting any repeating decimal to a fraction: 

Count the number of digits that repeat: the digits written under the bar. Write a fraction with the same number of 9's in its denominator and the repeating digits (without the bar) as the numerator. Then reduce if possible.

Rewriting repeating decimals as fractions: 

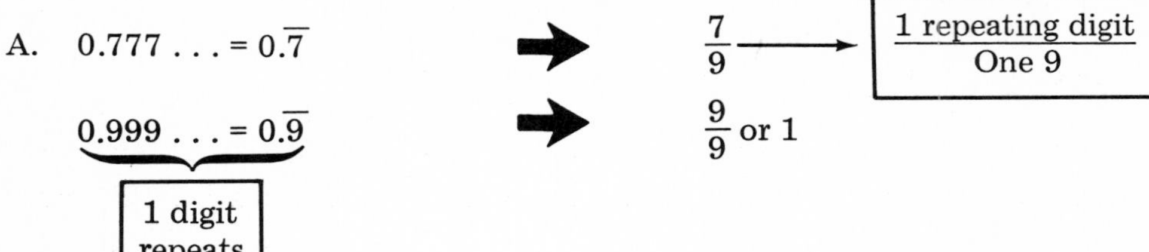

A. 0.777 . . . = 0.$\overline{7}$ → 7/9 → 1 repeating digit / One 9
0.999 . . . = 0.$\overline{9}$ → 9/9 or 1
1 digit repeats

B. $0.101010\ldots = 0.\overline{10}$ → $\frac{10}{99}$ → 2 repeating digits / Two 9's

$0.040404\ldots = 0.\overline{04}$ → $\frac{04}{99}$ or $\frac{4}{99}$

$0.272727\ldots = 0.\overline{27}$ → $\frac{27}{99}$ or $\frac{3}{11}$

2 digits repeat

C. 0.143143 . . . = 0.$\overline{143}$ → 143/999 → 3 repeating digits / Three 9's
0.006006 . . . = 0.$\overline{006}$ → 006/999 or 2/333
3 digits repeat

D. 0.10001000 . . . = 0.$\overline{1000}$ → 1,000/9,999 → 4 repeating digits / Four 9's
4 digits repeat

Rewriting repeating decimals as fractions: 

Complete the following:

A. $0.363636\ldots = 0.\overline{36} = \frac{\square}{99}$ or $\frac{4}{11}$

B. $0.019019\ldots = 0.\overline{019} = \dfrac{\bigcirc}{999}$

C. $0.10021002\ldots = 0.\overline{1002} = \dfrac{\triangle}{\text{⬡}}$ or $\dfrac{334}{3{,}333}$

Practice Exercises (52)

Rewrite each of the following as a fraction reduced to lowest terms:

1. $0.\overline{6}$
2. $0.\overline{15}$
3. $0.\overline{101}$
4. $0.\overline{7164}$

POST-TEST

1. Divide until a zero remainder is reached: $15.466 \div 2.2$
2. Round 394.197, correct to the nearest hundredth.
3. Divide by using the decimal-shift method: $18.5 \div 1{,}000$
4. Rewrite $\frac{3}{4}$ as a decimal, correct to the nearest hundredth.
5. Rewrite $0.\overline{12}$ as a fraction reduced to lowest terms.

LESSON 21
ADDITIONAL DECIMAL TOPICS

Objectives:

1. Write a decimal or whole number in scientific notation. (1-11)
2. Rewrite a number given in scientific notation in ordinary decimal notation. (12-15)
3. Multiply or divide large and small numbers by using scientific notation. (16-24)
4. Solve word problems that have values given in dollars, cents, and mills. (25-30)

Vocabulary:

Scientific Notation
Mill

PRE-TEST

1. Write the following numbers in scientific notation:
   (a) 17,000,000,000 (b) 0.00000821
2. Rewrite each of the following in ordinary decimal notation:
   (a) 3.05×10^7 (b) $4.1 \times \frac{1}{10^6}$
3. Perform each of the indicated operations using scientific notation:
   (a) $72{,}000{,}000 \times 0.03$ (b) $0.0000044 \div 11{,}000$
4. Calculate the amount of property tax due if the assessed value of the property is 2,500 dollars and the tax rate is 7.376 dollars per 100 dollars of the assessed value. (Round to the nearest cent.)

The use of very large and very small numbers is becoming more and more a part of the modern way of life. Each and every day people in complex societies find themselves faced with virtually unreadable numbers: (1)

The budget of the State of California for 1973 was approximately 9,800,000,000 dollars.

The distance that light travels in one year, which is called *one light-year*, is approximately 5,880,000,000,000 miles.

A single red cell of human blood contains approximately 270,000,000 hemoglobin molecules.

The diameter of a carbon atom is approximately 0.000000000001 meters.

A virus has a length of approximately 0.00000037 meters.

If reading very large and very, very small numbers is difficult, error-free computation with large and small numbers is virtually impossible. In fact, very few calculators are designed to do computations with 13-digit numbers such as appear in the previous frame. For this reason, *scientific notation*, one further decimal numeral form, is used to simplify computation with large and small numbers. (2)

Before discussing scientific notation, it is helpful to think about some of the ways that the number 3,007 can be named, for example. The decimal-shift methods of multiplication and division show that each of the following is equivalent to 3,007: (3)

300.7×10	$30{,}070 \times \frac{1}{10}$
30.07×10^2	$300{,}700 \times \frac{1}{10^2}$
3.007×10^3	$3{,}007{,}000 \times \frac{1}{10^3}$
$.3007 \times 10^4$	$30{,}070{,}000 \times \frac{1}{10^4}$
⋮	⋮

There are infinitely many ways to write the number 3,007 using only multiplication by powers of ten.

Scientific notation: (4)

When using scientific notation, all numbers are written as the product of a decimal numeral that is greater than (or equal) to 1 and less than 10 times some power of ten written in exponential form.

Of the equivalent forms of 3,007 given in the last frame, only one shows a product of a decimal between 1 and 10 times a power of 10:

$$\mathbf{3.007 \times 10^3}$$

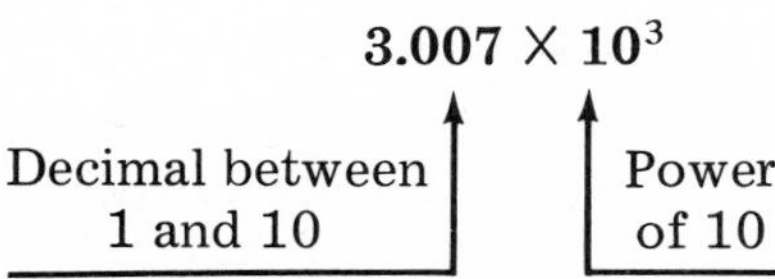

The number 3,007 is written in scientific notation as 3.007×10^3.

Scientific notation: ⑤

To write a large number such as 3,007 in scientific notation, the decimal-shift method of division and "Multiplication by One" are used:

"Why"

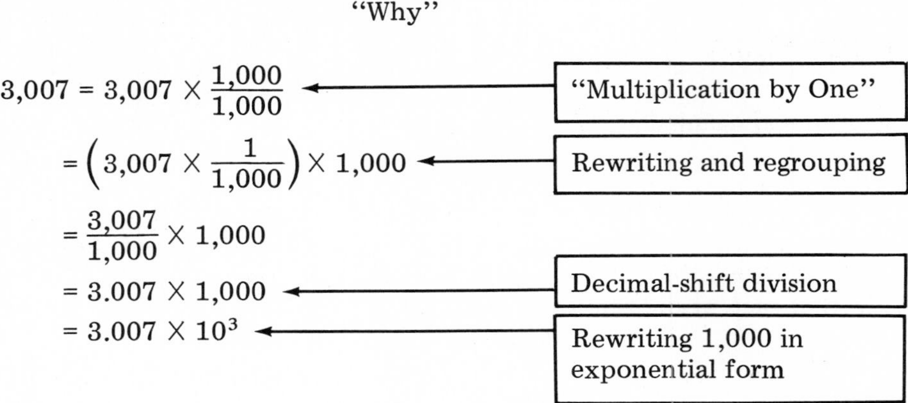

Notice that $\frac{1{,}000}{1{,}000}$ was chosen because it was necessary to shift the decimal point in 3,007.0 three places to the left to obtain a number with the same digits, but between 1 and 10.

Writing the number 3,007 in scientific notation: ⑥

"How"

Starting with the decimal numeral 3,007.0, shift the decimal point place-by-place to the <u>left</u> and determine if the resulting numeral is between 1 and 10.

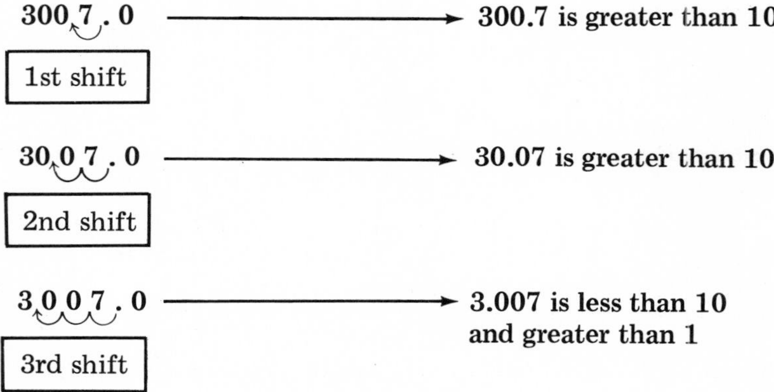

Then, since the decimal point was shifted three places to the left, representing a division by 1,000 or 10^3, show the multiplication of the resulting decimal numeral by the 3rd power of 10:

$$3{,}007.0 = 3.007 \times 10^3$$

3 shifts | 3rd power

Writing a small number in scientific notation: 0.0014 ⑦

"Why"

Using the decimal-shift method of multiplication by a power of ten and "Multiplication by One":

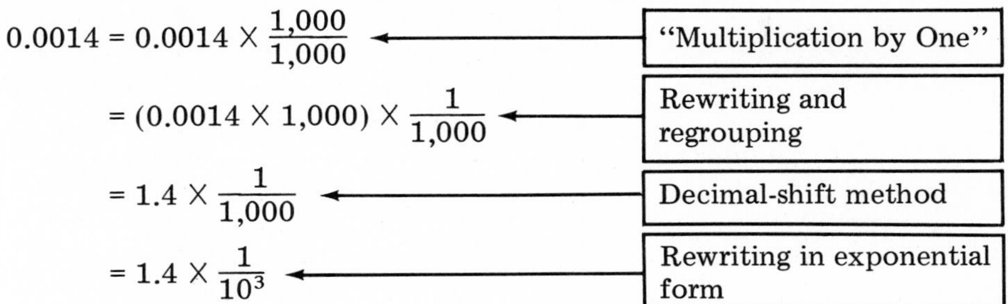

$$0.0014 = 0.0014 \times \frac{1{,}000}{1{,}000} \quad \text{“Multiplication by One”}$$

$$= (0.0014 \times 1{,}000) \times \frac{1}{1{,}000} \quad \text{Rewriting and regrouping}$$

$$= 1.4 \times \frac{1}{1{,}000} \quad \text{Decimal-shift method}$$

$$= 1.4 \times \frac{1}{10^3} \quad \text{Rewriting in exponential form}$$

Notice that $\frac{1{,}000}{1{,}000}$ was chosen because it was necessary to shift the decimal point in 0.0014 three places to the right to obtain a number with the same digits, but between 1 and 10.

Writing the number 0.0014 in scientific notation: ⑧

"How"

Starting with the decimal numeral 0.0014, shift the decimal point place-by-place to the *right* and determine if the resulting numeral is between 1 and 10:

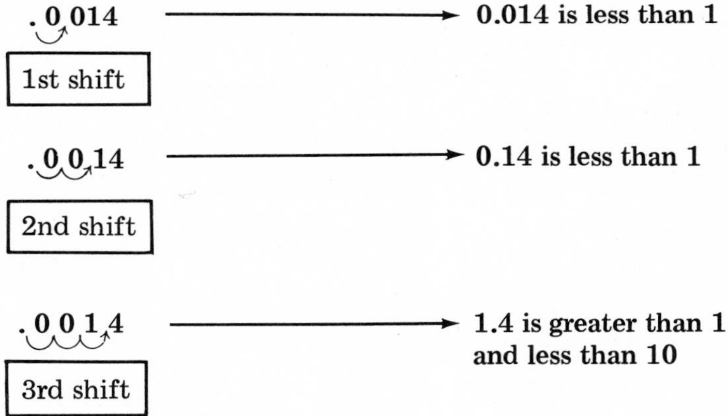

.0‸014 → **0.014 is less than 1**

1st shift

.0 0‸14 → **0.14 is less than 1**

2nd shift

.0 0 1‸4 → **1.4 is greater than 1 and less than 10**

3rd shift

Then, since a decimal point was shifted three places to the right, representing multiplication by 1,000 or 10^3, show the indicated division of the resulting decimal numeral by the 3rd power of 10 by showing the multiplication by the reciprocal of 10^3:

$$\mathbf{0.0014 = .001{,}4 \div 10^3 = 1.4 \times \frac{1}{10^3}}$$

Writing a number in scientific notation: 

To summarize:

To write a large number in scientific notation, shift the decimal point to the left and multiply by a power of ten with an exponent that indicates the number of places that the decimal point has been shifted.

For example.

$$2{,}000{,}000 = 2.000000. \; 0 \times 10^{6}$$

6 places to the left — 6th power

To write a small number in scientific notation, shift the decimal point to the right and multiply by the reciprocal of a power of ten with an exponent that indicates the number of places that the decimal point has been shifted.

For example,

$$0.0000301 = 0.00003.01 \times \frac{1}{10^{5}}$$

5 places to the right — 5th power

Writing numbers in scientific notation: 

Number		Scientific Notation
5,000,000	→	5.0×10^{6}
430,000,000	→	4.3×10^{8}
0.00007	→	$7.0 \times \frac{1}{10^{5}}$
0.0000000039	→	$3.9 \times \frac{1}{10^{9}}$

Writing numbers in scientific notation: (11)

Complete the following:

A. $9{,}800{,}000{,}000 = 9.8 \times 10^{\triangle}$

B. $5{,}880{,}000{,}000{,}000 = \bigcirc \times 10^{12}$

C. $0.00001 = 1.0 \times \frac{1}{10^{\square}}$

D. $0.00000037 = \hexagon \times \frac{1}{10^{7}}$

Rewriting numbers written in scientific notation: (12)

To rewrite a number given in scientific notation in ordinary decimal notation, simply carry out the indicated multiplication by the power of ten or the reciprocal of the power of ten.

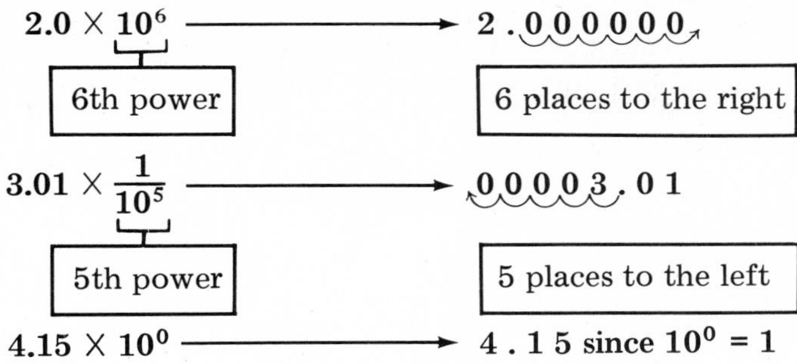

$4.15 \times 10^0 \longrightarrow 4.15$ since $10^0 = 1$

Rewriting in ordinary decimal notation: (13)

$6.037 \times 10^5 = 6\,.\,0\,3\,7\,0\,0$

5 places to the right

$7.2 \times 10^{12} = 7\,.\,2\,0\,0\,0\,0\,0\,0\,0\,0\,0\,0\,0$

12 places to the right

$6.037 \times \frac{1}{10^5} = 0\,0\,0\,0\,6.037$

5 places to the left

$7.2 \times \frac{1}{10^{12}} = 0\,0\,0\,0\,0\,0\,0\,0\,0\,0\,0\,7.2$

12 places to the left

Rewriting in ordinary decimal notation: (14)

Complete the following by correctly positioning the decimal point in the numeral on the right:

$8.01 \times 10^4 = 8\,0\,1\,0\,0\,0\,0$

$1.1 \times \frac{1}{10^5} = 0\,0\,0\,0\,0\,1\,1$

Practice Exercises

15

Write each number in scientific notation.

1. 3,000,000,000,000,000,000,000,000,000 candles are equivalent in light to the sun.
2. 635,000,000,000 different bridge hands can be dealt.
3. 0.000000006 meter is the width of a bacteria.
4. 0.00000825 meter is the diameter of a strand of human hair.

Write each number in ordinary decimal notation.

5. 5.32×10^7
6. 7.1×10^0
7. $4.3 \times \frac{1}{10^3}$
8. $1.2 \times \frac{1}{10^6}$

Computations with large and small numbers can often be simplified by using scientific notation. 16

For example, $2{,}100{,}000 \times 300$:

$$2{,}100{,}000 \times 300 = (2.1 \times 10^6) \times (3.0 \times 10^2)$$ ← Rewriting the factors in scientific notation

$$= (2.1 \times 3.0) \times (10^6 \times 10^2)$$ ← Regrouping

$$= 6.3 \times (10^6 \times 10^2)$$ ← Multiplying

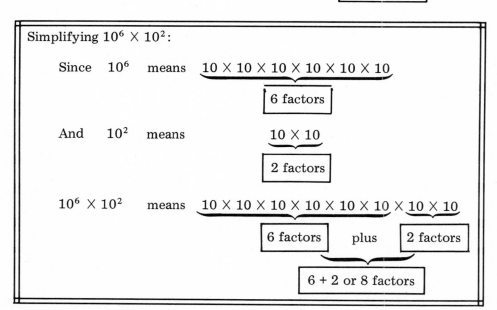

Simplifying $10^6 \times 10^2$:

Since 10^6 means $\underbrace{10 \times 10 \times 10 \times 10 \times 10 \times 10}_{\text{6 factors}}$

And 10^2 means $\underbrace{10 \times 10}_{\text{2 factors}}$

$10^6 \times 10^2$ means $\underbrace{10 \times 10 \times 10 \times 10 \times 10 \times 10}_{\text{6 factors}} \times \underbrace{10 \times 10}_{\text{2 factors}}$

6 factors plus 2 factors → 6 + 2 or 8 factors

Thus, $2{,}100{,}000 \times 300 = 6.3 \times (10^6 \times 10^2)$

$$= 6.3 \times 10^8$$ ← Adding the exponents

$$= 630{,}000{,}000$$ ← Rewriting

Computation using scientific notation: 0.00037 × 0.1004 (17)

$$0.00037 \times 0.1004 = \left(3.7 \times \frac{1}{10^4}\right) \times \left(1.004 \times \frac{1}{10^1}\right)$$ ← Writing in scientific notation

$$= \left(3.7 \times 1.004\right) \times \left(\frac{1}{10^4} \times \frac{1}{10^1}\right)$$ ← Regrouping

$$= 3.7148 \times \left(\frac{1}{10^4} \times \frac{1}{10^1}\right)$$ ← Multiplying

Simplifying $\frac{1}{10^4} \times \frac{1}{10^1}$:

$$\frac{1}{10^4} \times \frac{1}{10^1} = \frac{1 \times 1}{10^4 \times 10^1}$$

$$= \frac{1}{\underbrace{10 \times 10 \times 10 \times 10}_{\text{4 factors}} \times \underbrace{10}_{\text{1 factor}}}$$

4 factors + 1 factor

4 + 1 or 5 factors

Thus, $0.00038 \times 0.1004 = 3.7148 \times \left(\frac{1}{10^4} \times \frac{1}{10^1}\right)$

$$= 3.7148 \times \frac{1}{10^5}$$ ← Adding the exponents

$$= 0.000037148$$ ← Rewriting

Computation using scientific notation: 0.014 × 30,000 (18)

$$0.014 \times 30{,}000 = \left(1.4 \times \frac{1}{10^2}\right) \times \left(3.0 \times 10^4\right)$$ ← Writing in scientific notation

$$= (1.4 \times 3.0) \times \left(\frac{1}{10^2} \times 10^4\right)$$ ← Regrouping

$$= 4.2 \times \frac{10^4}{10^2}$$ ← Multiplying

Simplifying: $\frac{10^4}{10^2} = \frac{10 \times \not{10} \times \not{10} \times 10}{\not{10} \times \not{10}}$ ← 4 factors (numerator); ← 2 factors (denominator)

$$= 10 \times 10$$ ← 4 − 2 or 2 factors

$$= 10^2$$

Thus, $0.014 \times 30{,}000 = 4.2 \times 10^2$

$$= 420$$

Computation with scientific notation: 0.00501 × 20 (19)

$$0.00501 \times 20 = \left(5.01 \times \frac{1}{10^3}\right) \times \left(2.0 \times 10^1\right)$$ ← Writing in scientific notation

$$= (5.01 \times 2.0) \times \left(\frac{1}{10^3} \times 10^1\right)$$ ← Regrouping

$$= 10.02 \times \frac{10^1}{10^3}$$ ← Multiplying

Simplifying: $\frac{10^1}{10^3} = \frac{\overset{1}{\cancel{10}}}{10 \times \underset{1}{\cancel{10}} \times 10}$ ← 1 factor; ← 3 factors

$= \frac{1}{10 \times 10}$ ← 3 − 1 or 2 factors

$= \frac{1}{10^2}$

Thus, $0.00501 \times 20 = 10.02 \times \frac{1}{10^2}$

$= 0.1002$

Computation with scientific notation: (20)

Complete each of the following:

A. $3{,}000{,}000{,}000 \times 170 = (3.0 \times 10^9) \times (1.7 \times 10^2)$

$= (3.0 \times 1.7) \times (10^9 \times 10^2)$

$= 5.1 \times 10^{\square}$

$= 5.1\,0\,0\,0\,0\,0\,0\,0\,0\,0\,0$

$\square$ places

B. $0.0000115 \times 0.024 = \left(1.15 \times \frac{1}{10^5}\right) \times \left(2.4 \times \frac{1}{10^2}\right)$

$= (1.15 \times 2.4) \times \left(\frac{1}{10^5} \times \frac{1}{10^2}\right)$

$= 2.67 \times \frac{1}{10^{\triangle}}$

$= 0\,0\,0\,0\,0\,0\,0\,0\,2.\,76$

$\triangle$ places

C. $0.042 \times 7{,}100 = \left(4.2 \times \frac{1}{10^2}\right) \times (7.1 \times 10^3)$

$= (4.2 \times 7.1) \times \left(\frac{1}{10^2} \times 10^3\right)$

$= 29.82 \times 10^{\diamond}$

$= 29\,.\,8\,2$

$\diamond$ place

D. $0.0000001 \times 745 = \left(1.0 \times \frac{1}{10^7}\right) \times (7.45 \times 10^2)$

$$= (1.0 \times 7.45) \times \left(\frac{1}{10^7} \times 10^2\right)$$

$$= 7.45 \times \frac{1}{10^{\triangledown}}$$

$$= .0\,0\,0\,0\,7.45$$

$\triangledown$ places

Computation using scientific notation: (21)

Divide 7,500 by 150:

$$7{,}500 \div 150 = (7.5 \times 10^3) \div (1.5 \times 10^2)$$

$$= (7.5 \times 10^3) \times \left(\frac{1}{1.5 \times 10^2}\right)$$ ← Multiplying by the reciprocal of the divisor

$$= \frac{7.5 \times 10^3}{1.5 \times 10^2}$$

$$= \frac{7.5}{1.5} \times \frac{10^3}{10^2}$$ ← Rewriting

$$= 5 \times 10^1$$ ← $\frac{10^3}{10^2} = \frac{\not{10}^{\,1} \times \not{10}^{\,1} \times 10}{\not{10}_{\,1} \times \not{10}_{\,1}}$

$$= 50$$

Compute: 0.00036 ÷ 90 (22)

$$0.00036 \div 90 = 3.6 \times \frac{1}{10^4} \div (9.0 \times 10^1)$$ ← Rewriting each number in scientific notation

$$= \frac{3.6}{1} \times \frac{1}{10^4} \times \left(\frac{1}{9.0 \times 10^1}\right)$$ ← Multiplication by the reciprocal of the divisor

$$= \frac{3.6}{9 \times 10^4 \times 10^1}$$ ← Multiplying

$$= \frac{3.6}{9} \times \frac{1}{10^4 \times 10^1}$$ ← Rewriting

$$= 0.4 \times \frac{1}{10^4 \times 10^1}$$

$$= 0.4 \times \frac{1}{10^5}$$ ← $10^4 \times 10^1 = 10^5$ adding exponents

$$= 0.000004$$

Divide: 203.4 ÷ 0.006: 23

Complete the following:

$$203.4 \div 0.006 = (2.034 \times 10^2) \div \left(6.0 \times \frac{1}{10^3}\right)$$

$$= (2.034 \times 10^2) \div \left(\frac{6}{10^3}\right) \longleftarrow \boxed{\text{Multiplying}}$$

$$= (2.034 \times 10^2) \times \frac{10^3}{6} \longleftarrow \boxed{\text{Multiplying by the reciprocal of the divisor}}$$

$$= \frac{2.034 \times 10^2 \times 10^3}{6}$$

$$= \frac{2.034 \times 10^{\square}}{6} \longleftarrow \boxed{10^2 \times 10^3 = 10^{\square}}$$

$$= 0.339 \times 10^{\square} \longleftarrow \boxed{2.034 \div 6 = 0.339}$$

$$= 0\,.\,3\,3\,9\,0\,0$$

$\square$ places

Practice Exercises 24

Perform the indicated operation for each of the following using scientific notation:

1. 75,000,000 × 201,000
2. 0.00063 × 0.005
3. 60,200 × 0.031
4. 0.00413 × 240
5. 2,310,000 ÷ 6,000
6. 0.0007124 × 0.00004
7. 122,000 ÷ 0.0002
8. 0.000565 ÷ 500

The present-day methods of the business and financial communities require a working knowledge of decimal numerals. Since the cent is the lowest denomination in our monetary system, calculations performed must be rounded to the nearest cent or the nearest hundredth of a dollar. 25

In the world of finance, calculations are carried out to the nearest mill. A *mill* is 1 one-thousandth of a dollar or one-tenth of a cent. 

For example,

The Coast Unified School District taxes property at the rate of $\$4.23\underline{7}$ per \$100 of the assessed valuation.

One share of National Capital Fund is valued at $\$12.48\underline{2}$.

The Seal Beach City Council must raise the property tax rate $2.\underline{3}$ cents per \$100 of the assessed value.

The Partridge home is assessed at $8,500 for tax purposes. How much tax is paid to the Coast Unified School District? (Refer to the example in Frame 26.) (27)

How many hundreds of dollars valuation does the home have?

$$\$8,500 \div \$100 = 85$$

How much tax is due? Let T represent tax due.

Using a proportion:

Tax	$4.237	T
Valuation	1 Hundred	85 Hundreds

$$\frac{\$4.237}{1} = \frac{T}{85}$$

$$T = \$4.237 \times 85$$

$$T = \$360.145$$

$$T = \$360.15$$ ← Rounded to the nearest cent

Referring to the example in Frame 26: (28)

Ann Kelly wishes to sell 76 of her shares in National Capital Fund. How much will she receive for the sale of these shares? Let S represent the value of sold shares.

Using a proportion:

1 Share	76 Shares
$12.482	S

$$\frac{1}{\$12.482} = \frac{76}{S}$$

$$S = (\$12.482)\ (76)$$

$$S = \$948.632$$

$$S = \$948.63$$ ← Rounded to the nearest cent

Referring to the example in Frame 26: (29)

If the total assessed valuation for the property in the City of Seal Beach is $543,200,000, how much tax revenue will the raise produce?

How many hundreds of dollars of assessed valuation?

$$\$453,200,000 \div \$100 = \$4,532,000$$

How much tax revenue? Let X represent tax revenue.

$$\frac{\$0.023}{1} = \frac{X}{4,532,000}$$

$$X = (\$0.023)\ (\$4,532,000)$$

$$X = \$104,236.00$$

Practice Exercises

Solve each problem correct to the nearest cent.

1. A home assessed at \$11,250 and taxed at \$3.148 per \$100 assessed valuation has what amount of tax due?
2. A financial statement lists 325 shares of stock valued at \$8.456 per share. What is the total value of the stock?
3. If the tax rate is decreased \$0.084 per \$100 assessed valuation, what is the amount of reduction in tax for property assessed at \$15,875?

POST-TEST

1. Write the following numbers in scientific notation:
   (a) 923.4 (b) 0.00621
2. Rewrite each of the following in ordinary decimal notation:
   (a) 1.23×10^0 (b) $1.703 \times \frac{1}{10^4}$
3. Perform each of the indicated operations using scientific notation.
   (a) $61{,}000 \times 2{,}200$ (b) $0.0000816 \div 800{,}000$
4. Calculate the amount that can be realized from the sale of 32 shares of stock if each share is valued at \$22.352.

DRILL EXERCISES: UNIT 5

Rewrite each decimal either as a fraction or a mixed number reduced to lowest terms.

1. 3.62
2. 0.36
3. 71.47
4. 0.0062
5. 86.203
6. 16.72

Rewrite each decimal in expanded notation using the exponential forms of the powers of ten.

7. 45.82
8. 27.23
9. 1.468
10. 2.783
11. 0.0093
12. 0.4782

Name the place-value position of the underlined digits.

13. $507.87\underline{3}$
14. $0.\underline{4}782$
15. $3.407\underline{8}$

Rewrite each of the following decimals in word form:

16. 326.63
17. 0.371
18. 4.026
19. 420.002
20. 1,362.04
21. 10.010

Perform the indicated operations.

22. 26.3 + 7.4
23. 86.61 - 31.42
24. 16.721 + 3.6
25. 0.421 + 89.3
26. 3 - 0.421
27. 14.6 + 13.312 + 15.01
28. 0.0031 - 0.00047

29. $\begin{array}{r} 13.236 \\ 12.1 \\ +\ 4.70 \\ \hline \end{array}$

30. $\begin{array}{r} 27.3 \\ -\ 2.714 \\ \hline \end{array}$

31. $\begin{array}{r} 36.04 \\ -\ 3.77 \\ \hline \end{array}$

32. $\begin{array}{r} 15.74 \\ +14.97 \\ \hline \end{array}$

Multiply:

33. 36.2 × 0.3
34. 14.11 × 0.02
35. 26.8 × 0.0051
36. 0.003 × 2.4
37. 0.0621 × 0.21
38. 0.321 × 0.007

39. $\begin{array}{r} 12.60 \\ \times\ 2.13 \\ \hline \end{array}$

40. $\begin{array}{r} 124.100 \\ \times\quad 0.67 \\ \hline \end{array}$

41. Determine the value of $(0.02)^3$.
42. Determine the value of $(0.15)^2$.

Divide or multiply using the decimal-shift method.

43. 3.63 × 100
44. 3.26 ÷ 1,000
45. 87.6 ÷ 1,000
46. 0.0046 × 10

47. 16×100

48. $0.326 \div 10$

49. $1{,}236.42 \div 10{,}000$

50. $42.731 \times 10{,}000$

51. $0.0036 \times 1{,}000$

52. $0.6371 \div 100$

53. $432{,}179 \div 10{,}000$

54. $0.003261 \times 100{,}000$

Find each quotient, correct to the nearest tenth.

55. $3.6\overline{)7.31}$

56. $0.421\overline{)36.72}$

57. $0.037\overline{)0.4261}$

Find each quotient, correct to the nearest hundredth.

58. $8.2\overline{)367}$

59. $4.02\overline{)3.421}$

60. $0.021\overline{)6.23}$

Find each quotient, correct to the nearest thousandth.

61. $5.3\overline{)0.00362}$

62. $8.01\overline{)53.1}$

63. $0.0036\overline{)0.00473}$

Convert each fraction to a decimal, correct to the nearest thousandth.

64. $\frac{2}{3}$

65. $\frac{2}{7}$

66. $\frac{1}{8}$

67. $\frac{47}{3}$

68. $\frac{49}{8}$

69. $\frac{1}{5}$

70. $\frac{3}{5}$

71. $\frac{7}{8}$

72. $\frac{15}{16}$

Rewrite each repeating decimal as a fraction reduced to lowest terms.

73. $0.\overline{4}$

74. $0.\overline{6}$

75. $0.\overline{03}$

76. $0.\overline{07}$

77. $0.\overline{156}$

78. $0.\overline{012}$

Write each of the following numbers in scientific notation:

79. 3,400,000

80. 903,000,000

81. 0.00034

82. 0.0000904

83. 47.5

84. 865.4

Write each of the following in customary decimal notation:

85. 2.3×10^3

86. 4.6×10^2

87. $3.03 \times \frac{1}{10^4}$

88. $4.751 \times \frac{1}{10^3}$

Perform the indicated operations using scientific notation.

89. $56{,}000 \times 0.004$

90. $7{,}321 \times 0.02$

91. $0.0054 \div 600$

92. $770 \div 0.007$

Word problems

93. Toby Scott words for his father at the Recycle Cycle Shop for $3.80 an hour. For overtime he receives time and a half and on Saturday he earns double time. Find Toby's weekly earnings if he works 9 hours per day Monday through Friday and 3 hours on Saturday. (Consider 8 hours as a normal working day.)

94. Rhonda had $54.31 and then spent $13.25 for a record album, $6.27 for a gift, and $14.22 for gasoline. How much money does she have left?

95. Bill and Ann own a home assessed at $12,000. If for every $100 of assessed value they must pay $5.28 in taxes, how much is their tax bill?

96. If Robert Louis earns $263.78 for 40 hours of work, what is his hourly rate of pay? (Round to the nearest cent.)

97. If redwood compost costs $5.93 for 1 cubic yard, what is the cost of 2.5 cubic yards?

98. A manufacturer of precision parts has contracted to produce a steel pin 5.07 inches long. The buyer realizes that very few of the pins will measure exactly 5.07 inches and has agreed to buy the pins if they are either 0.001 inches too short or too long. This is called a *tolerance* of ±0.001. Would the buyer accept a pin that measures 5.0699? Would he accept a pin that measures 5.068?

99. If the assessed value of a home is $12,000 and the local school tax rate is increased from $1.78 per $100 to $1.975 per $100 of assessed valuation, by how much will the taxes on this home increase?

100. Larry's bicycle needs repair. The bicycle shop estimates that 5 hours of labor at $8.75 per hour and $27.35 in parts will put the bicycle in running order. A new bicycle would cost $78.95. How much will he save by getting the old bicycle repaired?

101. If an airplane uses 50.3 gallons of fuel per hour of flight, how many hours can the airplane fly on 193.5 gallons, correct to the nearest tenth of an hour.

102. If an automobile gets 15.4 miles to the gallon and the tank holds 22.3 gallons, how many miles can the car travel on a full tank?

103. Jerry spent $12.56 on round steak priced at $1.57 per pound. How many pounds did he buy?

1. The name of the place-value position of the underlined digit in the number 3,261.0$\underline{2}$536 is:
   (a) Tens
   (b) Tenths
   (c) Hundreds
   (d) Hundredths
   (e) None of the above
2. The number 6.0765 is written in expanded notation using the exponential forms of the powers of ten as:
   (a) $(6 \times 10^5) + (0 \times 10^4) + (7 \times 10^3) + (6 \times 10^2) + (5 \times 10^1)$
   (b) $(6 \times 10^0) + \left(0 \times \frac{1}{10^0}\right) + \left(7 \times \frac{1}{10^1}\right) + \left(6 \times \frac{1}{10^2}\right) + \left(5 \times \frac{1}{10^3}\right)$
   (c) $(6 \times 10^0) + \left(0 \times \frac{1}{10^1}\right) + \left(7 \times \frac{1}{10^2}\right) + \left(6 \times \frac{1}{10^3}\right) + \left(5 \times \frac{1}{10^4}\right)$
   (d) $(6 \times 10^1) + \left(0 \times \frac{1}{10^0}\right) + \left(7 \times \frac{1}{10^1}\right) + \left(6 \times \frac{1}{10^2}\right) + \left(5 \times \frac{1}{10^3}\right)$
   (e) None of the above
3. The number 582.0309 is written in word form as:
   (a) Five hundred eighty-two and three hundred nine ten-thousandths
   (b) Five hundred and eighty-two and three hundredths and nine ten-thousandths
   (c) Five hundred eighty-two and three hundred nine thousandths
   (d) Five hundred eighty-two and three hundredths nine ten-thousandths
   (e) None of the above
4. Find the sum: 0.6254 + 243 + 3.68
   (a) 686.5
   (b) 4.3297
   (c) 247.3054
   (d) 0.6865
   (e) None of the above
5. Find the difference: 3 – 2.6835
   (a) 1.3165
   (b) 0.3165
   (c) 1.4275
   (d) 1.6835
   (e) None of the above
6. Find the product: 23.16 × 0.34
   (a) 787.44
   (b) 78.744
   (c) 0.78744
   (d) 7.8744
   (e) None of the above
7. Find the product: 0.186 × 0.023
   (a) 0.4278
   (b) 0.04278
   (c) 0.004278
   (d) 4.278
   (e) None of the above

8. To multiply 25.432 by 10,000, the decimal is shifted:
   (a) 4 places to the right
   (b) 4 places to the left
   (c) 5 places to the right
   (d) 5 places to the left
   (e) None of the above

9. Divide until a zero remainder is reached: $0.06328 \div 4$
   (a) 0.1582
   (b) 0.01582
   (c) 0.001582
   (d) 1.582
   (e) None of the above

10. To divide 5.862 by 7.21, we rewrite:
   (a) 7.21 as 721 and 5.862 as 5,862
   (b) 7.21 as 721 and 5.862 as 586.2
   (c) 7.21 as 72.1 and 5.862 as 58.62
   (d) 7.21 as 7,210 and 5.862 as 5,862
   (e) None of the above

11. Round off 364.2074 to the nearest hundredth.
   (a) 364.207
   (b) 364.20
   (c) 364.208
   (d) 364.21
   (e) None of the above

12. Divide and round off the quotient to the nearest tenth: $0.1209 \div 0.21$
   (a) 0.5
   (b) 0.57
   (c) 0.6
   (d) 0.58
   (e) None of the above

13. When 7.89 is divided by 1,000, the decimal is shifted:
   (a) 3 places to the left
   (b) 3 places to the right
   (c) 4 places to the left
   (d) 4 places to the right
   (e) None of the above

14. Rewrite 4.64 as a fraction reduced to lowest terms.
   (a) $\frac{116}{25}$
   (b) $\frac{416}{25}$
   (c) $\frac{232}{50}$
   (d) $\frac{464}{100}$
   (e) None of the above

15. Rewrite $\frac{2}{7}$ as a decimal and round to the nearest thousandth.
   (a) 0.2857
   (b) 0.285
   (c) 0.2858
   (d) 0.286
   (e) None of the above

16. Mr. Sall owns 29 shares of CALCON stock that is valued at \$3.624 per share. What is the total value of his CALCON stock? Round your answer to the nearest cent.
(a) \$105.09
(b) \$105.10
(c) \$105.19
(d) \$105.20
(e) None of the above

17. Rewrite $0.\overline{21}$ as a fraction reduced to lowest terms.
(a) $\frac{21}{99}$
(b) $\frac{21}{999}$
(c) $\frac{7}{333}$
(d) $\frac{7}{33}$
(e) None of the above

18. Write 0.00267 in scientific notation.
(a) $0.267 \times \frac{1}{10^2}$
(b) 2.67×10^3
(c) $2.67 \times \frac{1}{10^3}$
(d) 26.7×10^4
(e) None of the above

19. Rewrite $4.17 \times \frac{1}{10^1}$ in ordinary decimal notation.
(a) 0.0417
(b) 0.417
(c) 41.7
(d) 4.17
(e) None of the above

20. Using scientific notation, find the value of $4{,}200{,}000 \times 0.00002$.
(a) 8.4×10^1
(b) 8.4×10^4
(c) $8.40 \times \frac{1}{10^1}$
(d) 2.1×10^1
(e) None of the above

LESSON 18:

Pre-test

1. $3\frac{1}{50}$
2. $(2 \times 10^1) + (3 \times 10^0) + \left(8 \times \frac{1}{10^1}\right) + \left(0 \times \frac{1}{10^2}\right) + \left(7 \times \frac{1}{10^3}\right) + \left(5 \times \frac{1}{10^4}\right)$
3. (a) Hundredths (b) Ten-thousandths
4. Four hundred seven and three hundred eight ten-thousandths

Practice Exercises

Frame 11

1. $\frac{1}{2}$
2. $\frac{1}{10{,}000}$
3. $7\frac{3}{25}$
4. $103\frac{3}{100}$
5. $1\frac{41}{100}$
6. $\frac{1}{4}$

Frame 14

1. $(5 \times 10^1) + (3 \times 10^0) + \left(8 \times \frac{1}{10}\right) + \left(3 \times \frac{1}{10^2}\right) + \left(2 \times \frac{1}{10^3}\right)$
2. $(9 \times 10^0) + \left(1 \times \frac{1}{10^1}\right) + \left(5 \times \frac{1}{10^2}\right) + \left(4 \times \frac{1}{10^3}\right) + \left(8 \times \frac{1}{10^4}\right)$
3. $\left(3 \times \frac{1}{10^1}\right) + \left(0 \times \frac{1}{10^2}\right) + \left(2 \times \frac{1}{10^3}\right) + \left(7 \times \frac{1}{10^4}\right)$
4. $\left(0 \times \frac{1}{10^1}\right) + \left(9 \times \frac{1}{10^2}\right) + \left(0 \times \frac{1}{10^3}\right) + \left(4 \times \frac{1}{10^4}\right)$

Frame 17

1. Hundredths
2. Ones
3. Ten-thousandths
4. Thousandths
5. Thousands
6. Hundred-thousandths

Frame 22

1. Seventy-one hundredths
2. Three and thirty-six ten-thousandths
3. One hundred and one-thousandth
4. Two hundred sixty-seven and seven tenths

Post-test

1. $\frac{1}{400}$
2. $(1 \times 10^1) + (7 \times 10^0) + \left(0 \times \frac{1}{10}\right) + \left(2 \times \frac{1}{10^2}\right) + \left(1 \times \frac{1}{10^3}\right)$
3. (a) Tenths (b) Thousandths
4. Nine hundred six and nine hundred six hundred-thousandths

LESSON 19:

Pre-test

1. (a) 177.32 (b) 21.689
2. (a) 3.329 (b) 0.0232
3. (a) 5.5272 (b) 0.03237
4. 0.01
5. 430,810

Practice Exercises

Frame 6

1. 49.211
2. 1.1034
3. 359.97
4. 37.2217

Frame 11

1. 13.731
2. 5.788
3. 546.4527
4. 0.575

Frame 19

1. 0.78
2. 1898.073
3. 1.2408
4. 0.0441
5. 0.00063
6. 0.0576

Frame 26

1. 2374.1
2. 6.71
3. 82.671
4. 3,687,000
5. 20
6. 73,600
7. 8,100

Post-test

1. (a) 259.65 (b) 2.209
2. (a) 15.755 (b) 0.312
3. (a) 108.452 (b) 0.0000486
4. 0.027
5. 34.91

LESSON 20:

Pre-test

1. 2.2
2. 3,742.073
3. 1.7815
4. 0.571
5. $\frac{1}{3}$

Practice Exercises

Frame 9

1. 31.4
2. 0.117
3. 0.0014
4. 1.701

Frame 13

1. 2,400
2. 122.8
3. 15
4. 5.79

Frame 24

1. 38.09
2. 0.038
3. 42.9
4. 5
5. 423.890
6. 0.0094

Frame 28

1. 8.333
2. 13.377
3. 0.035
4. 0.029

Frame 34

1. 0.01539
2. 0.0083
3. 17.937
4. 0.0831

Frame 41

1. 0.375
2. 0.833
3. 0.188
4. 0.412
5. 1.667
6. 0.750

Frame 52

1. $\frac{2}{3}$
2. $\frac{5}{33}$
3. $\frac{101}{999}$
4. $\frac{796}{1,111}$

Post-test

1. 7.03
2. 394.20
3. 0.0185
4. 0.75
5. $\frac{4}{33}$

LESSON 21:

Pre-test

1. (a) 1.7×10^{10} (b) $8.21 \times \frac{1}{10^6}$
2. (a) 30,500,000 (b) 0.0000041
3. (a) 2,160,000 (b) 0.0000000004
4. $184.40

Practice Exercises

Frame 15

1. 3×10^{27}
2. 6.35×10^{11}
3. $6 \times \frac{1}{10^9}$
4. $8.25 \times \frac{1}{10^6}$
5. 53,200,000
6. 7.1
7. 0.0043
8. 0.0000012

Frame 24

1. 15,075,000,000,000
2. 0.00000315
3. 1866.2
4. 0.9912
5. 385
6. 0.000000028496
7. 610,000,000
8. 0.00000113

Frame 30

1. $354.15
2. $2,748.20
3. $13.34

Post-test

1. (a) 9.234×10^2 (b) $6.21 \times \frac{1}{10^3}$
2. (a) 1.23 (b) 0.0001703
3. (a) 134,200,000 (b) 0.000000000102
4. $715.26

Answers To Drill Exercises: Unit 5

1. $3\frac{31}{50}$
2. $\frac{9}{25}$
3. $71\frac{47}{100}$
4. $\frac{31}{5,000}$
5. $86\frac{203}{1,000}$
6. $16\frac{18}{25}$
7. $(4 \times 10) + (5 \times 10^0) + \left(8 \times \frac{1}{10^1}\right) + \left(2 \times \frac{1}{10^2}\right)$
8. $(2 \times 10) + (7 \times 10^0) + \left(2 \times \frac{1}{10^1}\right) + \left(3 \times \frac{1}{10^2}\right)$
9. $(1 \times 10^0) + \left(4 \times \frac{1}{10^1}\right) + \left(6 \times \frac{1}{10^2}\right) + \left(8 \times \frac{1}{10^3}\right)$
10. $(2 \times 10^0) + \left(7 \times \frac{1}{10^1}\right) + \left(8 \times \frac{1}{10^2}\right) + \left(3 \times \frac{1}{10^3}\right)$
11. $\left(0 \times \frac{1}{10^1}\right) + \left(0 \times \frac{1}{10^2}\right) + \left(9 \times \frac{1}{10^3}\right) + \left(3 \times \frac{1}{10^4}\right)$
12. $\left(4 \times \frac{1}{10^1}\right) + \left(7 \times \frac{1}{10^2}\right) + \left(8 \times \frac{1}{10^3}\right) + \left(2 \times \frac{1}{10^4}\right)$
13. Thousandths
14. Tenths
15. Ten-thousandths
16. Three hundred twenty-six and sixty-three hundredths
17. Three hundred seventy-one thousandths
18. Four and twenty-six thousandths
19. Four hundred twenty and two thousandths
20. One thousand three hundred sixty-two and four hundredths
21. Ten and ten thousandths
22. 33.7
23. 55.19
24. 20.321
25. 89.721
26. 2.579
27. 42.922
28. 0.00263
29. 30.036
30. 24.586
31. 32.27
32. 30.71
33. 10.86
34. 0.2822
35. 0.13668
36. 0.0072
37. 0.013041
38. 0.002247
39. 26.838
40. 83.147
41. 0.000008
42. 0.0225
43. 363. or 363
44. 0.00326
45. 0.0876
46. 0.046
47. 1,600. or 1,600
48. 0.0326
49. 0.123642
50. 427,310
51. 3.6
52. 0.006371
53. 43.2179

54. 326.1
55. 2.0
56. 87.2
57. 11.5
58. 44.76
59. 0.85
60. 296.67
61. 0.001
62. 6.629
63. 1.314
64. 0.667
65. 0.286
66. 0.125
67. 15.667
68. 6.125
69. 0.200
70. 0.600
71. 0.875
72. 0.938
73. $\frac{4}{9}$
74. $\frac{2}{3}$
75. $\frac{1}{33}$
76. $\frac{7}{99}$
77. $\frac{52}{333}$
78. $\frac{4}{333}$
79. 3.4×10^6
80. 9.03×10^8
81. $3.4 \times \frac{1}{10^4}$
82. $9.04 \times \frac{1}{10^5}$
83. 4.75×10^1
84. 8.654×10^2
85. 2,300
86. 460
87. 0.000303
88. 0.004751
89. 224
90. 146.42
91. 0.000009
92. 110,000
93. \$203.30
94. \$20.57
95. \$633.60
96. \$6.59 per hour
97. \$14.83
98. Yes, No
99. \$23.40
100. \$7.85
101. 3.8 hours
102. 343.42 miles
103. 8 pounds

Answers To Self-Test: Unit 5

1. (d) 2. (c) 3. (a) 4. (c) 5. (b) 6. (d) 7. (c)
8. (a) 9. (b) 10. (b) 11. (d) 12. (c) 13. (a) 14. (a)
15. (d) 16. (b) 17. (d) 18. (c) 19. (b) 20. (a)

UNIT SIX

Percent

Lesson 22	THE MEANING OF PERCENT
Lesson 23	SOLVING PERCENT PROBLEMS
Lesson 24	APPLICATIONS OF PERCENT

LESSON 22
THE MEANING OF PERCENT

Objectives:

1. Rewrite a percent as a fraction or mixed number. (1–10)
2. Rewrite a fraction or mixed number as a percent. (11–16)
3. Rewrite a percent as a decimal. (17–22)
4. Rewrite a decimal as a percent. (23–28)
5. Complete a Percent-Fraction-Decimal Triangle. (29–32)
6. Arrange combinations of percents, fractions, and decimals in order of increasing value. (33–36)

Vocabulary:

Percent

PRE-TEST

1. Rewrite each percent as a fraction or mixed number.
   (a) $8\frac{1}{2}\%$ (b) 225%
2. Rewrite each fraction or mixed number as a percent.
   (a) $2\frac{3}{4}$ (b) $\frac{3}{8}$
3. Rewrite each percent as a decimal.
   (a) 0.5% (b) 3.14%
4. Rewrite each decimal as a percent.
   (a) 0.003 (b) 7.58
5. Complete the Percent-Fraction-Decimal Triangle for 3%.

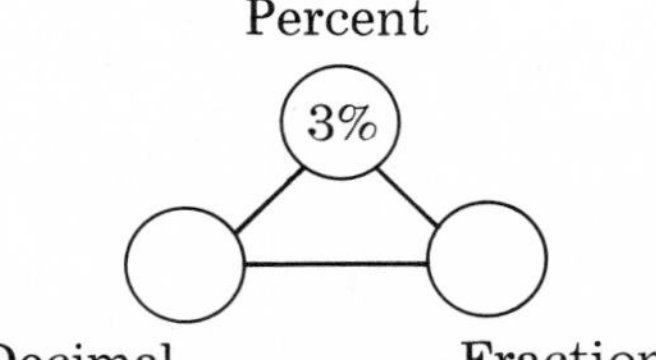

6. Arrange the following in order of increasing value: 3.1, 300%, and $\frac{8}{3}$.

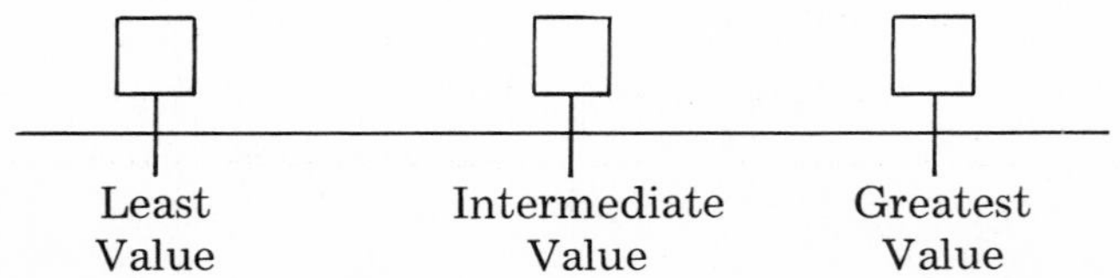

Any person who pays taxes, buys a house or automobile on time, or borrows money must understand the meaning of "percent." The term *percent* comes from the Latin phrase *per centum*, which means "by the hundred." The symbol for percent is %. (1)

The expression 14% means 14 parts of 100 equal parts of a whole or $14 \times \frac{1}{100}$. Thus, 14% means fourteen one-hundredths. This percent can be expressed in three different, but equivalent ways, all meaning fourteen one-hundredths: (2)

14%	$\frac{14}{100}$	0.14
Percent	**Fraction**	**Decimal**

The percent symbol may be used with whole numbers, fractions, and mixed numbers, but always with the same meaning. (3)

$$15\% = \frac{15}{100}$$

$$100\% = \frac{100}{100}$$

$$1\% = \frac{1}{100}$$

$$7.5\% = \frac{7.5}{100}$$

$$12\frac{1}{2}\% = \frac{12\frac{1}{2}}{100}$$

$$\frac{3}{4}\% = \frac{\frac{3}{4}}{100}$$

Since one meaning of $\frac{15}{100}$ is 15 parts of a whole divided into 100 equal parts, the following relationship is always true: (4)

$$15\% = 15 \times \frac{1}{100}$$

$$\blacktriangle\% = \blacktriangle \times \frac{1}{100}$$

This relationship is used to rewrite percent expressions.

Rewriting a percent: (5)

Before calculations with percents can be performed, the percent expression must be rewritten as either a fraction, a mixed numeral, or a decimal. The following examples will show how to rewrite a percent as a fraction.

Rewriting 175% as a fraction: ⑥

$$175\% = 175 \times \frac{1}{100} \longleftarrow \boxed{\text{Definition of percent}}$$

$$= \frac{\overset{1}{\cancel{5}} \times \overset{1}{\cancel{5}} \times 7}{4 \times \underset{1}{\cancel{5}} \times \underset{1}{\cancel{5}}} \longleftarrow \boxed{\text{Multiplying}}$$

$$= \frac{7}{4} \text{ or } 1\frac{3}{4}$$

The equivalent fraction is always rewritten as a mixed number and is reduced if necessary.

Rewriting $12\frac{1}{2}$% as a fraction: ⑦

Complete the following:

$$12\frac{1}{2}\% = 12\frac{1}{2} \times \frac{\bigcirc}{\square}$$

$$= \frac{25}{2} \times \frac{\bigcirc}{\square}$$

$$= \frac{\overset{1}{\cancel{25}} \times 1}{2 \times \underset{4}{\cancel{100}}}$$

$$= \frac{1}{8}$$

Rewriting 4.5% as a fraction: ⑧

Complete the following:

$$4.5\% = 4.5 \times \frac{\bigcirc}{\square}$$

$$= 4\frac{1}{2} \times \frac{\bigcirc}{\square} \longleftarrow \boxed{4.5 = 4\frac{5}{10} = 4\frac{1}{2}}$$

$$= \frac{9}{2} \times \frac{\bigcirc}{\square}$$

$$= \frac{9 \times 1}{2 \times 100}$$

$$= \frac{\triangle}{\hexagon}$$

Rewriting $\frac{5}{16}\%$ as a fraction: (9)

Complete the following:

$$\frac{5}{16}\% = \frac{5}{16} \times \frac{\bigcirc}{\square}$$

$$= \frac{\overset{1}{\cancel{5}} \times 1}{16 \times \underset{20}{\cancel{100}}}$$

$$= \frac{1}{\hexagon}$$

Practice Exercises (10)

Rewrite each percent as a fraction or mixed number and reduce to lowest terms.

1. 56%
2. $2\frac{1}{2}\%$
3. 8.25%
4. 150%
5. $5\frac{1}{3}\%$
6. $\frac{1}{2}\%$

Rewriting a fraction as a percent: (11)

It is often necessary to write a fraction as a percent. This can also be accomplished by applying the definition and using multiplication by 1 or $\frac{100}{100}$.

Rewriting $\frac{1}{2}$ as a percent: (12)

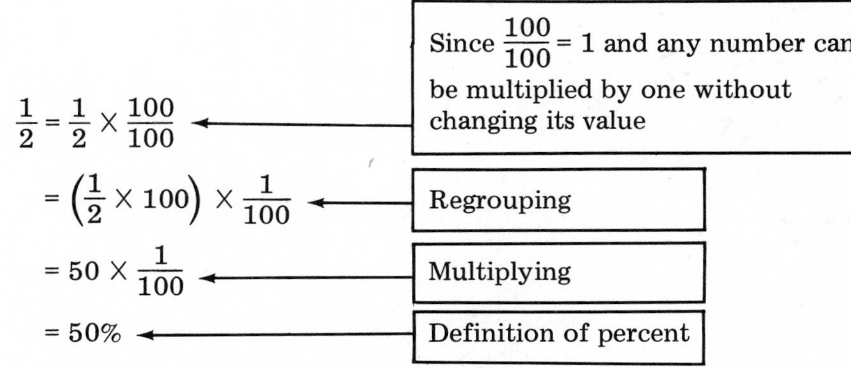

$\frac{1}{2} = \frac{1}{2} \times \frac{100}{100}$
Since $\frac{100}{100} = 1$ and any number can be multiplied by one without changing its value
$= \left(\frac{1}{2} \times 100\right) \times \frac{1}{100}$
Regrouping
$= 50 \times \frac{1}{100}$
Multiplying
$= 50\%$
Definition of percent

Rewriting $\frac{2}{3}$ as a percent: (13)

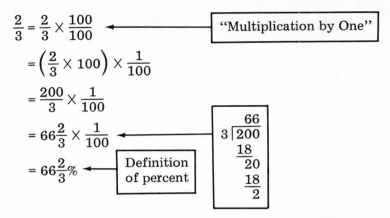

$$\frac{2}{3} = \frac{2}{3} \times \frac{100}{100}$$ ← "Multiplication by One"

$$= \left(\frac{2}{3} \times 100\right) \times \frac{1}{100}$$

$$= \frac{200}{3} \times \frac{1}{100}$$

$$= 66\frac{2}{3} \times \frac{1}{100}$$ ← $3\overline{)200}$ = 66, 18, 20, 18, 2

$$= 66\frac{2}{3}\%$$ ← Definition of percent

Rewriting $2\frac{1}{4}$ as a percent: (14)

$$2\frac{1}{4} = \frac{9}{4}$$ ← Rewriting $2\frac{1}{4}$ as a fraction

$$= \frac{9}{4} \times \frac{100}{100}$$ ← "Multiplication by One"

$$= \left(\frac{9}{4} \times 100\right) \times \frac{1}{100}$$

$$= (9 \times 25) \times \frac{1}{100}$$

$$= 225 \times \frac{1}{100}$$

$$= 225\%$$ ← Definition of percent

Rewriting $1\frac{3}{4}$ as a percent: (15)

Complete the following:

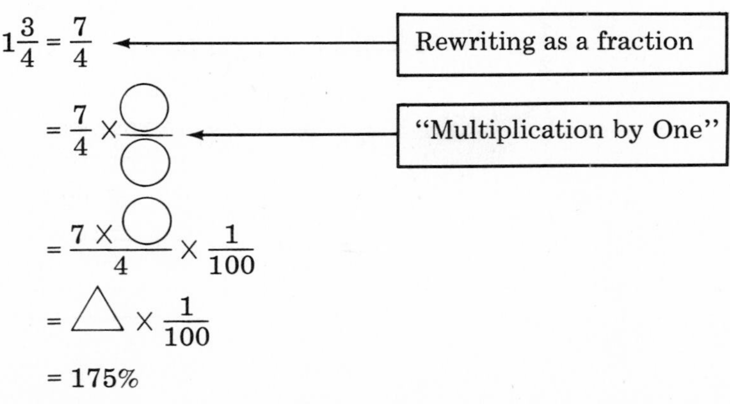

$$1\frac{3}{4} = \frac{7}{4}$$ ← Rewriting as a fraction

$$= \frac{7}{4} \times \frac{\bigcirc}{\bigcirc}$$ ← "Multiplication by One"

$$= \frac{7 \times \bigcirc}{4} \times \frac{1}{100}$$

$$= \triangle \times \frac{1}{100}$$

$$= 175\%$$

Practice Exercises ⑯

Rewrite each fraction as a percent.

1. $\frac{2}{3}$
2. $\frac{3}{2}$
3. $\frac{1}{9}$
4. $\frac{4}{5}$
5. $6\frac{1}{2}$
6. $\frac{5}{6}$

Rewriting a percent as a decimal: ⑰

To rewrite a percent as a decimal, first rewrite the percent as a fraction and then rewrite the fraction in the equivalent decimal form.

Rewriting 5% as a decimal: ⑱

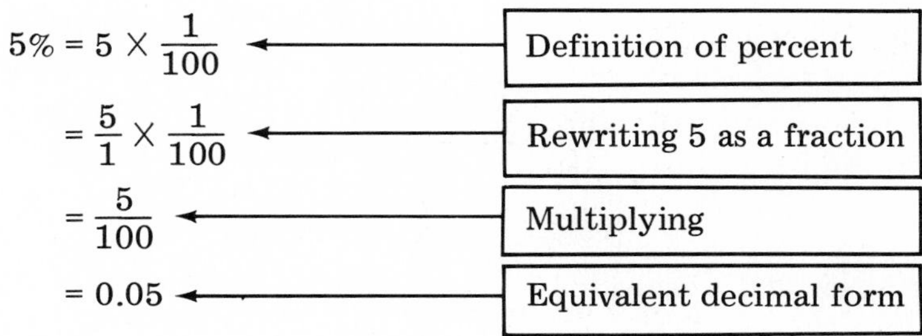

Rewriting $3\frac{1}{3}\%$ as a decimal, correct to the nearest hundredth: ⑲

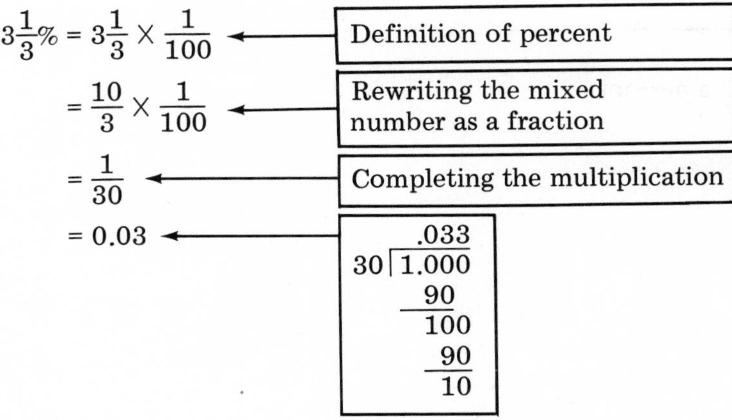

$3\frac{1}{3}\% = 3\frac{1}{3} \times \frac{1}{100}$
Definition of percent
$= \frac{10}{3} \times \frac{1}{100}$
Rewriting the mixed number as a fraction
$= \frac{1}{30}$
Completing the multiplication
$= 0.03$
.033
30 | 1.000
90
100
90
10

Rewriting 183% as a decimal: (20)

Complete the following:

$$183\% = 183 \times \frac{\bigcirc}{\square}$$

$$= \frac{183}{1} \times \frac{\bigcirc}{\square}$$

$$= \frac{183}{\square}$$

$$= 1.83 \longleftarrow$$ Using the decimal-shift method for division by a power of ten

---

Rewriting 0.25% as a decimal: (21)

Complete the following:

$$0.25\% = 0.25 \times \frac{\bigcirc}{\square}$$

$$= \frac{0.25}{\square}$$

$$= 0.0025 \longleftarrow$$ Using the decimal-shift method for division by a power of ten

---

Practice Exercises 

Rewrite each of the following as a decimal, correct to the nearest thousandth:

1. 28%
2. 0.2%
3. 166%
4. $5\frac{3}{4}\%$

Rewriting decimals as percents: 

To rewrite a decimal as a percent follow the same procedure as is used in changing a fraction to a percent. (Review Frame 11 of this lesson, if necessary.)

Rewriting 0.06 as a percent: 

$0.06 = 0.06 \times \frac{100}{100}$	"Multiplication by One"
$= (0.06 \times 100) \times \frac{1}{100}$	Regrouping
$= 6 \times \frac{1}{100}$	Using decimal-shift method for multiplication by a power of ten
$= 6\%$	Definition of percent.

Rewriting $0.03\frac{1}{3}$ as a percent: 

$0.03\frac{1}{3} = 0.03\frac{1}{3} \times \frac{100}{100}$	"Multiplication by One"
$= \frac{3\frac{1}{3}}{100} \times \frac{100}{100}$	$0.03\frac{1}{3}$ is read $3\frac{1}{3}$ hundredths
$= \left(\frac{3\frac{1}{3}}{100} \times 100\right) \times \frac{1}{100}$	
$= 3\frac{1}{3} \times \frac{1}{100}$	
$= 3\frac{1}{3}\%$	Definition of percent

Rewriting 0.7 as a percent: 26

Complete the following:

$$0.7 = 0.7 \times \frac{\square}{\square}$$

$$= \left(0.7 \times \square\right) \times \frac{1}{\square}$$

$$= 70 \times \frac{1}{\square}$$

$$= 70\%$$

Rewriting 1.67 as a percent: (27)

Complete the following:

$$1.67 = 1.67 \times \frac{\square}{\square}$$

$$= \left(1.67 \times \square\right) \times \frac{1}{\square}$$

$$= \triangle \times \frac{1}{\square}$$

$$= 167\%$$

Practice Exercises

Rewrite each decimal as a percent. Use decimal notation.

1. 0.3
2. 0.025
3. $0.06\frac{1}{4}$
4. 1.25

Percent-Fraction-Decimal Triangle:

To help visualize the equivalence of percents, fractions, and decimals, the following triangular pattern is useful.

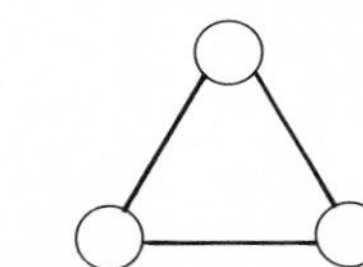

This pattern is called a *Percent-Fraction-Decimal Triangle*, or the *PFD Triangle.*

The PFD Triangle for 75% is shown below.

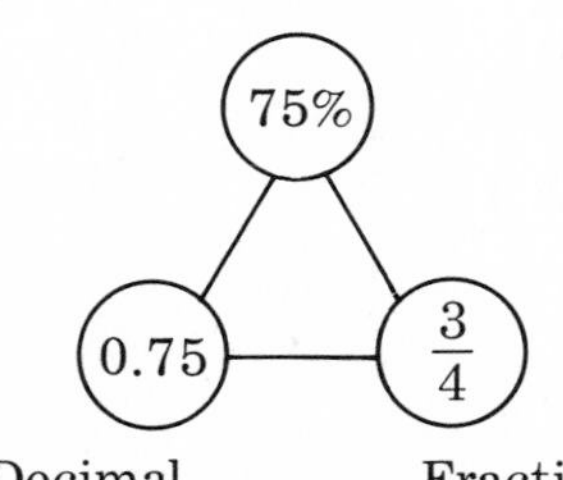

Rewriting 135% using the PFD Triangle: 

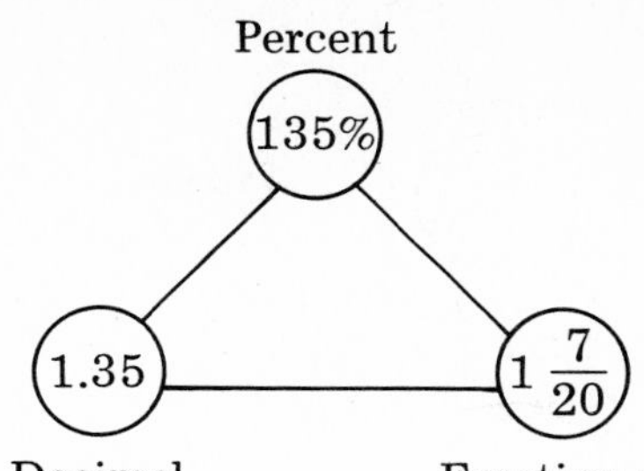

$$135\% = 135 \times \frac{1}{100}$$
$$= \frac{135}{100}$$
$$= 1.35$$

$$135\% = 135 \times \frac{1}{100}$$
$$= \frac{135}{100}$$
$$= \frac{\overset{1}{\cancel{5}} \times 27}{\underset{1}{\cancel{5}} \times 20}$$
$$= \frac{27}{20} \text{ or } 1\frac{7}{20}$$

Completing the PFD Triangle for 0.083: 

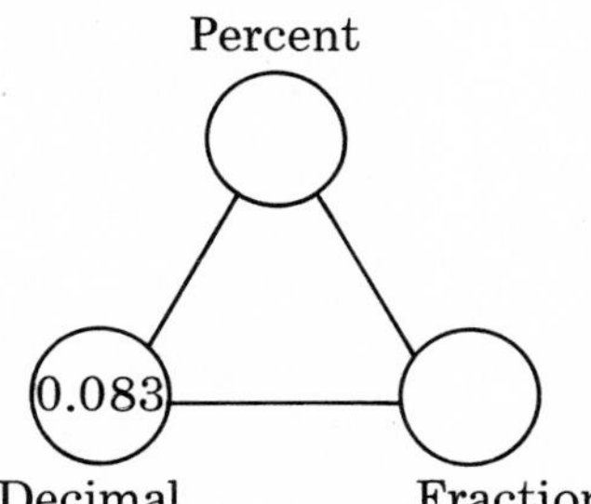

Determining the equivalent percent:

$$0.083 = 0.083 \times \frac{\square}{\square}$$
$$= \left(0.083 \times \square\right) \times \frac{1}{\square}$$
$$= 8.3 \times \frac{1}{\square}$$
$$= \bigcirc$$

Determining the equivalent fraction:

$$0.083 = \frac{\text{⬡}}{\triangle}$$

Thus, $0.083 = 8.3\%$ and $0.083 = \frac{83}{1{,}000}$.

Completing the PFD Triangle for $5\frac{1}{8}$: 

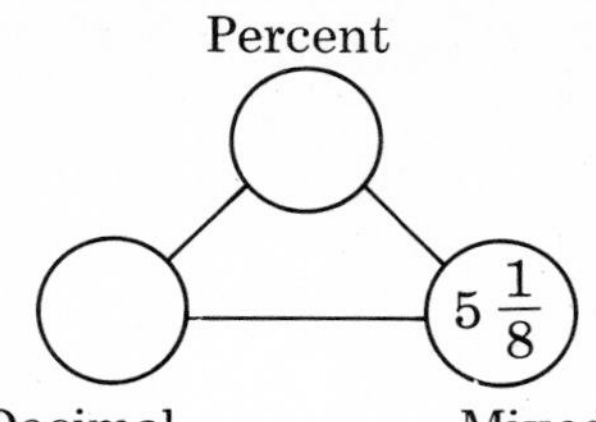

Determining the equivalent percent:

$$5\frac{1}{8} = 5\frac{1}{8} \times \frac{\square}{\square}$$

$$= \frac{41}{8} \times \frac{\square}{\square}$$

$$= \left(\frac{41}{8} \times \square\right) \times \frac{1}{\square}$$

$$= 512\frac{1}{2} \times \frac{1}{\square}$$

$$= ⬡$$

Determining the equivalent decimal, correct to the nearest thousandth:

$$5\frac{1}{8} = \frac{\triangle}{8}$$

$$8\overline{)41.000}$$

Thus $5\frac{1}{8} = 512\frac{1}{2}\%$ and $5\frac{1}{8} = 5.125$.

Which is larger: 1.5 or 200%? 250% or 3? 4 or 400%? (33)

To answer these questions, a number scale is very useful.

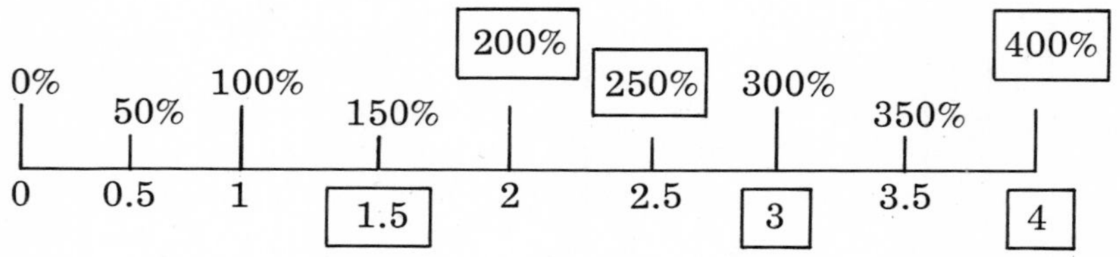

When you compare the relative value of any two numbers, locate each on the number scale. The number to the right will always be the larger number. If two numbers occupy the same position on the scale, they are equal to each other. 200% is to the right of 1.5; thus, 200% is larger than 1.5.

250% is to the left of 3; thus, 250% is smaller than 3. Note that 400% and 4 occupy the same position; thus, they are equal to each other.

When you compare the relative values of percents, fractions, and decimal numbers, rewrite the percents and fractions as decimals so that they can be arranged in order of increasing value.  34

Arrange 150%, 0.98, and $\frac{12}{5}$ in order of increasing value:

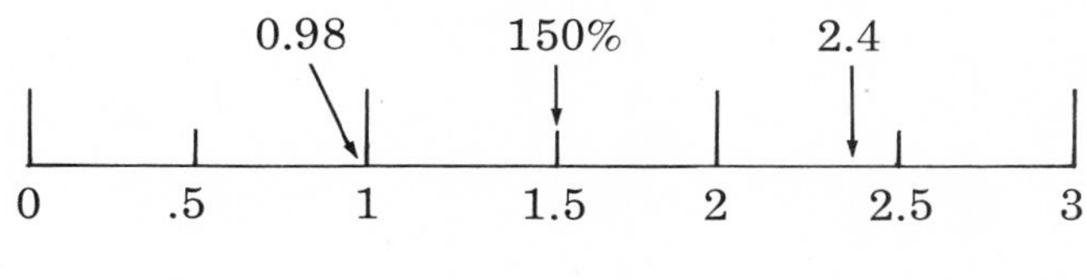

$$150\% = 150 \times \frac{1}{100} = \frac{150}{100} = 1.5 \qquad \text{and} \qquad \frac{12}{5} = \frac{12}{5} \times \frac{2}{2} = \frac{24}{10} = 2.4$$

Now locate the decimals 1.5, 2.4, and 0.98 on the number scale:

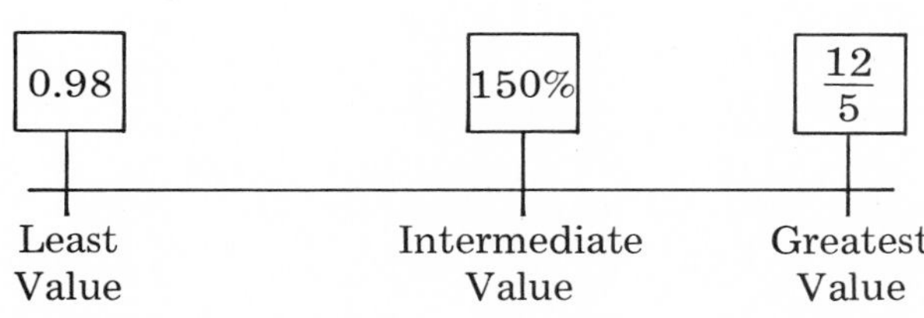

0.98 — Least Value; 150% — Intermediate Value; $\frac{12}{5}$ — Greatest Value

---

Arrange the following in order of increasing value: 200%, 0.53, and $\frac{7}{3}$  35

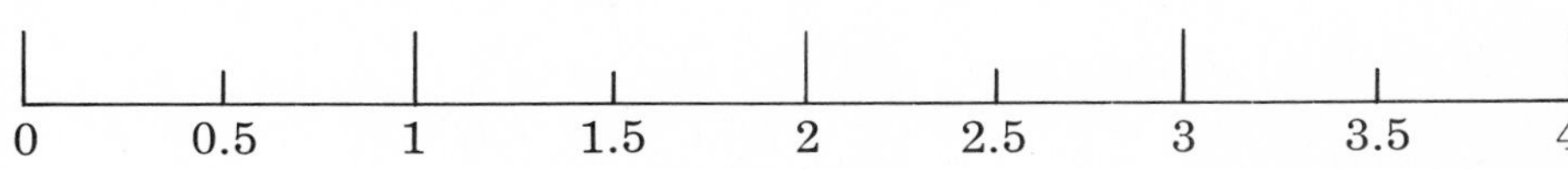

Complete the following:

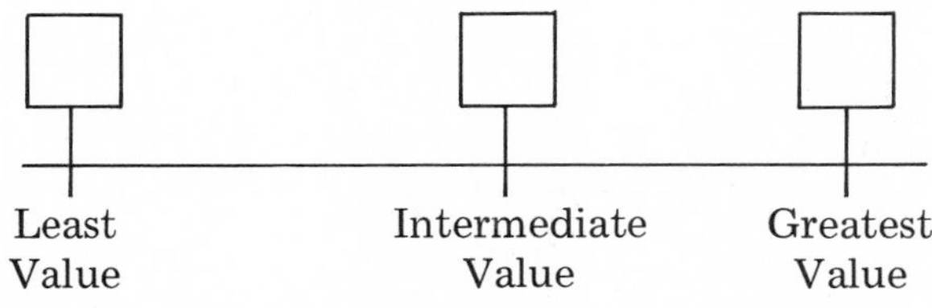

Practice Exercises

Complete each PFD Triangle:

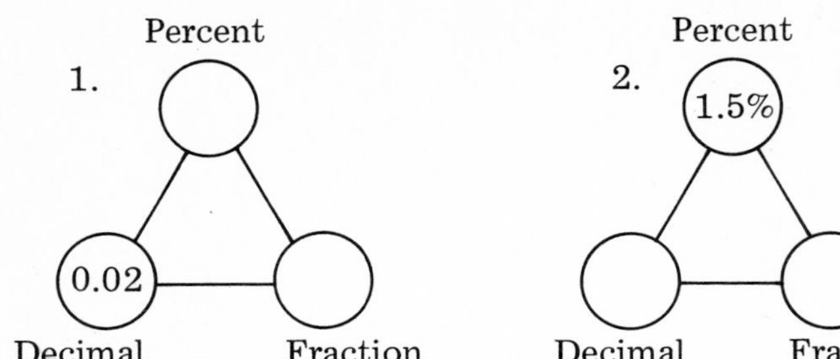

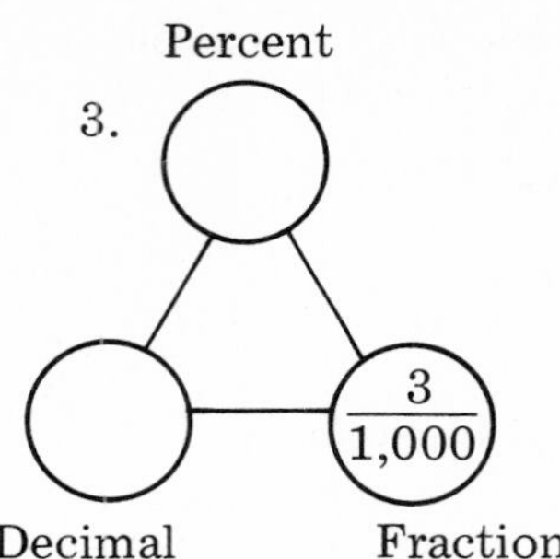

4. Arrange the following in order of increasing value: 0.02, 1.5%, and $\frac{3}{1,000}$.

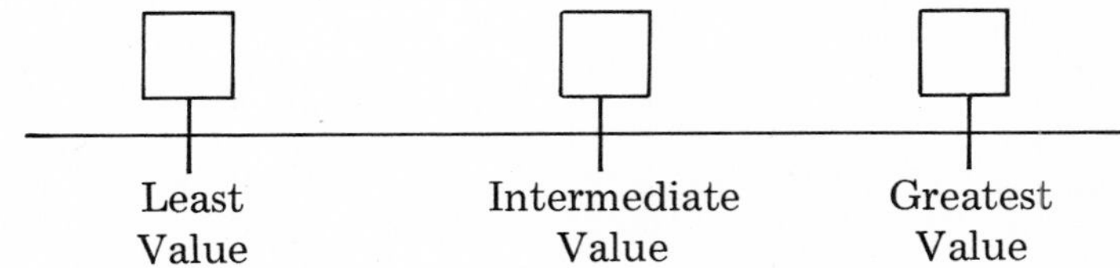

POST-TEST

1. Rewrite each percent as a fraction or mixed number.
   (a) $15\frac{3}{4}\%$ (b) 160%
2. Rewrite each fraction or mixed number as a percent.
   (a) $3\frac{1}{5}$ (b) $\frac{5}{8}$
3. Rewrite each percent as a decimal.
   (a) 0.04% (b) 2.37%
4. Rewrite each decimal as a percent.
   (a) 0.058 (b) 2.77
5. Complete the PFD Triangle for 14%.

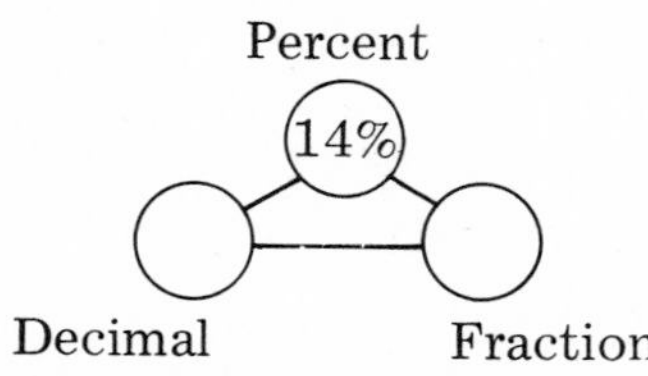

6. Arrange the following in order of increasing value: 111%, $\frac{5}{4}$, and 1.01

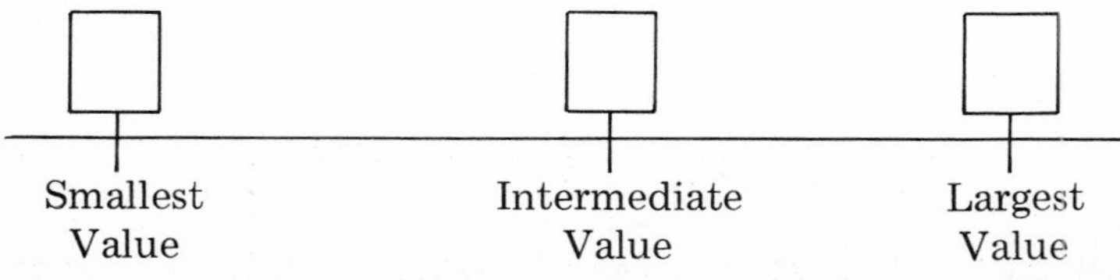

LESSON 23
SOLVING PERCENT PROBLEMS

Objective:

1. Translate a percent problem into an equation and solve. (1–17)

Vocabulary:

Percentage
Rate
Base
Percentage Formula

PRE-TEST

1. Translate each of the following percent problems into an equation and solve:
   (a) What number is $5\frac{1}{2}\%$ of \$68.20, correct to the nearest cent?
   (b) $\frac{3}{4}\%$ of what is 2.5?
   (c) 4.8 is what percent of 1.2?
   (d) If Nicolette correctly answered 45 out of 60 questions on her midterm exam, what percent score did she earn?
   (e) If 85% of the patients treated for a particular disease are cured and 700 people seek treatment, how many people are cured?

Any statement involving percent can be translated into an equation. (1)

For example,

50% of 60 is 30 translates as $50\% \times 60 = 30$. Notice that "of" is translated as "×" and "is" as "=." Since 50% is 0.50 or $\frac{1}{2}$ and since "of" means multiply, 50% of 60 means to multiply one-half times sixty. Thus, 30 is $\frac{1}{2}$, or 50%, of 60.

Translating problems involving percent: (2)

Translating a percent statement into an equation is extremely useful in solving problems involving percent in which some essential part is unknown.

Following are examples of translations of percent problems:

(a) **50% of some number is 30:**

$$50\% \times B = 30$$

"Of" is replaced by "×." "Some number" is replaced by B. "Is" is replaced by =.

(b) **What number is 50% of 60?**

$$P = 50\% \times 60$$

P replaces "what number."

(c) **30 is what percent of 60?**

$$30 = R \times 60$$

R means "what percent?"

Solving percent problems: What number is 15% of 600? ③

First, make an estimate of the answer, keeping in mind that 15% of 600 means $\frac{15}{100}$ of 600. Would the answer be closer to 100 or to 900? Which seems more reasonable? 900 seems too large since only $\frac{15}{100}$ of 600 is needed.

Translate into an equation and solve:

$$P = 15\% \times 600$$ ⟵ *P* replaces "what number"

$$P = \frac{15}{\cancel{100}_{1}} \times \frac{\cancel{600}^{6}}{1}$$

$$P = 90$$

Thus, 90 is 15% of 600.

Solving percent problems: What number is $3\frac{1}{2}\%$ of 35? ④

Estimating first:

Since $3\frac{1}{2}\% = \frac{7}{2}\% = \frac{7}{200}$, $3\frac{1}{2}\%$ of 35 has to be much smaller than 35. Let *P* replace "what number." Then, translating and solving:

$$P \text{ is } 3\tfrac{1}{2}\% \text{ of } 35$$

$$P = 3\tfrac{1}{2}\% \times 35$$

$$P = \frac{7}{\cancel{200}_{40}} \times \frac{\cancel{35}^{7}}{1}$$

$$P = \frac{49}{40} \text{ or } 1\frac{9}{40}$$

Thus, $1\frac{9}{40}$ is $3\frac{1}{2}\%$ of 35. Notice that $1\frac{9}{40}$ is much smaller than 35.

Solving percent problems: 30% of 50 is what number? (5)

First estimating,

Since 30% is approximately $\frac{1}{3}$, an answer that is close to $\frac{1}{3}$ of 50 is reasonable. Certainly 150 would be too large and 1.5 would be too small.

Translate into an equation and solve, letting P replace "what number":

Complete the following:

$$30\% \times 50 = P$$

$$\frac{\square}{100} \times 50 = P$$

$$\bigcirc = P$$

Thus, 30% of 50 is 15. This is reasonably close to our estimate of $\frac{1}{3}$ of 50.

Solving percent problems: 15 is what percent of 40? (6)

To estimate the answers, ask the question: 15 is what part of 40? It is easy to see that 15 is approximately $\frac{1}{3}$ of 40. Since $\frac{1}{3}$ is approximately 0.33, then $\frac{1}{3}$ is close to 33%.

Translate into an equation and solve, letting R replace "what percent":

$$15 = R \text{ of } 40$$

$$15 = R \times 40$$

$$\frac{15}{40} = \frac{R \times 40}{40}$$ ← Divide both sides of the equation by 40.

$$0.375 = R$$

$$37.5\% = R$$ ← 0.375 must be changed to a equivalent percent since R is "what percent."

or

$$37\tfrac{1}{2}\% = R$$

Thus, 15 is 37.5% of 40. The estimate of 33% was too small, but it is useful in preventing errors such as 3.75% or 375%, for example.

Solving percent problems: 78 is 25% of what number? (7)

Estimating:

Since 25% = $\frac{1}{4}$, then 78 is $\frac{1}{4}$ of some number. Should this number be larger or smaller than 78? Of course, it should be larger, and four times larger since 78 is $\frac{1}{4}$ of this number. An estimate of 300 sounds reasonable.

Translate into an equation and solve, letting B replace "what number."

$$78 = 25\% \times B$$

$$78 = \frac{1}{4} \times B$$

$$4 \times 78 = \left(4 \times \frac{1}{4}\right) \times B$$

$$312 = B$$

Thus, 78 is 25% of 312, which is close to the estimate of 300. A method of checking the answer is to find 25% of 312.

$$\begin{aligned} 25\% \text{ of } 312 &= 25\% \times 312 \\ &= \frac{1}{4} \times \frac{312}{1} \\ &= \frac{312}{4} \\ &= 78 \end{aligned}$$

Percentage formula: 8

In the preceding examples the variables P, R, and B have been used. Choosing a variable is usually up to the individual, but there was a reason for choosing these particular variables.

Consider the statement:

78 = 25% of 312

78 is a part of 312 and is called the *percentage.* The 25% is called the *rate* and it indicates the relative size of the percentage, 78, compared to 312, which is the total or *base*. If P stands for percentage, R for rate, and B for base, then all problems in this lesson can be translated into the following equation, which is called the *percentage formula.*

Percentage = Rate × Base

$$P = R \times B$$

Identify the percentage, rate, and base in the following sentence: 

30% of 80 is 24

The rate, R, is 30%. The total or base, B, is 80. The part or percentage, P, is 24.

Translate into an equation and solve: 

$8\frac{1}{2}\%$ of 200 is what number?

Complete the following:

$$8.5\% \times 200 = P$$

$$\bigcirc \times \square = P$$

$$17 = P$$

Translate into an equation and solve: What percent of 16 is 12? (11)

Complete the following:

$$R \times \square = \bigcirc$$

$$R = \frac{\bigcirc}{\square}$$

$$R = .75$$

$$R = \hexagon$$

Be sure to change the answer to percent when solving for "what rate."

Translate into an equation and solve: 2.5% of what number is 20? (12)

Complete the following:

$$2.5\% \times B = 20$$

$$\square \times B = 20$$

$$B = \frac{20}{\square}$$

$$B = 800$$

Practice Exercises (13)

Translate each problem into an equation and solve.

1. What number is 2% of 60?
2. 30 dollars is what percent of 60 dollars?
3. 20% of what number is 800?
4. 0.5 is what percent of 0.4?
5. 115% of 70 is what number?
6. 0.4% of what number is 3.6?

Word problems: 14

How many questions were answered correctly if a student received a grade of 80% on a 20-question biology exam?

The total number correct is a percentage of the total number of questions.

Let P be the number of questions answered correctly. The base (total number of questions) is 20 and the rate is 80%.

$$P \text{ is } 80\% \text{ of } 20$$

$$P = (0.80)\,(20)$$

$$P = 16$$

The carpenter's union negotiated an 8% salary increase for its members. If Dave Amos, a carpenter, receives an increase of 64 dollars per month, what did Dave Amos earn each month before the increase? 

Complete the following:

Raise is 8% of previous salary

$$\$64 = 8\% \times B$$

$$\$64 = \square \times B$$

$$\frac{\$64}{\square} = \frac{\square \times B}{\square}$$

$$\$\bigcirc = B$$

A poll taker sent out 800 copies of a questionnaire. If 680 were returned completed, what percent of the copies were returned? 

Complete the following:

Number returned is what percent of the number sent?

$$680 = R \times \square$$

$$\frac{680}{\square} = \frac{R \times \square}{\square}$$

$$\triangle\,\% = R$$

Practice Exercises

Translate each word problem into an equation and solve.

1. How much is 9% of $1.64? (Correct to the nearest cent)
2. What percent of 5.5 is 1.1?
3. A tank is filled to 65% of its capacity of 1,200 gallons. How many gallons are in the tank?
4. The number 2.59 is 35% of what number?
5. A dart team won 18 games and lost 6 games. What percent of the games played did the team lose?
6. Mr. Wilcox saved $1.25 by buying fertilizer at a price reduction of 10%. What was the regular price?

POST-TEST

1. Translate each percent problem into an equation and solve.
   (a) What number is $4\frac{1}{2}$% of $75.20? (Correct to the nearest cent.)
   (b) 0.8% of what number is 7.3?
   (c) 5.4 is what percent of 0.9?
   (d) If 3% of the automobile parts manufactured by a company do not meet the quality control standards, and if 15,000 parts were manufactured last month, how many parts were rejected?
   (e) Of the 912 registered voters in a precinct, only 600 voted in the last municipal election. What percent of the registered voters turned out to vote? (Correct to the nearest whole percent.)

LESSON 24
APPLICATIONS OF PERCENT

Objectives:

1. Solve tax problems. (1–8)
2. Solve commission problems. (9–13)
3. Determine a percent of increase or decrease. (14–18)
4. Solve discount problems. (19–22)
5. Determine a percent of profit or loss. (23–26)
6. Solve interest problems. (27–33)

Vocabulary:

Sales Tax	Percent of Discount
Social Security Tax	Percent of Profit
Federal Income Tax	Percent of Loss
Commission	Interest
Percent of Increase	Principal
Percent of Decrease	Service Charge

PRE-TEST

1. Find John Rubenstein's monthly take-home pay if he earns 980 dollars and has $5\frac{1}{2}\%$ social security tax deducted and 12% federal income tax withheld.
2. Betty McIntosh sells magazine subscriptions and earns a 35% commission. If she sold 420 dollars worth of subscriptions, how much commission did she earn?
3. The population of Fortune City was 35,800 last year and is 30,500 this year. Calculate the rate of decrease in population. (Correct to the nearest tenth of a percent.)
4. A gallon of paint originally sold for $9.50. What is the new price if it is discounted 20%?
5. The local market pays 50 cents for a loaf of whole wheat bread and sells the bread for 53 cents. What is the rate of profit?
6. The Blackstone family had a 240-dollar unpaid balance last month on its bank credit card. If the bank charges $1\frac{1}{2}\%$ interest per month on the unpaid balance, what was the interest charge last month?

Applications of percent arise in many everyday situations. Some of these are listed below and will be discussed in this lesson. 1

Taxes	Commission
Profit–Loss	Increase-Decrease
Interest	Discount

In most applications the percentage formula is used. For example, 2

Percentage = Rate × Base

Sales tax = Rate × Amount of purchase

Commission = Rate × Amount of sales

Taxes: 3

Sales tax is levied on the amount of the purchase.

How much sales tax is levied on a purchase of a toaster marked $17.50 if the sales tax rate is 6%?

Sales tax is 6% of $17.50.

Sales tax = (0.06) ($17.50)

Sales tax = $1.05

Taxes: 4

The *total payment* is the purchase price plus the sales tax.

Total payment = $17.50 + $1.05

= $18.55

Total payment is 106% of the price of the toaster since $\frac{\$18.55}{\$17.50}$ is equal to 1.06

Taxes: 5

Social security tax was levied at the rate of $5\frac{1}{2}\%$ of a person's income and deducted from his salary.

Bill Fong earned $7.40 per hour. What amount was deducted for social security tax?

Social security tax was $5\frac{1}{2}\%$ of the income.

Social security tax = (0.055) ($7.40) per hour

Social security tax = $0.407 or $0.41 per hour

Taxes: 

Federal income tax is also deducted from a person's income. The tax rate varies according to his income and certain other factors. Suppose Bill Fong's income tax bracket required that 14% of his income be withheld. How much income tax was withheld per hour?

Income tax is 14% of income.

Income tax = (0.14) ($7.40) per hour

Income tax = $1.036 or $1.04 per hour

Taxes: 

What was Bill Fong's take-home pay for an 8-hour work day?

Take-home pay per hour was income less Social Security tax and income tax.

Take-home pay per hour = $7.40 – ($0.41 + $1.04)

= $7.40 – $1.45

= $5.95

Take-home pay per day = 8 × $5.95

= $47.60

Practice Exercises 

Solve each tax problem.

1. What is the total payment for a television set priced at $128.77 if the sales tax rate is 5%? (Correct to the nearest cent.)
2. Martha earned $5.50 per hour. If she paid $5\frac{1}{2}$% social security tax and 18% income tax, what was her take-home pay for every 40-hour work week?

Commission: 

A salesperson may earn a commission based upon the amount of sales. The amount of *commission* is based upon the rate of commission and the amount of the sale.

Percentage = Rate × Base

Commission = Rate × Sales

Commission: 

Frank Matinick is a real estate salesman. He earns 3% of his sales, and the real estate company which listed the home for sale also receives 3%. Suppose Frank sells the Adams' home, which was originally listed for sale by the Apex Realty Company. If he sells the home for $32,000, what is his commission?

The commission is rate times sales.

$$C = 3\% \times 32{,}000$$
$$C = (0.03)\ (\$32{,}000)$$
$$C = \$960$$

Commission: (11)

How much will the Adams family pay in total real estate commissions?

The total real estate commission is 6% of \$32,000.

Total real estate commission = (0.06) (\$32,000)
= \$1,920

The total real estate commission is the salesman's commission plus the company's commission.

Commission: (12)

Maria works as a saleslady in a furniture store. She earns a salary of 90 dollars per week, plus a 12% commission based on her weekly sales. If Maria sells \$1,340 dollars worth of furniture this week, what is her total weekly income?

Total income is salary plus commission.

Total income = \$90.00 + (12%) (\$1,340)
= \$90.00 + (0.12) (\$1,340)
= \$90.00 + \$160.80
= \$250.80

Practice Exercises (13)

Solve each problem.

1. Marv Snow sells automobiles and earns a 5% commission. Last month he sold 7 cars for a total sales amount of 16,100 dollars. How much commission did Marv earn?
2. Mercedes is a waitress in the "Flickering Lite" cafe. Her tips average 8% of the cost of the food that she serves. She earns 50 dollars a week plus tips. If she served 840 dollars worth of food last week, how much was her income for the week?

Percent of increase–percent of decrease: (14)

Percent of increase is calculated from the amount of increase and the original amount.

Percentage = Rate × Base

Amount increase = Percent of increase × Original amount

Percent of decrease is calculated from the amount of decrease and the original amount.

Percentage = Rate X Base

Amount decrease = Percent of decrease X Original amount

Percent of increase: (15)

The number of automobile accidents in Orange County for one year was 4,914. The year before it was 4,200. What was the percent of increase in the number of accidents?

Original amount = 4,200

New amount = 4,914

Amount of increase = 4,914 – 4,200

= 714

714 is what percent of 4,200?

$$714 = R \times 4{,}200$$

$$\frac{714}{4{,}200} = R$$

$$0.17 = R$$

$$17\% = R$$

The number of automobile accidents increased 17% this year. Or, we could also say the number of automobile accidents this year are 117% of those in the previous year, since $\frac{4{,}914}{4{,}200} = 1.17$.

Percent of decrease: 

Mr. Wallace bought a Cadillac for 8,000 dollars. Later he sold it for 7,500 dollars. What was the percent of decrease?

Original amount = 8,000 dollars

New amount = 7,500 dollars

Amount of decrease = 8,000 dollars – 7,500 dollars

= 500 dollars

500 dollars is what percent of 8,000 dollars?

$$\$500 = R \times \$8{,}000$$

$$\frac{\$500}{\$8{,}000} = R$$

$$0.0625 = R$$

$$6.25\% = R$$

The car decreased 6.25% in value. Or, we could say the selling price is 93.75% of the purchase price, since $\frac{\$7{,}500}{\$8{,}000} = 0.9375$

Percent of increase:

In 1965 the price of ground beef was 50 cents per pound. In 1975 the price of ground beef increased to 90 cents per pound. What is the percent of increase to the nearest whole percent?

Original amount = 50 cents
New amount = 90 cents

Amount of increase = 90 cents − 50 cents = 40 cents

40 cents is what percent of 50 cents?

$$40¢ = R \times 50¢$$

$$\frac{40¢}{50¢} = \frac{R \times 50¢}{50¢}$$

$$0.8 = R$$

$$80\% = R$$

Practice Exercises

Find the percent of increase or decrease.

1. A company earned 180,000 dollars in 1969 and 234,000 dollars in 1970. Find the percent of increase in earnings.
2. Mr. Jones earned 8,000 dollars last year and 7,600 dollars this year. Find the percent of decrease in earnings.

Percent of discount:

Percent of discount is calculated from an amount of discount and an original selling price.

Percentage = Rate × Base

Amount of discount = Percent of discount × Original selling price

Discount price:

A coat originally marked at 45 dollars is discounted 20%. What is the discount price?

The discount price is the original price less the amount discounted.

Discount price = \$45 − (0.20) (\$45)
= \$45 − \$9
= \$36

Note that the discount price is 80% of the original selling price, since $\frac{36}{45} = 0.8$.

Percent of discount: 21

A store advertises 20% to 50% off on all merchandise. A sweater, originally marked at \$15, is marked at \$10.50. Calculate the rate of discount.

The amount of discount is a percent of the original selling price.

$$\$15.00 - \$10.50 = R\ (\$15)$$

$$\$4.50 = R\ (\$15)$$

$$\frac{\$4.50}{\$15} = R$$

$$0.30 = R$$

$$30\% = R$$

Practice Exercises

Solve the following discount problems:

1. A suit selling for 150 dollars was discounted to 90 dollars. What was the rate of discount?
2. A set of golf clubs was discounted 30%. If the clubs had originally sold for 170 dollars, what was the discount price?

Percent of profit–percent of loss:

Percent of profit or *percent of loss* on business sales may be calculated using by the merchant's cost as the base (sometimes the selling price is used as the base).

Percentage = Rate X Base

Profit = Rate X Cost

Loss = Rate X Cost

Percent of profit:

A used car dealer buys a car for 700 dollars which is to be sold at a 10% profit. Compute the amount of profit and the selling price.

The profit is 10% of the cost.

Profit = (0.10) (\$700)

Profit = \$70

The selling price is the cost plus profit.

Selling price = \$700 + \$70

Selling price = \$770

The selling price is 110% of the dealer's cost.

Percent of loss: 

A home valued at 50,000 dollars is damaged by an earthquake. The resulting value of the home is 29,000 dollars. What is the percent loss?

The loss is a percent of the value.

$$\$50{,}000 - \$29{,}000 = R\ (\$50{,}000)$$
$$\$21{,}000 = R\ (\$50{,}000)$$
$$\frac{\$21{,}000}{\$50{,}000} = R$$
$$0.42 = R$$
$$42\% = R$$

Practice Exercises 

Calculate the percent of profit or the percent of loss.

1. If a 4,000-dollar car depreciates 27% in value, what is it worth?
2. What percent of profit, to the nearest tenth of a percent, is earned by selling merchandise that originally costs 43 dollars for 57 dollars? (Use the merchant's cost as the base.)

Simple interest: 

Interest is the payment for the use of money which has either been borrowed or saved. The amount of money borrowed or saved is called the *principal.*

Interest = Rate × Principal × Time (in years)

$$I = R \times P \times T$$

Find the interest earned if 500 dollars is invested at 6% per year for 1 year.

$$\text{Interest} = (6\%)\ (\$500)\ (1)$$
$$I = (0.06)\ (\$500)\ (1)$$
$$I = \$30$$

Simple interest: 

Find the amount of interest earned if the 500 dollars are invested at 6% per year for 4 years.

$$I = R \times P \times T \text{ (in years)}$$
$$I = (6\%)\ (\$500)\ (4)$$
$$I = (0.06)\ (\$500)\ (4)$$
$$I = \$120$$

When using the simple interest formula, 

$$I = R \times P \times T$$

Time, T, must be expressed in *years* as shown below:

$$6 \text{ months} = \frac{6}{12} \text{ or } \frac{1}{2} \text{ year}$$

$$30 \text{ days} = \frac{30}{360} \text{ or } \frac{1}{12} \text{ year}$$

When you compute interest, use 360 days = 1 business year and 30 days = 1 business month.

Simple interest: 

David Steen borrowed 75 dollars for 60 days at 18% interest per year. How much will he pay back?

The amount paid back is the principal plus interest.

$$\text{Amount paid back} = \$75 + (18\%)\ (\$75)\left(\frac{60}{360}\right) \longleftarrow \boxed{60 \text{ days} = \frac{60}{360} \text{ years}}$$

$$= \$75 + (0.18)\ (\$75)\left(\frac{1}{6}\right)$$

$$= \$75 + \$2.25$$

$$= \$77.25$$

Simple interest: 

Calculate the interest earned on 800 dollars at 6% per year invested for 9 months.

$$\text{Interest} = (0.06)\ (\$800)\left(\frac{9}{12}\right) \longleftarrow \boxed{9 \text{ months} = \frac{9}{12} \text{ years}}$$

$$= \$36$$

Carrying charge: 

A credit card *carrying charge* can be calculated in the same way as simple interest.

The Premier department store charges an 18% per year carrying charge on the unpaid balance resulting from credit purchases. The White family purchased a refrigerator for 348 dollars and charged the total purchase on its credit card. How much carrying charge is levied at the end of the first month?

Carrying charge = Rate × Principal × Time

$$\text{Carrying charge} = (0.18)\ (\$348)\ \left(\frac{1}{12}\right)$$

$$= \$5.22$$

Practice Exercises

Find the interest on each of the following:

1. 1,000 dollars at 8% for 9 months
2. 400 dollars at 7.75% for 90 days
3. 7,200 dollars at 11% for 4 years
4. 500 dollars at $6\frac{1}{4}$% for 2 years

POST-TEST

1. Sammy Lee earned 12,500 dollars last year. If $5\frac{1}{2}$% social security tax was deducted and 14% federal income tax was withheld, how much did he take home?
2. John Cranston earns 55 dollars per week plus a 15% commission on his sales of sewing machines. If he sold 870 dollars worth of sewing machines, how much would his weekly income be?
3. The number of deer in the local mountains has increased from 350 to 420 in the last 5 years. What percent increase is this?
4. A pair of pants was discounted 3 dollars from a price of 16 dollars. Calculate the rate of discount.
5. A storekeeper paid 85 cents each for pens that contained scented ink. Because the pens didn't sell, the storekeeper sold the pens for 50 cents each. What was the storekeeper's percent of loss, to the nearest tenth of a percent?
6. How much interest will a 1,500-dollar principal earn at $6\frac{1}{2}$% for 30 months?

DRILL EXERCISES: UNIT 6

Convert the following percents to decimals, correct to the nearest thousandth:

1. 10%
2. 25%
3. 135%
4. 75%
5. $\frac{1}{4}\%$
6. 275%
7. 5%
8. 325%
9. $7\frac{1}{2}\%$
10. $66\frac{2}{3}\%$
11. 750%
12. 6%
13. 0.75%
14. $2\frac{1}{2}\%$
15. 32.5%
16. 0.02%

Convert the following percents to fractions or mixed numbers in lowest terms:

17. 50%
18. 30%
19. 5%
20. 36%
21. 74%
22. 96%
23. $37\frac{1}{2}\%$
24. $6\frac{1}{4}\%$
25. $16\frac{2}{3}\%$
26. $8\frac{1}{3}\%$
27. 160%
28. $162\frac{1}{2}\%$
29. $266\frac{2}{3}\%$

Convert the following to percents using mixed numbers or fractional notation where appropriate.

30. $\frac{9}{100}$
31. $\frac{1}{4}$
32. $\frac{1}{5}$
33. $\frac{1}{10}$
34. $\frac{3}{25}$
35. $\frac{3}{20}$
36. $\frac{1}{8}$
37. $\frac{2}{9}$
38. $\frac{5}{6}$
39. $\frac{21}{35}$
40. $\frac{6}{7}$

Convert the following to percents using decimal notation and rounding to the nearest tenth of a percent.

41. 0.28
42. 0.65

43. 0.043

44. 0.0051

45. 0.4

46. 0.3

47. 5.00

48. 7

49. 1.37

50. 1.9

51. $0.62\frac{1}{2}$

52. $0.07\frac{3}{4}$

53. $1.16\frac{2}{3}$

54. 0.875

55. 0.004

56. $0.00\frac{1}{2}$

57. $0.00\frac{2}{3}$

Write an equation for each of the following and solve:

58. Find 46% of 13.
59. The number 33 is 18% of what number?
60. What number is 21% of 76?
61. The number 45 is what percent of 72?
62. 76% of what number is 76?
63. Find 234% of 59.
64. 56 is 153% of what number?
65. What number is 43.1% of 47?
66. What percent of 48 is 754?
67. 71 is 0.4% of what number?
68. 100 is $4\frac{1}{2}$% of what number?
69. 30 is $5\frac{1}{4}$% of what number?
70. Mr. Jones owns 51% of a company worth 1,600,000 dollars. What is the value of his holdings?
71. One ruler is 15 inches long; another ruler is 12 inches long. What percent of the length of the longer ruler is the shorter ruler?
72. Property taxes are based on 23.4% of the market value of a home. The tax base for a house is \$4,387.50. What is its market value?
73. The population of a town is 237,000. This is 160% of its population 20 years ago. What was its population 20 years ago rounded to the nearest thousand?
74. A family spends 22% of its monthly income for housing. If its monthly income is 960 dollars, how much does the family spend for housing?
75. A meter stick is 109.4% as long as a yard stick. How many inches long is a meter stick? (1 yard = 36 inches.)
76. A National League baseball team wins 101 of 162 games played. What percent of its games did the team win, correct to the nearest tenth of a percent?
77. 24-carat gold is pure gold. What percent gold is 14-carat gold?
78. A bronze alloy contains 4.9% zinc by weight. How much can be made using 10 pounds of zinc?
79. Brian receives a 10-dollar per week raise. This is equivalent to an 8% increase. What was his pay before the raise?

80. If sales tax is $4\frac{1}{2}\%$, find the sales tax on a purchase of $14.85, correct to the nearest cent.

81. In a shipment of light bulbs, 3% were deficient. If the number of deficient light bulbs was 108, what was the total number of light bulbs in the shipment?

82. Last year it cost Jim $25.80 for books. This year his books cost $38.70. What was the percent of increase?

83. A bowler scored 150 in the first game and 165 in the second game. What was his percent of increase?

84. The average age of American men at first marriage changed from 25.9 years in 1900 to 22.8 in 1965. For women, the change was from 21.9 to 20.6.
   (a) What was the percent of decrease for males, correct to the nearest tenth of a percent?
   (b) What was the percent of decrease for females, correct to the nearest tenth of a percent?

85. Linda weighed 120 pounds last year. She went on a diet and 6 months later she weighed only 98 pounds. What is the percent of decrease in her weight, correct to the nearest whole percent?

86. The gross national product changed from 589.2 billion dollars in 1963 to 713.9 billion dollars in 1966. What is the percent of increase, correct to the nearest tenth of a percent?

87. Find the rate of decrease if a man's pay falls from $3.75 per hour to $3.50 per hour.

88. In baking a large cake, the amount of flour was increased from 2 cups to 15 cups. What percent of increase was this?

89. What is the interest on 700 dollars, correct to the nearest cent, at the rate of 5% for:
   (a) 1 year
   (b) 10 months
   (c) 1 month
   (d) 2.5 years
   (e) 45 days

90. What is the interest on $240 for 8 months at the rate of:
   (a) 7%
   (b) 3.5%
   (c) 6%
   (d) $4\frac{3}{4}\%$.

Self Test: Unit 6

1. Express 6.7% as a decimal, correct to the nearest thousandth.
   (a) 6.7
   (b) 0.0067
   (c) 0.067
   (d) 670
   (e) None of the above

2. Express $3\frac{3}{4}$ as a percent.
   (a) $3\frac{3}{4}\%$
   (b) 375%
   (c) 0.0375%
   (d) 3.75%
   (e) None of the above

3. Express 0.0033 as a percent.
   (a) 0.0033%
   (b) 33%
   (c) 0.033%
   (d) 0.33%
   (e) None of the above

4. Express $37\frac{1}{2}\%$ as a fraction in lowest terms.
   (a) $\frac{1}{3}$
   (b) $\frac{7}{16}$
   (c) $\frac{2}{7}$
   (d) $\frac{3}{8}$
   (e) None of the above

5. What number is 24% of 3.8?
   (a) 0.0912
   (b) 91.2
   (c) 9.12
   (d) 912
   (e) None of the above

6. 106% of 80 is what number?
   (a) 84.8
   (b) 848
   (c) 80.48
   (d) 8.48
   (e) None of the above

7. Find $5\frac{3}{4}\%$ of $7\frac{1}{5}$.
   (a) 4.14
   (b) 0.0414
   (c) 0.414

(d) 41.4
(e) None of the above

8. What percent of 0.48 is 0.30?
(a) 0.625%
(b) $62\frac{1}{2}\%$
(c) 0.0625%
(d) $\frac{5}{8}\%$
(e) None of the above

9. 13% of what number is 52?
(a) 6.76
(b) 4
(c) 676
(d) 400
(e) None of the above

10. Which one of the following represents the smallest value?
(a) 83%
(b) 0.01
(c) 8.3
(d) $\frac{8}{9}$
(e) 0.8%

11. John's test score was 80%. If he answered 24 questions correctly, how many questions were on the test?
(a) 30
(b) 28
(c) 32
(d) 34
(e) None of the above

12. A $5\frac{1}{2}\%$ annual raise is granted G. W. C. teachers. If Mr. Reynold's salary is 12,300 dollars, what is the amount of his raise for 1 year?
(a) $6.77
(b) $67.65
(c) $6,765.00
(d) $676.50
(e) None of the above

13. Sally has reduced her weight from 148 pounds to 126 pounds. Compute the percent of weight lost, correct to the nearest whole percent.
(a) 14%
(b) 86%
(c) 15%
(d) 85%
(e) None of the above

14. Roberto invests 335 dollars in a savings account for 1 year at 6% annual interest. At the end of the year what is the total amount in his savings account?
(a) $335.00
(b) $355.10
(c) $20.10
(d) $600.00
(e) None of the above

15. Mike's earnings are 724 dollars per month. The social security tax of $4\frac{1}{2}\%$ of his earnings is deducted from his salary. Determine the amount deducted for social security each month.
(a) $3.26
(b) $325.80
(c) $0.33
(d) $30.08
(e) None of the above

16. A new automobile depreciates approximately 8% of its value when driven off the car dealer's lot after being sold. The Millers purchased a new car for 3,720 dollars. What is the car's depreciation in value due to being sold and, thus, becoming a "used" car?
(a) $297.60
(b) $29.76
(c) $2,976.00
(d) $2.98
(e) None of the above

17. Sales tax will be levied upon gasoline. If a gallon of gasoline costs 62 cents, what will the cost be, including the sales tax of 5%? (Round to the nearest cent.)
(a) 63 cents
(b) 64 cents
(c) 65 cents
(d) 66 cents
(e) None of these

18. A gallon of paint originally sold for $12.95. What is the new price if it is discounted 15%? (Correct to the nearest cent.)
(a) $12.80
(b) $11.01
(c) $194.00
(d) $1.94
(e) None of the above

19. Approximately 0.2% of the ball-point pens made by the Speedrite Company are defective. If $2\frac{1}{2}$ million pens are made, approximately how many are defective?
(a) 5,000
(b) 50,000
(c) 500
(d) 500,000
(e) None of the above

20. Suppose that a nation's population doubles every 50 years. What is the percent of increase of population after 100 years if originally the population is 10,000,000.
(a) 400%
(b) 200%
(c) 600%
(d) 300%
(e) None of the above

LESSON 22:

Pre-test

1. (a) $\frac{17}{200}$ (b) $\frac{9}{4}$ or $2\frac{1}{4}$
2. (a) 275% (b) $37\frac{1}{2}\%$
3. (a) 0.005 (b) 0.0314
4. (a) 0.3% (b) 758%
5.

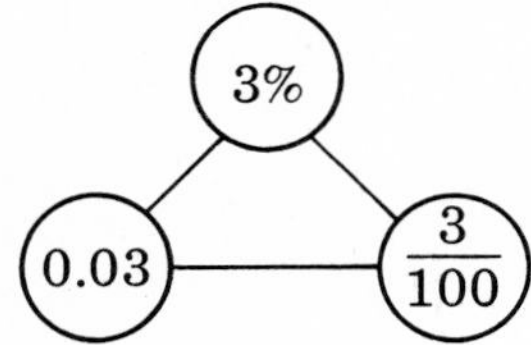

6. Least value: $\frac{8}{3} = 2.\overline{6}$
   Intermediate value: 300% = 3.0
   Greatest value: 3.1

Practice Exercises

Frame 10

1. $\frac{14}{25}$ 2. $\frac{1}{40}$ 3. $\frac{33}{400}$ 4. $\frac{3}{2}$ or $1\frac{1}{2}$ 5. $\frac{4}{75}$ 6. $\frac{1}{200}$

Frame 16

1. $66\frac{2}{3}\%$ 2. 150% 3. $11\frac{1}{9}\%$ 4. 80% 5. 650% 6. $83\frac{1}{3}\%$

Frame 22

1. 0.280 2. 0.002 3. 1.660 4. 0.058

Frame 28

1. 30% 2. 2.5% 3. 6.25% 4. 125%

Frame 36

1.

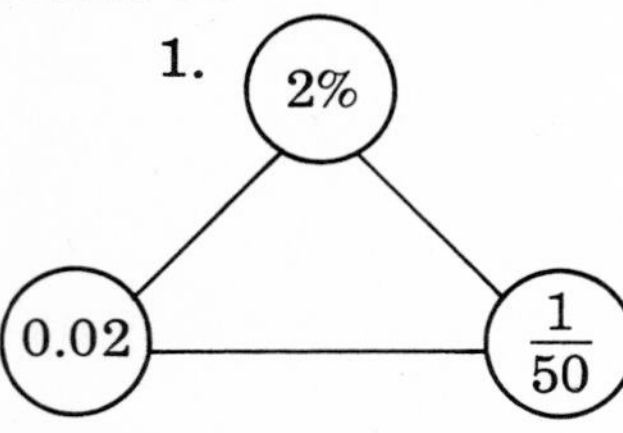

2. 1.5% — 0.015 — $\frac{3}{200}$

3.

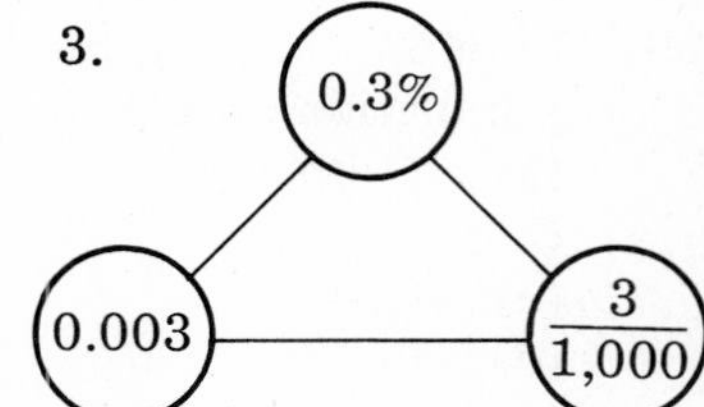

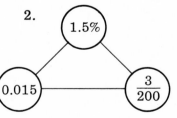

4. Least value: $\frac{3}{1{,}000} = 0.003$
   Intermediate value: 1.5% = 0.015
   Greatest value: 0.02 = 0.020

Post-test

1. (a) $\frac{63}{400}$ (b) $\frac{8}{5}$ or $1\frac{3}{5}$
2. (a) 320% (b) $62\frac{1}{2}\%$

3. (a) 0.0004 (b) 0.0237

4. (a) 5.8% (b) 277%

5.

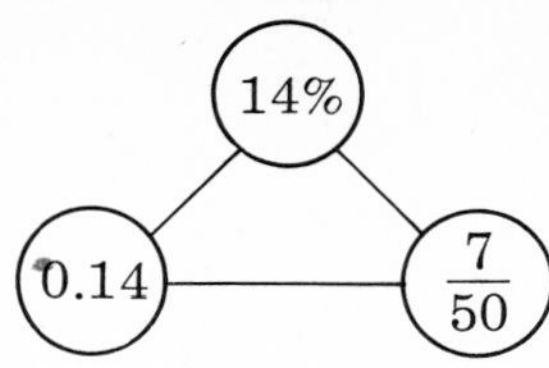

6. Least value: 1.01
Intermediate value: 111% = 1.11
Greatest value: $\frac{5}{4} = 1.25$

LESSON 23:

Pre-test

1. (a) $P = 5\frac{1}{2}\% \times \$68.20; P = \$3.75$
   (b) $\frac{3}{4}\% \times B = 2.5; B = 333.\overline{3}$ or $333\frac{1}{3}$
   (c) $4.8 = R \times 1.2; R = 400\%$
   (d) $45 = R \times 60; R = 75\%$
   (e) $P = 85\% \times 700; P = 595$ people

Practice Exercises

Frame 13

1. $P = 2\% \times 60; P = 1.2$
2. $\$30 = R \times \$60; R = 50\%$
3. $20\% \times B = 800; B = 4{,}000$
4. $0.5 = R \times 0.4; R = 125\%$
5. $115\% \times 70 = P; P = 80.5$
6. $0.4\% \times B = 3.6; B = 900$

Frame 17

1. $P = 9\% \times \$1.64; P = \$.15$
2. $R \times 5.5 = 1.1; R = 20\%$
3. $65\% \times 1{,}200 = P; P = 780$
4. $2.59 = 35\% \times B; B = 7.4$
5. $6 = R \times 24; R = 25\%$
6. $\$1.25 = 10\% \times B; B = \$12.50$

Post-test

1. (a) $P = 4\frac{1}{2}\% \times \$75.20; P = \$3.38$
   (b) $0.8\% \times B = 7.3; B = 912.5$
   (c) $5.4 = R \times 0.9; R = 600\%$
   (d) $P = 3\% \times 15{,}000; P = 450$ parts
   (e) $600 = R \times 912; R = 66\%$

LESSON 24:

Pre-test

1. \$808.50
2. \$147.00
3. 14.8%
4. \$7.60
5. 6%
6. \$3.60

Practice Exercises

Frame 8

1. \$135.21
2. \$168.30

Frame 13

1. \$805.00
2. \$117.20

Frame 18

1. 30%
2. 5%

Frame 22

1. 40% 2. $119.00

Frame 26

1. $2,920.00 2. 32.6%

Frame 33

1. $60.00 2. $7.75 3. $3,168.00 4. $62.50

Post-test

1. $10,062.50 2. $185.50 3. 20%
4. 18.75% 5. 41.2% 6. $243.75

Answers To Drill Exercises: Unit 6

1. 0.100
2. 0.250
3. 1.350
4. 0.750
5. 0.003
6. 2.750
7. 0.050
8. 3.250
9. 0.075
10. 0.667
11. 7.500
12. 0.060
13. 0.008
14. 0.025
15. 0.325
16. 0.000
17. $\frac{1}{2}$
18. $\frac{3}{10}$
19. $\frac{1}{20}$
20. $\frac{9}{25}$
21. $\frac{37}{50}$
22. $\frac{24}{25}$
23. $\frac{3}{8}$
24. $\frac{1}{16}$
25. $\frac{1}{6}$
26. $\frac{1}{12}$
27. $1\frac{3}{5}$
28. $1\frac{5}{8}$
29. $2\frac{2}{3}$
30. 9%
31. 25%
32. 20%
33. 10%
34. 12%
35. 15%
36. $12\frac{1}{2}\%$
37. $22\frac{2}{9}\%$
38. $83\frac{1}{3}\%$
39. 60%
40. $85\frac{5}{7}\%$
41. 28%
42. 65%
43. 4.3%
44. 0.5%
45. 40%
46. 30%
47. 500%
48. 700%
49. 137%
50. 190%
51. 62.5%
52. 7.8%
53. 116.7%
54. 87.5%
55. 0.4%
56. 0.5%
57. 0.7%
58. $P = 46\% \times 13; P = 5.98$
59. $18\% \times B = 33; B = 183\frac{1}{3}$
60. $P = 21\% \times 76; P = 15.96$
61. $72 \times R = 45; R = 62.5\%$
62. $76\% \times B = 76; B = 100$
63. $234\% \times 59 = P; P = 138.06$
64. $56 = 153\% \times B; B = 36\frac{92}{153}$
65. $P = 43.1\% \times 47; P = 20.257$
66. $R \times 48 = 754; R = 1{,}570\frac{5}{6}\%$
67. $71 = 0.4\% \times B; B = 17{,}750$
68. $100 = 4\frac{1}{2}\% \times B; B = 2{,}222\frac{2}{9}$
69. $30 = 5\frac{1}{4}\% \times B; B = 571\frac{3}{7}$
70. $P = 51\% \times \$1{,}600{,}000; P = \$816{,}000$
71. $R \times 15 = 12; R = 80\%$
72. $23.4\% \times B = \$4{,}387.50; B = \$18{,}750$
73. $160\% \times B = 237{,}000; B = 148{,}000$
74. $22\% \times \$960 = P; P = \$211.20$
75. $109.4\% \times 36$ inches $= P; P = 39.384$ inches
76. $R \times 162 = 101; R = 62.3\%$
77. $R \times 24 = 14; R = 58\frac{1}{3}\%$
78. $4.9\% \times B = 10; B = 204\frac{4}{49}$
79. $8\% \times B = \$10.00; B = \$125.00$

80. $4\frac{1}{2}\% \times \$14.85 = P; P = \$.67$

81. $3\% \times B = 108; B = 3{,}600$

82. 50%

83. 10%

84. (a) 12.0% (b) 5.9%

85. 18%

86. 21.2%

87. $6\frac{2}{3}\%$

88. 650%

89. (a) $\$700 \times 5\% \times 1 = \$35.00$

(b) $\$700 \times 5\% \times \frac{10}{12} = \$29.17$

(c) $\$700 \times 5\% \times \frac{1}{12} = \$2.92$

(d) $\$700 \times 5\% \times 2.5 = \$87.50$

(e) $\$700 \times 5\% \times \frac{45}{360} = \$4.38$

90. (a) $\$240 \times 7\% \times \frac{8}{12} = \$11.20$

(b) $\$240 \times 3.5\% \times \frac{8}{12} = \$5.60$

(c) $\$240 \times 6\% \times \frac{8}{12} = \$9.60$

(d) $\$240 \times 4\frac{3}{4}\% \times \frac{8}{12} = \$7.60$

Answers To Self Test: Unit 6

1. (c)	2. (b)	3. (d)	4. (d)	5. (e)	6. (a)	7. (c)
8. (b)	9. (d)	10. (e)	11. (a)	12. (d)	13. (c)	14. (b)
15. (e)	16. (a)	17. (c)	18. (b)	19. (a)	20. (d)	

UNIT SEVEN

Measurement

Lesson 25	AN INTRODUCTION TO MEASUREMENT
Lesson 26	METRIC MEASURES
Lesson 27	OTHER METRIC MEASURES
Lesson 28	APPLICATIONS OF MEASUREMENTS

LESSON 25
AN INTRODUCTION TO MEASUREMENT

Objectives:

1. Use dimension analysis to convert units of measure in the English system. (1–17)
2. Add or subtract denominate numbers. (18–25)
3. Multiply or divide denominate numbers. (26–32)

Vocabulary:

Denominate Number
Dimension Analysis
Conversion Factor

PRE-TEST

1. Convert 3 hours to seconds by using dimension analysis.
2. Perform the following operations:
   (a) Add 2 pints 1 cup 7 ounces and 5 pints 1 cup 4 ounces.
   (b) Subtract 2 pints 1 cup 7 ounces from 5 pints 1 cup 4 ounces.
   (c) $3\frac{1}{2}$ feet $\times \frac{5}{8}$.
   (d) (3 pounds 10 ounces) $\div$ 4.

Once a society progresses beyond the more primitive levels of trade, there is a need for a system of measurements. 1

BARTERING

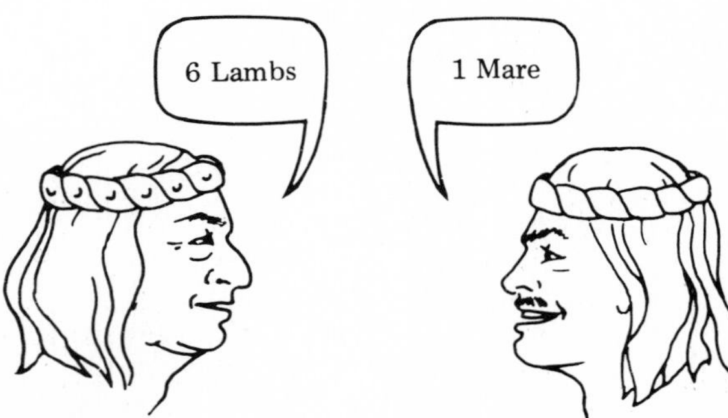

Each man is satisfied since each agrees on the value of commodities he wishes to exchange.

Each man has doubts about the value of the commodities he wishes to exchange.

A *system of standard measures* is necessary in any advanced society to facilitate commerce. It must meet the following requirements: (2)

The basic units of measure must be clearly defined.

The basic units must be standardized in such a way so that they can be easily reproduced.

The units of measure must meet the needs of the persons within the society that accepts them as standard.

All systems of measurements are invented or evolve out of the inventions of man as determined by the specific needs of society.

The English system of measures, long used in this and other English-speaking countries of the world, is a classic example of evolved units of measure meeting diversified needs. The basic units reflect the cultural influences of many early societies. (3)

The origin and evolution of some early measures: the cubit

The cubit is familiar to many people because of Biblical references to this unit of measure: "The length of the ark shall be three hundred cubits, the breadth of it fifty cubits, and the height of it thirty cubits."

A cubit was originally the length of a man's forearm from the tip of his middle finger to the tip of the elbow.

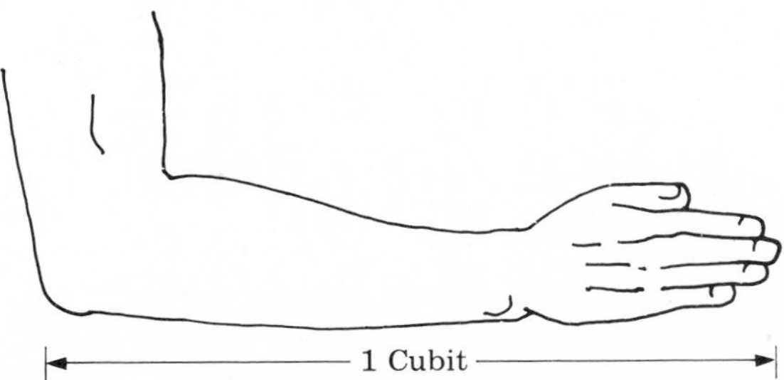

Since this measure varied according to the stature of the man, the Egyptians attempted to standardize this unit some time around 3,000 B.C. by using stone or wooden standard cubits.

The Egyptian cubit was approximately 21 inches long and was subdivided into *palms* and *digits*.

EGYPTIAN

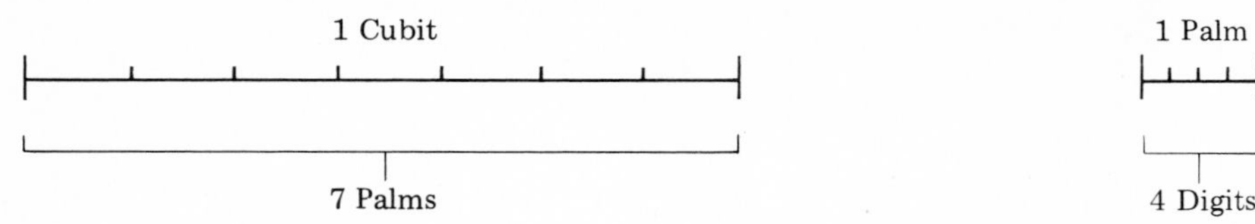

The Romans adopted the cubit and introduced this measure of length into the areas of Europe that fell under their influence. The Roman standard cubit differed from the Egyptian standard cubit. 5

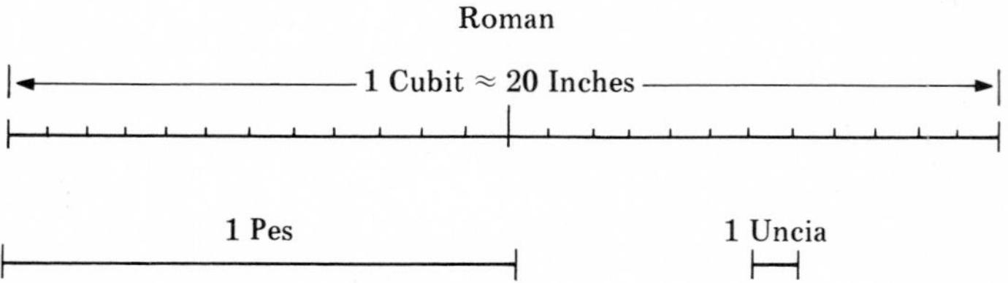

Notice that there are 12 *unciae* in 1 *pes.*

From the Roman measures, the following English measures evolved:

uncia ⟶ **inch**

pes ⟶ **foot**

The yard, 3 feet, replaced the cubit in England. The yard was "standardized" during the reign of Henry the First as the distance between the tip of the king's nose and the end of his middle finger.

The origins of some other English system measures reflect their haphazard development: 

LENGTHS

Statute mile	5,280 feet	From the Latin *mille passus*, 1,000 paces or, approximately, 5,000 feet.
Nautical mile (*knot*)	6,076.10333 feet	Early sailors measured the speed of their ships by allowing a rope knotted into 47 feet 3 inch lengths to fall into the water for 28 seconds. Each knot that was pulled into the wake of the ship in the 28 seconds represented 1 knot or 1 nautical mile per hour.

Fathom	6 feet	Depth of water measured by ropes knotted into 6-foot lengths; the distance from finger tip to finger tip of the extended arms of a sailor.
		AREAS
Acre	160 square rods	The area a man could plow in one day with a yoke (a pair) of oxen.
		WEIGHT
Grain	$\frac{1}{7{,}000}$ pound (avoirdupois)	Early Babylonians used grain seeds as a standard measure of weight.
Ounce	$\frac{1}{16}$ pound (avoirdupois)	Also from the Roman *uncia*; the *uncia* was also a basic monetary unit with a value determined by weight.
		TIME
Minutes	$\frac{1}{60}$ of an hour	From the Latin *pars minuta prima*, the first small part, or the first subdivision.
Seconds	$\frac{1}{360}$ of an hour	*Pars minuta secunda*, the second small part.

It is difficult to believe that any one person knows all the standard units of measure, and their relationships, in the English system. There is further confusion generated by the fact that more than one standard unit bears the same name. For example, compare the following: 

Dry Capacity	to	**Liquid Capacity**
1 pint = 33.6 cubic inches		1 pint = 28.875 cubic inches
1 quart = 2 pints		1 quart = 2 pints
1 peck = 8 quarts		1 gallon = 4 quarts

Weight—Avoirdupois	to	**Weight—Apothecary or Troy**
1 ounce = 437.5 grains		1 ounce = 480 grains
1 pound = 16 ounces		1 pound = 12 ounces

Only because of their familiarity will English system units be used for the remainder of this lesson to show the unit conversion method and operations with denominate numbers. Learning the method of conversion is important; learning the basic units is not. The necessary tables of measures in the English system are given in the Appendix.

Denominate numbers: (8)

Measurements of all kinds are usually given as *denominate numbers.* A *denominate number* consists of two parts: a number and a unit of measure.

Number	*Unit of Measure*	*Denominate Number*
3	days	3 days
55	miles per hour	55 miles per hour
12	ounces	12 ounces
100	feet	100 feet

Converting denominate numbers: (9)

The primary method for converting denominate numbers to other, equivalent denominate numbers with different units of measure is the *dimension analysis* method.

For example, how many minutes are there in 2 days?

2 days = ? minutes

Converting denominate numbers: 

The following are taken from the Tables of Measurements in the Appendix and will be used to make the conversion from days to minutes:

1 day = 24 hours

1 hour = 60 minutes

Converting denominate numbers: 

Conversion factors are formed from the table values:

1 day = 24 hours

$$\frac{1\text{ day}}{1\text{ day}} = \frac{24\text{ hours}}{1\text{ day}} \leftarrow \boxed{\text{Dividing by 1 day}}$$

$$1 = \frac{24\text{ hours}}{1\text{ day}}$$

Think: $\frac{1}{1} = 1$ as if the units of measure could be "cancelled," even though cancellation actually applies only to numbers.

$\frac{24\text{ hours}}{1\text{ day}}$ is called a **conversion factor**

A number of days can be multiplied by this conversion factor to convert to an equivalent number of hours. Notice that this is comparable to using "Multiplication by One."

Converting denominate numbers: (12)

There are two conversion factors that can be derived from each individual table entry.

From the Table	*Conversion Factor*	*Use*
1 day = 24 hours	$\frac{24 \text{ hours}}{1 \text{ day}}$	Converting days to hours
	$\frac{1 \text{ day}}{24 \text{ hours}}$	Converting hours to days
1 hour = 60 minutes	$\frac{60 \text{ minutes}}{1 \text{ hour}}$	Converting hours to minutes
	$\frac{1 \text{ hour}}{60 \text{ minutes}}$	Converting minutes to hours

Converting denominate numbers: 

To decide which conversion factors to use to convert 2 days to minutes, think of "cancelling" units:

$$\frac{2\ \cancel{\text{days}}}{1} \times \frac{?\ \cancel{\text{hours}}}{?\ \cancel{\text{days}}} \times \frac{?\ \text{minutes}}{?\ \cancel{\text{hours}}}$$

Days "cancel" days and hours "cancel" hours, leaving minutes as the only unit of measure, the desired result. The appropriate numbers from the tables are then placed in the conversion factor:

$$\frac{2\ \cancel{\text{days}}}{1} \times \frac{24\ \cancel{\text{hours}}}{1\ \cancel{\text{day}}} \times \frac{60\ \text{minutes}}{1\ \cancel{\text{hour}}}$$

or

$$2 \times 24 \times 60 \text{ minutes} = 2{,}880 \text{ minutes}$$

Convert 17 yards to inches: 

17 yards = ? inches

Think: $\frac{17\ \cancel{\text{yards}}}{1} \times \frac{?\ \cancel{\text{feet}}}{?\ \cancel{\text{yard}}} \times \frac{?\ \text{inches}}{?\ \cancel{\text{foot}}}$

From the table: 1 yard = 3 feet and 1 foot = 12 inches

$$17\ \cancel{\text{yards}} \times \frac{3\ \cancel{\text{feet}}}{1\ \cancel{\text{yard}}} \times \frac{12\ \text{inches}}{1\ \cancel{\text{foot}}}$$

or

$$17 \times 3 \times 12 \text{ inches} = 612 \text{ inches.}$$

Convert 60 miles per hour to feet per second: 15

$$60 \text{ miles per hour can be written as } 60\frac{\text{miles}}{\text{hour}}.$$

This problem differs from the other examples in this lesson in that it combines two different types of conversions in one problem:

The conversion of a unit of length (miles to another unit of length (feet); and the conversion of a unit of time (hours) to another unit of time (seconds).

The following table entries will be used:

To convert from miles to feet	5,280 feet = 1 mile
To convert from hours to seconds	1 hour = 60 minutes
	1 minute = 60 seconds

Then,

$\frac{5{,}280 \text{ feet}}{1 \text{ mile}}$ is the conversion factor used to convert miles to feet,

$\frac{1 \text{ hour}}{60 \text{ minutes}}$ is used to convert minutes to hours, and

$\frac{1 \text{ minute}}{60 \text{ seconds}}$ is used to convert seconds to minutes.

So,

$$60\frac{\text{miles}}{\text{hour}} = 60\frac{\cancel{\text{miles}}}{\cancel{\text{hour}}} \times \frac{5{,}280 \text{ feet}}{1 \cancel{\text{mile}}} \times \frac{1 \cancel{\text{hour}}}{60 \cancel{\text{minutes}}} \times \frac{1 \cancel{\text{minute}}}{60 \text{ seconds}}$$

$$= \frac{\overset{1}{\cancel{60}} \times 5{,}280}{\underset{1}{\cancel{60}} \times 60} \frac{\text{feet}}{\text{second}} \text{ or } 88 \frac{\text{feet}}{\text{second}}$$

Convert 10 pounds (avoirdupois) to ounces: 

10 pounds = ? ounces

$$\textbf{Think:} \quad \frac{10 \cancel{\text{pounds}}}{1} \times \frac{? \text{ ounces}}{? \cancel{\text{pounds}}}$$

Complete the following by using the tables, if necessary, to determine the number of ounces per pound:

$$\frac{10\ \cancel{\text{pounds}}}{1} \times \frac{\square\ \text{ounces}}{\triangle\ \cancel{\text{pounds}}}$$

or

$$10 \times \square\ \text{ounces} = \bigcirc\ \text{ounces}$$

Practice Exercises (17)

Convert each of the following to equivalent denominate numbers with the specified unit of measure.

1. 6 gallons to pints
2. 2.9 tons to pounds
3. $\frac{7}{8}$ yards to feet
4. 46 ounces to pounds

Arithmetic operations may be performed on denominate numbers or on combinations of denominate and non-denominate numbers. Addition and subtraction must only be performed on two or more denominate numbers *with identical units of measure.* 

Addition of denominate numbers: (19)

7 days	12 hours	45 minutes
+3 days	16 hours	21 minutes

Notice that the units of measure are aligned, just as place-value positions are aligned: days are added to days; hours are added to hours; and minutes are added to minutes.

Adding:

7 days	12 hours	45 minutes
+3 days	16 hours	21 minutes
10 days	28 hours	66 minutes

Rewriting:

10 days 28 hours 66 minutes

is equivalent to

10 days + (1 day 4 hours) + (1 hour 6 minutes)

or

11 days 5 hours 6 minutes

Since 66 minutes is greater than the next unit of measure of time (1 hour = 60 minutes), "carry" 60 minutes as 1 hour; and since 28 hours is greater than the next unit of measure (24 hours = 1 day), "carry" 24 hours as 1 day.

Add 2 yards 2 feet 7 inches and 2 feet 9 inches: 20

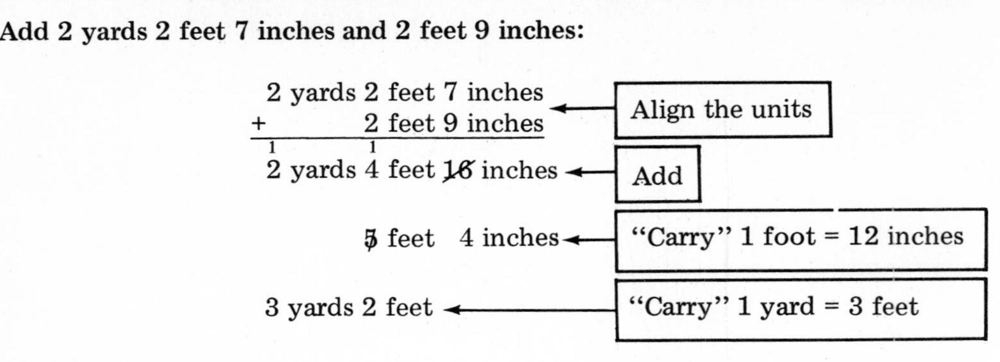

Thus, 2 yards 4 feet 16 inches = 3 yards 2 feet 4 inches.

Add 3 gallons 2 quarts and 3 quarts 1 pint: 21

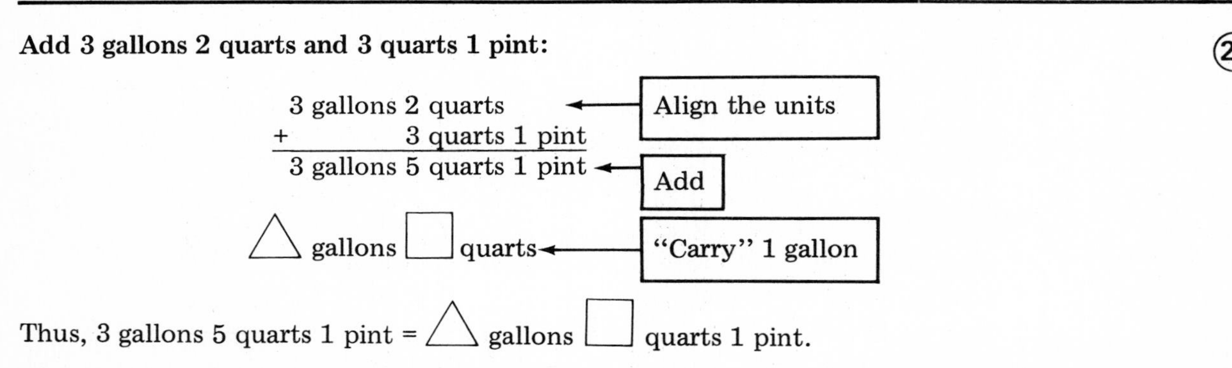

Thus, 3 gallons 5 quarts 1 pint = △ gallons □ quarts 1 pint.

Subtraction of denominate numbers: 22

Just as a form of "carrying" is used in addition of denominate numbers, a form of "borrowing" must be used in some subtractions.

For example, subtract 3 days 16 hours 21 minutes from 7 days 12 hours 45 minutes.

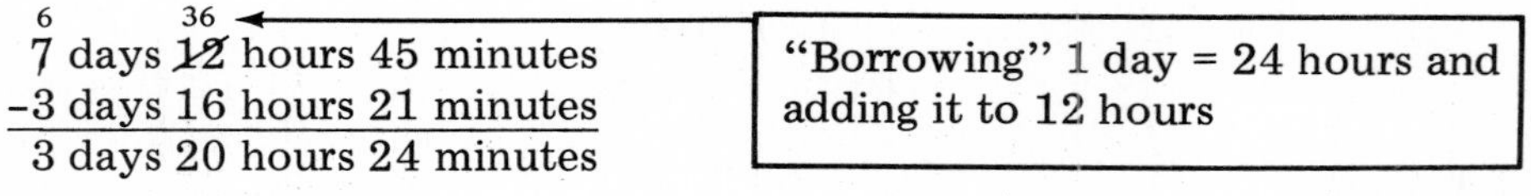

6 36
7 days 12 hours 45 minutes
−3 days 16 hours 21 minutes
3 days 20 hours 24 minutes
"Borrowing" 1 day = 24 hours and adding it to 12 hours

Subtract 4 pounds 11 ounces from 7 pounds: 23

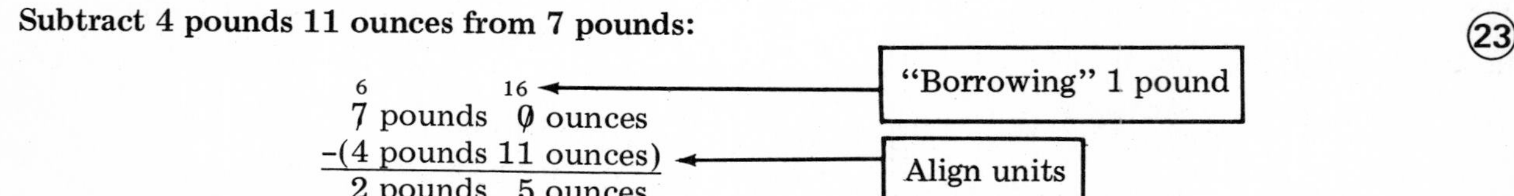

Subtract 13 hours 23 minutes from 1 day 6 hours 15 minutes: 24

Complete the following (in two steps):

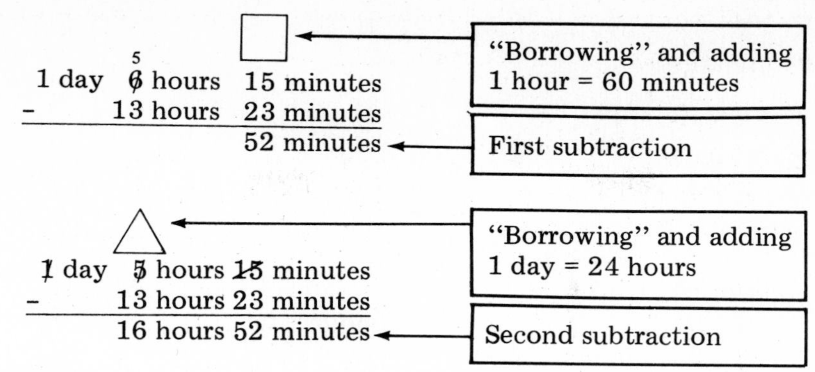

Practice Exercises 

Find each sum or difference.

1. 5 feet 6 inches
 +3 feet 9 inches

2. 8 weeks 6 days 17 hours
 +3 weeks 5 days 23 hours

3. 3 gallons 1 quart
 −1 gallon 3 quarts

4. 53 minutes 41 seconds
 −47 minutes 53 seconds

A denominate number can be multiplied or divided by a non-denominate number. 

For example, multiply 3 feet 9 inches by 7:

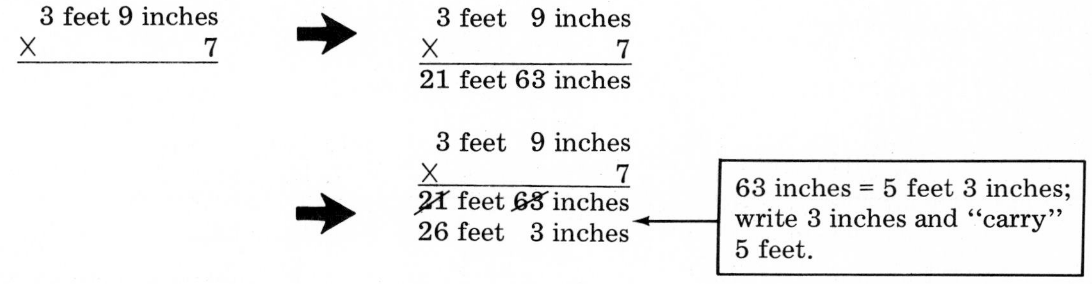

Multiply 1 gallon 3 quarts 1 pint by 3: (27)

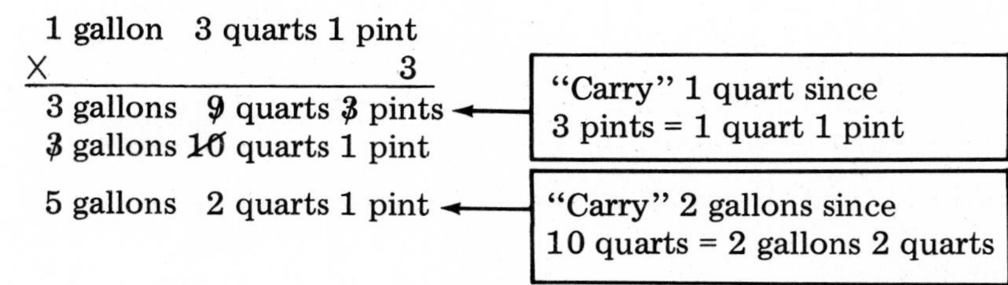

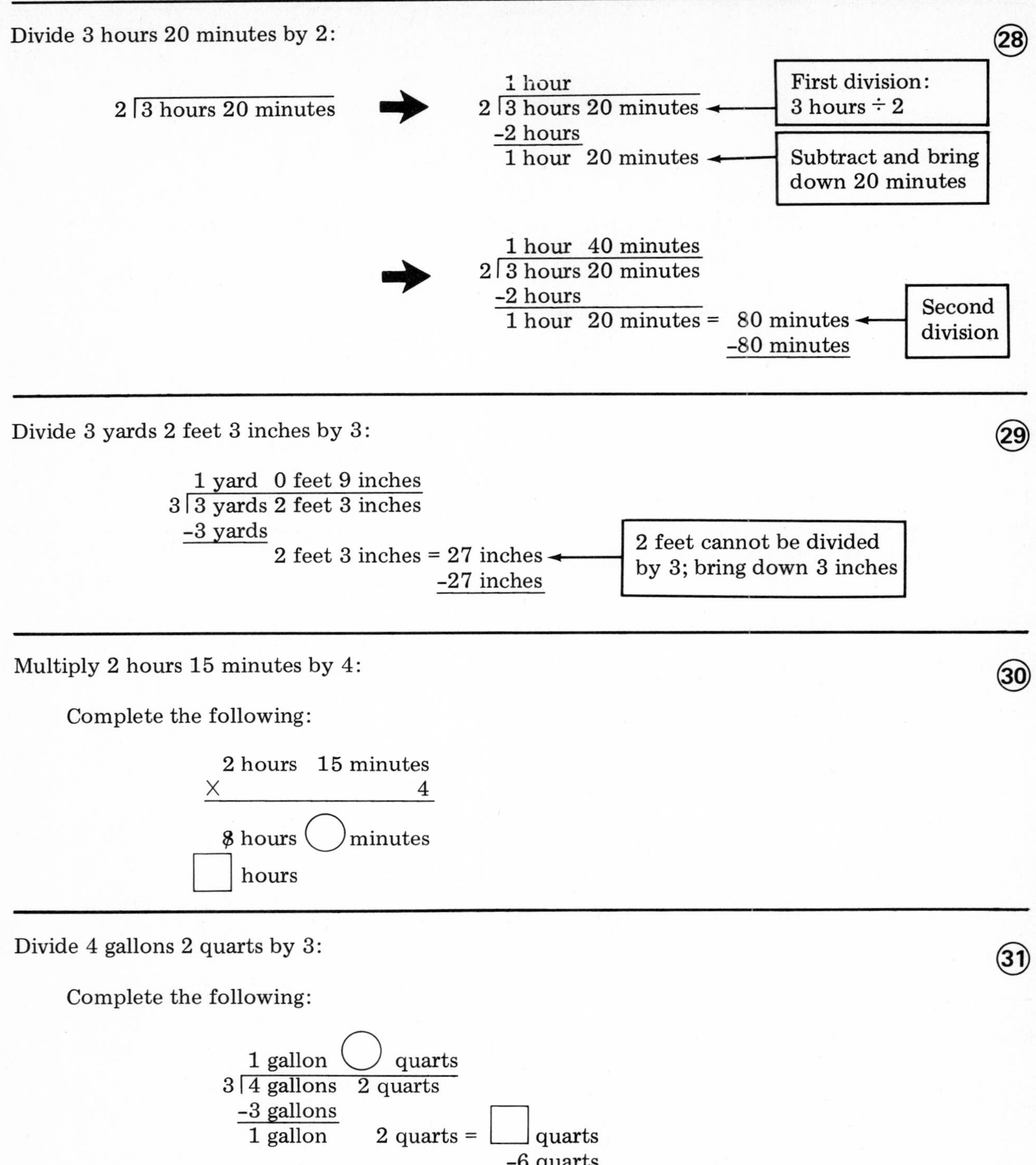

Divide 3 hours 20 minutes by 2: (28)

```
2 ⟌3 hours 20 minutes

     1 hour
2 ⟌3 hours 20 minutes   ← First division: 3 hours ÷ 2
  -2 hours
   1 hour  20 minutes   ← Subtract and bring down 20 minutes

     1 hour  40 minutes
2 ⟌3 hours 20 minutes
  -2 hours
   1 hour  20 minutes =  80 minutes  ← Second division
                        -80 minutes
```

Divide 3 yards 2 feet 3 inches by 3: (29)

```
     1 yard  0 feet 9 inches
3 ⟌3 yards 2 feet 3 inches
  -3 yards
           2 feet 3 inches = 27 inches  ← 2 feet cannot be divided by 3; bring down 3 inches
                            -27 inches
```

Multiply 2 hours 15 minutes by 4: (30)

Complete the following:

```
  2 hours   15 minutes
×                    4
  8 hours  ◯ minutes
  □ hours
```

Divide 4 gallons 2 quarts by 3: (31)

Complete the following:

```
     1 gallon   ◯  quarts
3 ⟌4 gallons  2 quarts
  -3 gallons
   1 gallon    2 quarts = □ quarts
                         -6 quarts
```

Practice Exercises

1. Multiply 11 gallons 1 quart by 15.
2. Multiply 7 yards 2 feet 7 inches by 4.
3. Divide 3 gallons 3 quarts by 5.
4. Divide 9 hours 14 minutes 44 seconds by 2.

POST-TEST

1. Convert 7 days to minutes by using dimension analysis.
2. Perform the following operations:
   (a) Add 5 yards 2 feet 9 inches and 7 yards 2 feet 8 inches.
   (b) Subtract 5 yards 2 feet 9 inches from 7 yards 2 feet 8 inches.
   (c) Multiply: (3.1 inches) (0.6).
   (d) Divide: (5 days 3 hours) ÷ 3.

LESSON 26
METRIC MEASURES

Objectives:

1. Name and abbreviate seven common units of measure in the SI system for each of the following: length, weight, and volume or capacity. (1–18)
2. Convert metric units of length, weight, and volume or capacity to other units of the same measure. (19–31)
3. Convert metric measures of length, weight, and volume or capacity to equivalent English units of measure. (32–34)
4. Convert English measures of length, weight, and volume or capacity to equivalent metric units of measure. (32–34)

Vocabulary:

SI System (Metric System)

Meter

Gram

Liter

Metric Prefixes and Abbreviations:

- Milli- (m)
- Centi- (c)
- Deci- (d)
- Deka- (da)
- Hecto- (h)
- Kilo- (k)

Metric Ton

PRE-TEST

1. Correctly complete the following:
   (a) 1 kilometer = ________ meters.
   (b) Centigrams is abbreviated ______ .
2. Convert 278 milliliters to liters.
3. Convert 120 kilometers to miles, correct to the nearest mile.
4. Convert 7 quarts to liters.

A moment's time spent examining any tables of measures for the English system should convince anyone that this system has evolved with little rhyme or reason. There are so few similarities between the numbers of subdivisions for the various units that it almost seems inappropriate to call the English measures a "system." When England changed over to the SI system in the 1960's, the United States became the last major world power using the English system of measures. 1

The *SI system*, the International System of Units, commonly called the *Metric System*, was invented by a commission of French scientists in the late eighteenth century. 

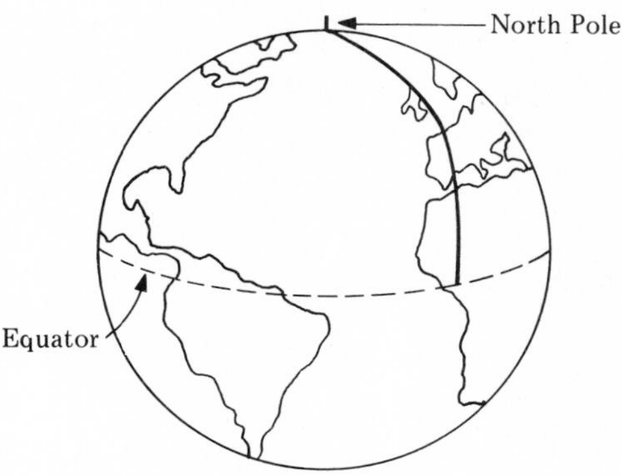

The *meter*, the basic unit used to define all other units of measure in the SI system, was originally established as one ten-millionth of the distance from the equator to the North Pole.

The meter is now defined as a specific multiple of the wavelength of the orange-red line in the spectrum of krypton-86 gas under specific conditions. While this fact may mean very little to most people using the meter, it does mean that the exact measure of one meter can be readily reproduced anywhere in the world, eliminating the necessity for a standard meter bar, as was used in the past. 3

Metric lengths: the meter 

The *meter*, abbreviated m (no period), is the basic unit for all measures of length. The yard, like all measures in use in the United States, is legally defined in terms of the meter.

$$\textbf{1 yard} = \textbf{0.9144 meter}$$
$$\textbf{1 meter} \approx \textbf{1.094 yards}$$

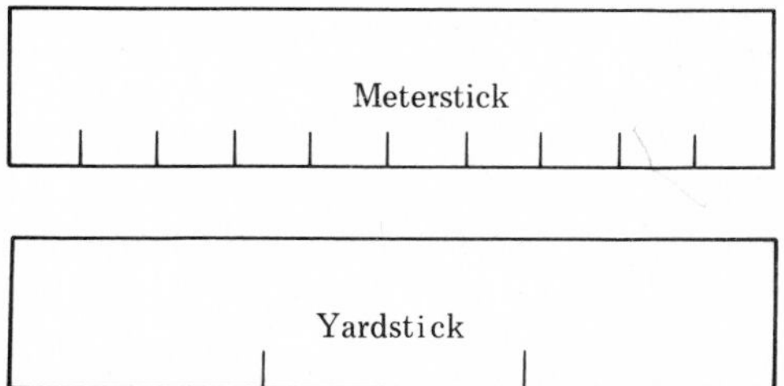

One meter is slightly longer than one yard.

The monetary system in the United States is similar to the SI system of measures. For purposes of comparison between the subdivisions and multiples of the meter and common currency, 1 dollar will be considered as the "basic unit of U.S. money." 5

Measures of length smaller than the meter: the decimeter 6

If 1 meter is divided into 10 equal parts, each of these 10 subdivisions is 1 *decimeter*, abbreviated *dm* (no period is used).

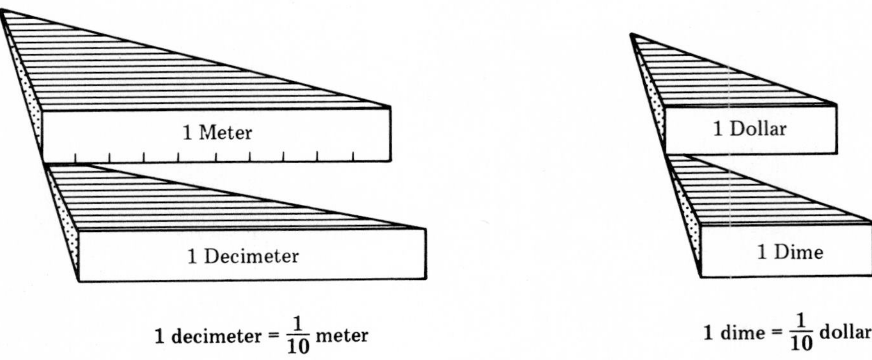

1 decimeter = $\frac{1}{10}$ meter   1 dime = $\frac{1}{10}$ dollar

Think of 1 dime as 1 "decidollar"; the prefix *deci* (d) means "one-tenth of."

Measures of length smaller than the meter: the centimeter 7

If 1 decimeter is divided into 10 equal parts (or 1 meter is divided into 100 equal parts), each of these subdivisions is called 1 *centimeter*, abbreviated *cm* (no period).

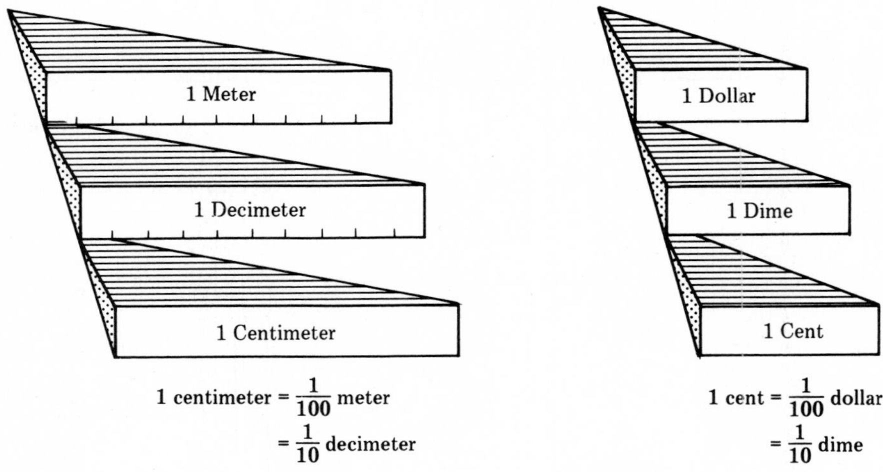

1 centimeter = $\frac{1}{100}$ meter
= $\frac{1}{10}$ decimeter

1 cent = $\frac{1}{100}$ dollar
= $\frac{1}{10}$ dime

Think of 1 cent as 1 "centidollar"; the prefix *centi* (c) means "one-hundredth of."

Measure of length smaller than the meter: the millimeter 8

If 1 centimeter is divided into 10 equal parts (or 1 meter is divided into 1,000 equal parts), each of these subdivisions is called 1 *millimeter*, abbreviated mm (no period).

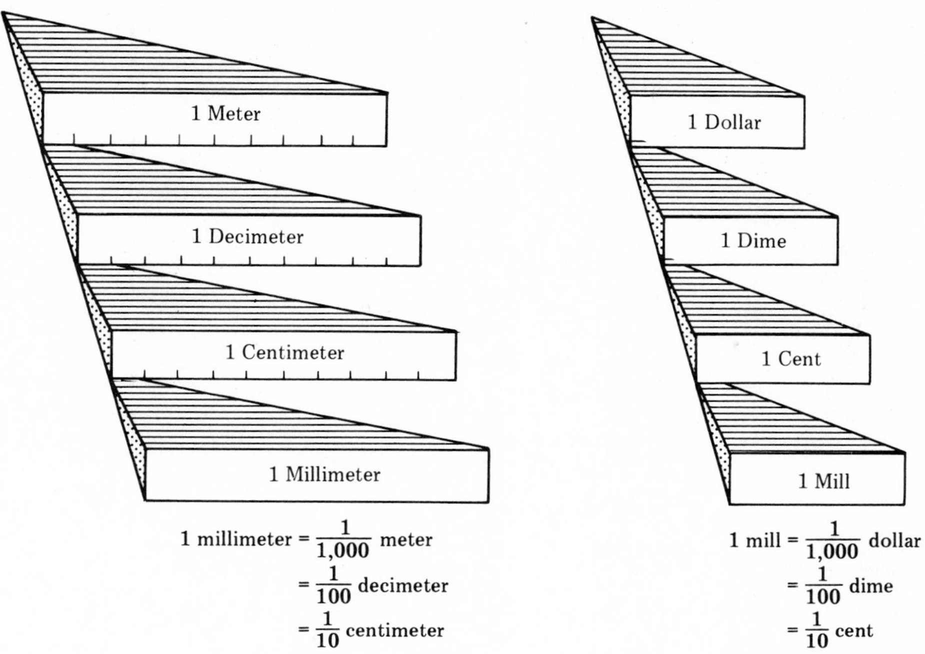

$$1 \text{ millimeter} = \frac{1}{1{,}000} \text{ meter} = \frac{1}{100} \text{ decimeter} = \frac{1}{10} \text{ centimeter}$$

$$1 \text{ mill} = \frac{1}{1{,}000} \text{ dollar} = \frac{1}{100} \text{ dime} = \frac{1}{10} \text{ cent}$$

Think of 1 mill as 1 "millidollar"; the prefix *milli* (m) means "one-thousandth of."

The prefixes for metric measures of length smaller than the millimeter are given in the Appendix.

Measures of length larger than the meter: the dekameter 9

If 10 meters are laid end to end, the measure of this length is 1 *dekameter*, abbreviated dam (no period).

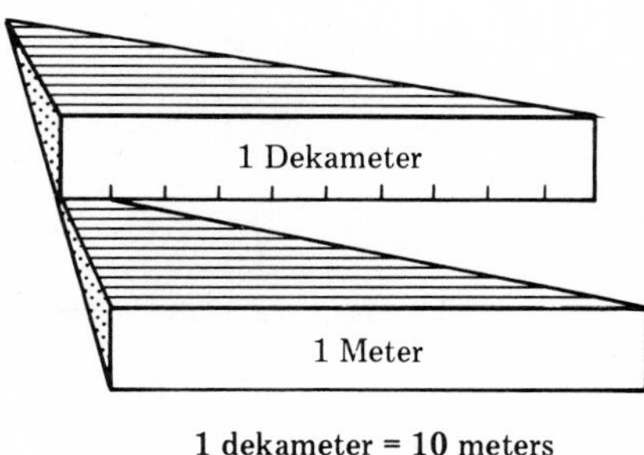

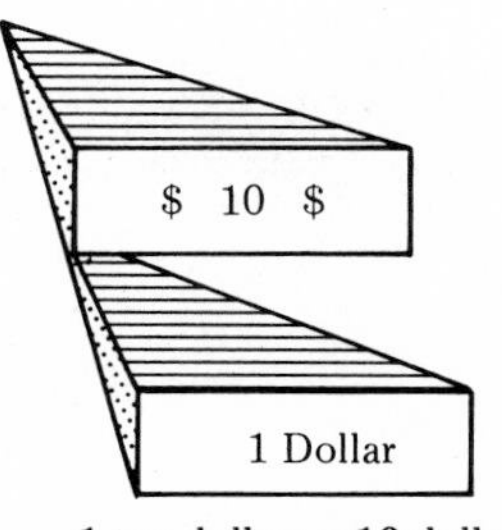

Think of 1 ten-dollar bill as 1 "dekadollar"; the prefix *deka* (da) means "ten times." (This prefix may also be spelled "deca.")

Measures of length larger than the meter: the hectometer

If 10 dekameters (or 100 meters) are laid end to end, the measure of this length is 1 *hectometer*, abbreviated hm (no period).

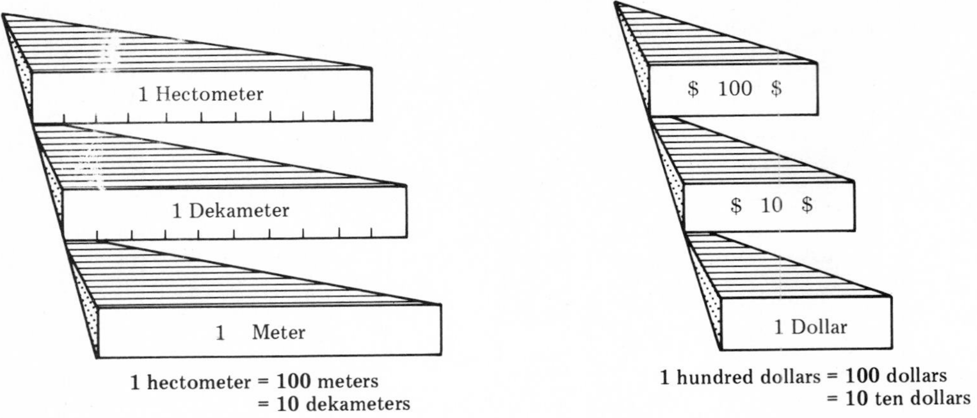

Think of 1 one-hundred dollar bill as 1 "hectodollar"; the prefix *hecto* (h) means "100 times."

Measures of length larger than the meter: the kilometer

If 10 hectometers (or 1,000 meters) are laid end to end, the measure of this length is 1 *kilometer*, abbreviated km (no period).

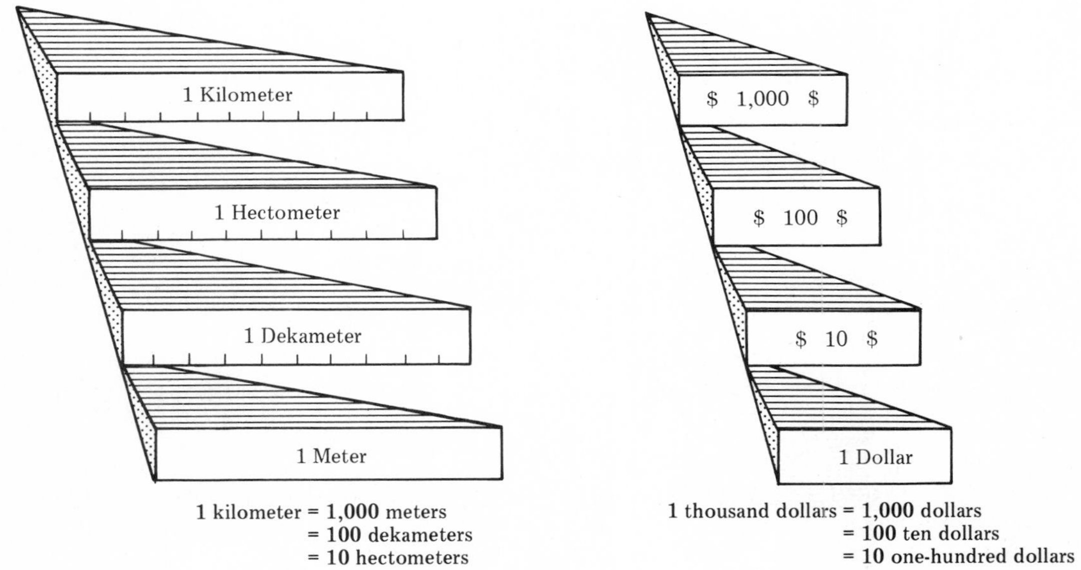

Think of 1 one-thousand dollar bill as 1 "kilodollar"; the prefix *kilo* (k) means "one thousand times."

The prefixes for metric measures of length larger than the kilometer are given in the Appendix.

Measures of length: 

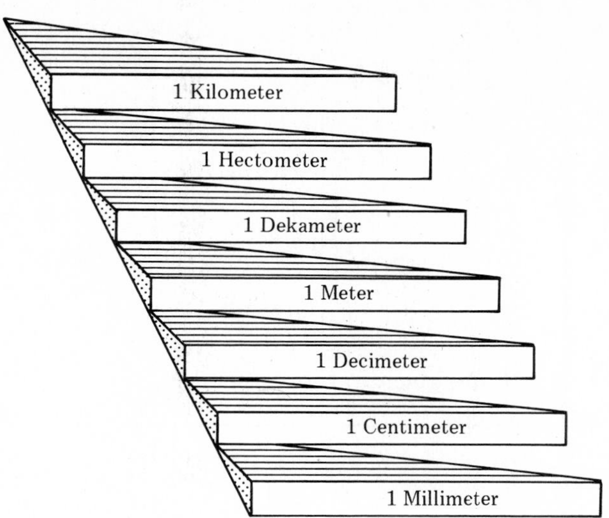

1 kilometer (km) = 1,000 meters

1 hectometer (hm) = 100 meters

1 dekameter (dam) = 10 meters

1 Meter

1 decimeter (dm) = $\frac{1}{10}$ meter

1 centimeter (cm) = $\frac{1}{100}$ meter

1 millimeter (mm) = $\frac{1}{1,000}$ meter

Common measures of length: 

The most commonly used measures of length in the SI system are the meter, the millimeter, the centimeter, and the kilometer. These metric units of length are usually used for the corresponding English units shown below.

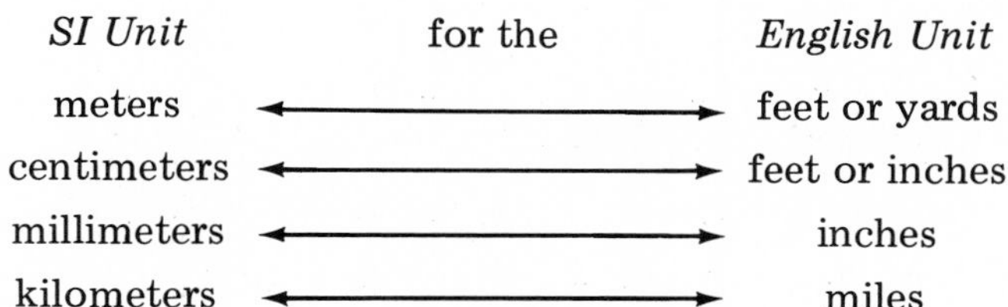

SI Unit	for the	*English Unit*
meters	⟷	feet or yards
centimeters	⟷	feet or inches
millimeters	⟷	inches
kilometers	⟷	miles

If a person's height is 5 feet, he is about 1.5 meters tall. The speed limit may be written as 88 kilometers per hour, but you won't be exceeding 55 miles per hour. A bust size of 91 centimeters for Miss 36-24-35 sounds all right, but what about buying a pair of men's pants in size 100 centimeters by 80 centimeters?

Metric weights: 

The *gram*, abbreviated g (no period), is the base unit for weights in the SI system. Other metric weights are related to the gram exactly as metric lengths are related to the meter. The weight measure names are formed by attaching the same prefixes used for length to the base unit, gram.

1 kilogram (kg)	=	**1,000 grams**
1 hectogram (hg)	=	**100 grams**
1 dekagram (dag)	=	**10 grams**
	1 Gram	
1 decigram (dg)	=	$\frac{1}{10}$ **gram**
1 centigram (cg)	=	$\frac{1}{100}$ **gram**
1 milligram (mg)	=	$\frac{1}{1,000}$ **gram**

The prefixes for other metric measures of weight are given in the Appendix.

Common measures of weight: 15

The most commonly used weight measures in the SI system are the gram, the kilogram, and the metric ton (t). (1 metric ton = 1,000 kilograms.)

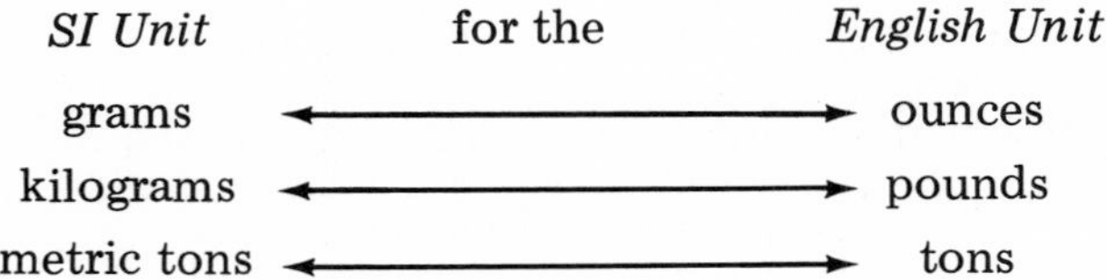

The weight of a paper clip is approximately 1 gram. A large loaf of bread weighs approximately 1 kilogram. A compact car weighs approximately 1 metric ton.

Metric measures of volume or capacity:  16

The *liter*, abbreviated l (no period), is the base unit for volumes or capacities. Metric measures of volume are named in exactly the same way as are measures of length and weight.

1 kiloliter (kl)	=	**1,000 liters**
1 hectoliter (hl)	=	**100 liters**
1 dekaliter (dal)	=	**10 liters**
	1 Liter	
1 deciliter (dl)	=	$\frac{1}{10}$ **liter**
1 centiliter (cl)	=	$\frac{1}{100}$ **liter**
1 milliliter (ml)	=	$\frac{1}{1,000}$ **liter**

The prefixes for other metric measures of volume are given in the Appendix.

Common measures of volume or capacity: (17)

The most commonly used volume and capacity measures in the SI system are the liter and the milliliter.

SI Unit	for the	*English Unit*
liter	⟷	pints, quarts, and gallons
milliliter	⟷	ounces

One teaspoon is approximately 5 milliliters; one quart of milk is slightly less than 1 liter of milk.

Practice Exercises (18)

Fill in the blanks to correctly complete the following:

1. 10 kilograms = ________ hectograms.
2. Milliliters is abbreviated ______.
3. 72 km per hour means 72 ______________ per hour.
4. 1,000 grams = 1 ______________.
5. 10 centimeters = __________ decimeters.
6. 2.5 l means 2.5 ____________ .
7. Hectograms is abbreviated ______.
8. 24.2 dam means 24.2 ______________ .

Writing metric measures: (19)

To show how metric measures are written, it is useful to again think of the comparison with money.

For example,

1 dollar   6 dimes   3 cents

is written

$1.63

1 meter   6 decimeters   3 centimeters

is written

1.63m

Decimal notation is used to write all metric measures.

Converting metric units: 

Since all measures in the metric system can be derived from a base unit by multiplying or dividing by powers of ten, conversions are much easier than in the English system.

For example,

$$\mathbf{2.54\ hectometers = ?\ meters}$$

Since 1 hectometer = 100 meters, the necessary conversion factor is

$$\frac{100\ \text{m}}{1\ \text{hm}}$$

Then,

$$2.54\ \text{hm} = 2.54\ \cancel{\text{hm}} \times \frac{100\ \text{m}}{1\ \cancel{\text{hm}}}$$

$$= 2.54 \times 100\ \text{m}$$

$$= 254.\ \text{m}$$

Since unit conversions in the SI system involve multiplication and division by powers of ten, it may be helpful to review the decimal-shift method in Unit 5.

Convert 2.54 centimeters to meters: 

Since 100 centimeters = 1 meter, the conversion factor is $\frac{1\ \text{m}}{100\ \text{cm}}$.

Then,

$$2.54\ \text{cm} = 2.54\ \cancel{\text{cm}} \times \frac{1\ \text{m}}{100\ \cancel{\text{cm}}}$$

$$= 2.54 \times \frac{1\ \text{m}}{100}$$

$$= \frac{2.54}{100}\ \text{m}$$

$$= .0254\ \text{m}$$

Converting metric units: 

Consider the examples in the last two frames:

$$\mathbf{2.54\ hm = 2.54 \times 100\ m}$$

$$\mathbf{2.54\ cm = \frac{2.54}{100}\ m}$$

Notice in the first example that converting from the larger measure, hectometers, to the smaller measure, meters, is done by multiplying by 100 or 10^2. In the second example, converting from a smaller measure, centimeters, to a larger measure, meters, is done by dividing by 100 or 10^2.

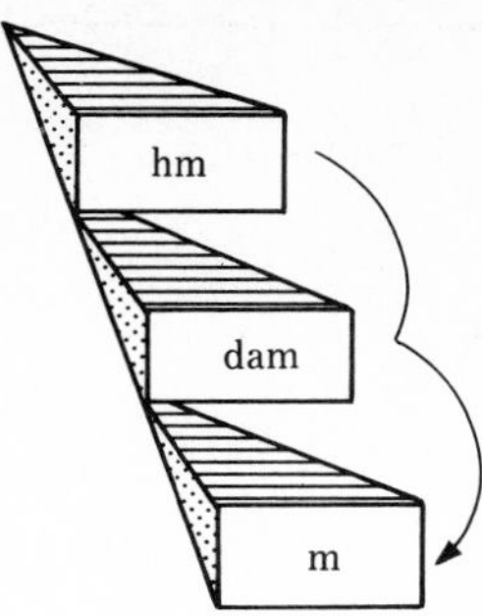

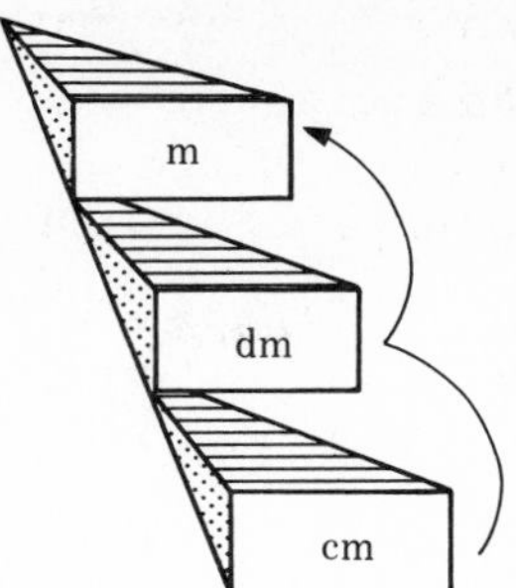

Hectometers to meters is *two* steps down the scale of measures of length; multiply by 10^2.

Centimeters to meters is *two* steps up the scale of measures of length; divide by 10^2.

Converting metric units: 

Since all base metric measures have equivalent subdivisions and multiples and since they are all denoted by the same prefixes, the following "step" scale can be used for any conversions of units within the same measures.

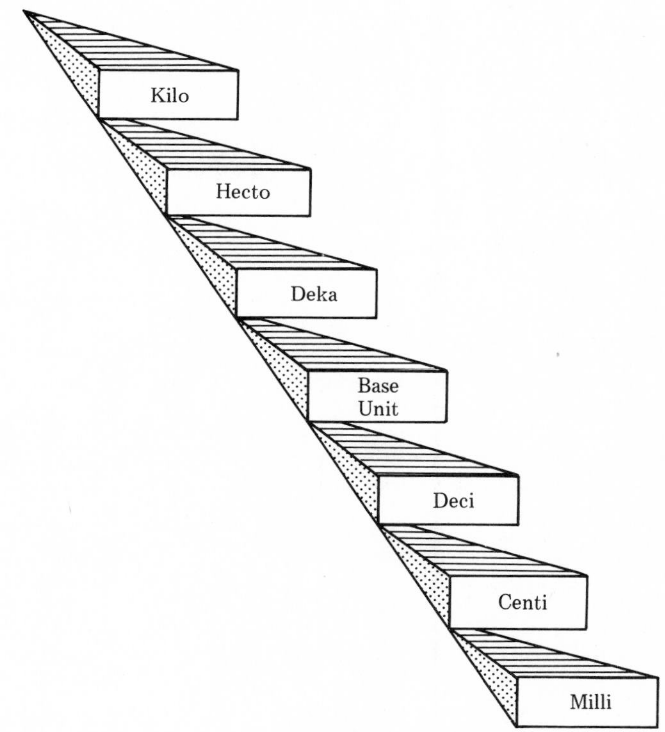

Convert 341 milligrams to grams: 

First count *up* the number of "steps" from "milli" to "base":

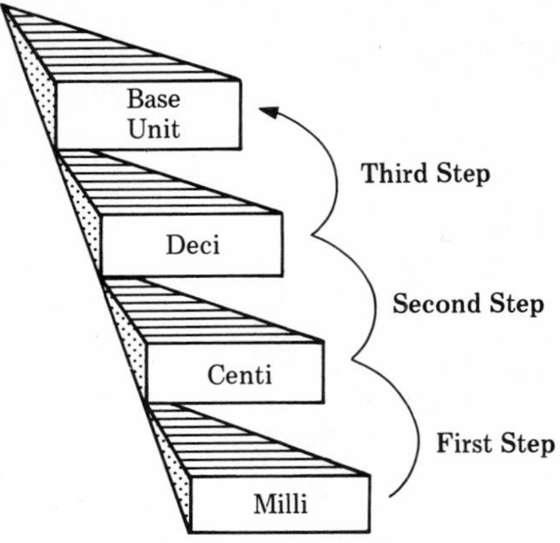

Since the measure in grams is three "steps" *up*, divide by 10^3 or 1,000:

341 mg ➡ $\frac{341}{1,000}$ g ➡ 0.341 g

Convert 0.234 kiloliters to liters: 

First count *down* the number of "steps" from "kilo" to "base":

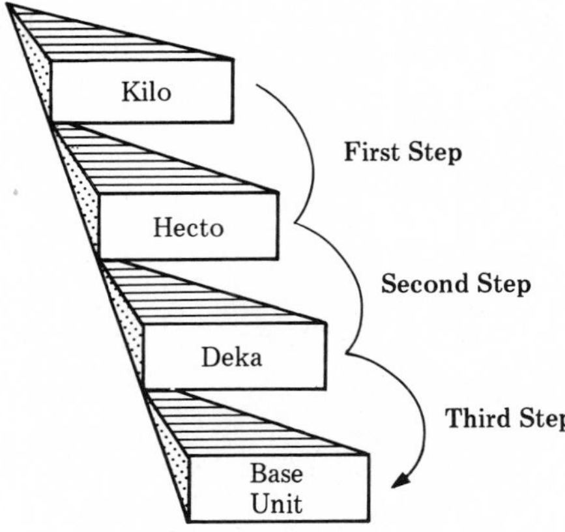

Since the measure in liters is three "steps" *down*, multiply by 10^3 or 1,000:

0.234 kl ➡ **0.234 × 1,000 l** ➡ 234. l

To convert from a smaller unit to a larger unit of metric measure: 

> *Divide* by the power of ten in which the exponent counts the number of "steps" *up* from the smaller to the larger measure.

To convert from a larger unit to a smaller unit of metric measure:

Multiply by the power of ten in which the exponent counts the number of "steps" *down* from the larger to the smaller measure.

Convert 720 decimeters to hectometers: (27)

"Deci" to "hecto" is *3* "steps" *up*:

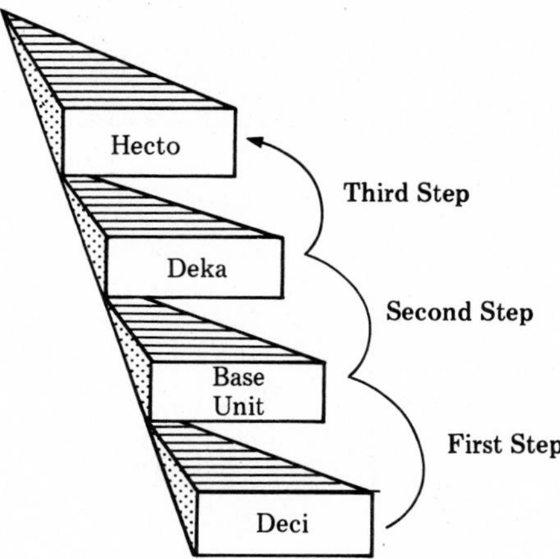

Divide by 10 to the *3*rd power:

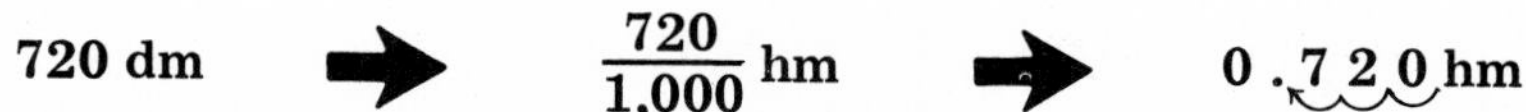

Convert 23.4 dekagrams to decigrams:  (28)

"Deka" to "deci" is *2* "steps" *down*:

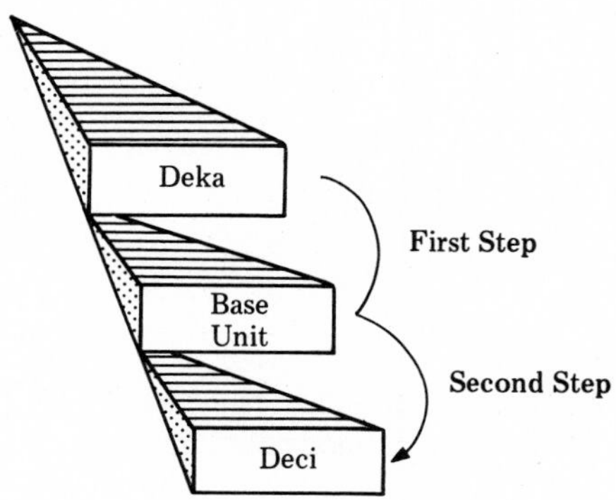

Multiply by 10 to the *2*nd power:

Convert 11.3 liters to hectoliters: 

Complete the following:

"Base" to "hecto" is ☐ "steps" up.
(Refer to Frame 24.)

$$11.3\ \text{l} = \frac{11.3}{10^{\square}}\ \text{hl}$$
$$= 0\,.113\ \text{hl}$$

Convert 13.2 kilometers to meters: 

Complete the following:

"Kilo" to "base" is ☐ "steps" down.
(Refer to Frame 24.)

$$13.2\ \text{km} = 13.2 \times 10^{\square}\ \text{m}$$
$$= 13200.\ \text{m}$$

The metric measure 13,200 m may appear in print as

13 200 m

The international agreement is to use spaces instead of commas to aid in reading large and small numbers.

Practice Exercises 

Convert each of the following to the indicated unit of measure:

1. 342 centimeters to decimeters.
2. 13.7 kilograms to grams.
3. 530 milliliters to liters.
4. 6.2 dekameters to millimeters.
5. 820 centigrams to hectograms.
6. 12.3 kiloliters to dekaliters.

English/metric conversions: 

Equivalent metric-English measures are given in Table 6 in the Appendix.

A. Convert 60 miles to kilometers.

From the table: 1 mile = 1.609 kilometers

Using dimension analysis:

$$60 \text{ miles} = 60\ \cancel{\text{mi}} \times \frac{1.609 \text{ km}}{1\ \cancel{\text{mi}}}$$

$$= 60 \times 1.609 \text{ km}$$

$$= 96.54 \text{ km}$$

B. Convert 4.3 kilograms to pounds.
From the table: 1 pound = 0.4536 kilograms

$$4.3 \text{ kilograms} = 4.3\ \cancel{\text{kg}} \times \frac{1 \text{ lb}}{0.4536\ \cancel{\text{kg}}}$$

$$= \frac{4.3}{0.4536} \text{ lb}$$

= 9.48 pounds, correct to the nearest hundredth of a pound

English/metric conversions: (33)

Complete the following:

A. Convert $2\frac{1}{2}$ quarts to liters.

From the table: 1 quart = $\square$ liter

$$2.5 \text{ qt} = 2.5\ \cancel{\text{qt}} \times \frac{\square\ 1}{1\ \cancel{\text{qt}}}$$

$$= 2.5 \times \square\ 1$$

$$= 2.375 \text{ l}$$

Note that $2\frac{1}{2}$ was rewritten as 2.5 since metric units are expressed in decimal notation.

B. Convert 30 centimeters to inches.

From the table: 1 inch = $\bigcirc$ centimeters

$$30 \text{ centimeters} = 30\ \cancel{\text{cm}} \times \frac{1 \text{ in.}}{\bigcirc\ \cancel{\text{cm}}}$$

$$= \frac{30}{\bigcirc} \text{ in.}$$

= 11.8 inches, correct to the nearest tenth of an inch

Practice Exercises

(34)

Convert each of the following to the indicated unit of measure.

1. $30\frac{1}{2}$ inches to centimeters.
2. 190 milliliters to quarts. (Hint: first convert milliliters to liters.)
3. 730 miles to kilometers.
4. 22.68 grams to ounces (avoirdupois).
5. 0.3 quarts to liters.
6. 10.1 meters to yards, correct to the nearest yard.

POST-TEST

1. Correctly complete the following:

   (a) $\frac{1}{1{,}000}$ liter = ______________ milliliter.

   (b) Kilometers is abbreviated ______ .

2. Convert 24 liters to milliliters.
3. Convert 3.7 kilograms to pounds, correct to the nearest pound.
4. Convert 15 yards to meters, correct to the nearest meter.

LESSON 27
OTHER METRIC MEASURES

Objectives:

1. Determine a rectangular surface area in a specified metric measure given the dimensions in either English or metric measures. (1–10)
2. Solve a word problem that has metric measures of rectangular surface areas (11–17)
3. Determine the volume or capacity of a rectangular object in a specified metric measure given the dimensions in either English or metric measures. (18–24)
4. Solve a word problem that has metric measures of volume for rectangular objects. (25–28)
5. Convert temperatures from degrees Celsius to degrees Fahrenheit or from degrees Fahrenheit to degrees Celsius. (29–33)

Vocabulary:

Length	Depth
Unit Scale	Height
Width	Volume (or Capacity)
Area	Cubic Units of Measure
Square Units of Measure	Degrees Celsius
Hectare	Degrees Fahrenheit

PRE-TEST

1. What is the surface area, in square centimeters, of a rectangular television screen that measures 17 inches by 14 inches? (Correct to the nearest square centimeter.)
2. If 1 can of paint will cover 90 square meters, how many cans of paint will it take to paint the two side walls in a theater if each wall is 6.5 meters high and 20 meters long? (Round to the nearest whole can.)
3. What is the capacity, in cubic meters, of a box 20 centimeters by 15 centimeters by 10 centimeters?
4. If topsoil costs $2.00 per cubic meter, what will it cost to cover an area 15 meters long and 10 meters wide to a depth of 0.5 meters with topsoil?
5. Convert each of the following: (Correct to the nearest tenth of a degree.)
   (a) 72°F to degrees Celsius.
   (b) 37°C to degrees Fahrenheit.

In the last lesson, metric measures of length were discussed. 1

The length of this line is 10 centimeters.

What does it mean to say, "The length of this line is 10 centimeters?"

Length: 2

One way to think of *length* is to say that it measures the "distance between."

The length of this line is a measure of the distance between the end bars.

The length of a room is the distance between two walls.

The length of a football field is the distance between the goal posts.

Length: 3

To "measure a length" means that a *unit scale* has been chosen and a comparison has been made to determine how many of these units can be fitted onto the length being measured.

The unit scale chosen to measure this line is 1 centimeter.

1 Centimeter

1 2 3 4 5 6 7 8 9 10

1 Centimeter

Since ten 1-centimeter units can be fitted between the end bars, the length of the line is 10 centimeters.

Measures of length alone do not provide the answers to many commonplace problems. 

How much carpeting is needed to cover the floor of a room that is 8 meters long and 6 meters wide?

How much grass seed is needed to plant a lawn 30 meters long and 20 meters wide?

Before questions such as these can be answered, it is necessary to know something about *measures of area.*

Area: 

To decide how much carpeting is needed to cover the floor of a room that is 8 meters long and 6 meters wide, and how it is to be measured, a sketch of the room is useful.

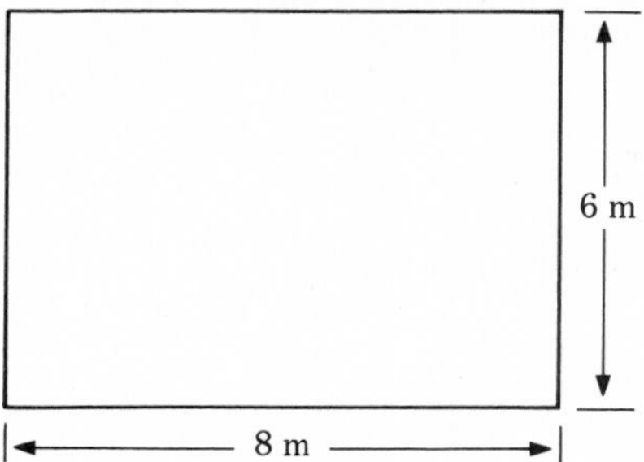

Notice that the measures of the *length* and the *width* of the room are *both* measures of length: the length, 8 meters, is the distance between the two walls that are farthest apart; the width, 6 meters, is the distance between the two walls that are closest to each other.

Area: 

Even though the room dimensions, 8 m by 6 m, are given in measures of length, the question that must be answered is what is the total floor surface of the room?

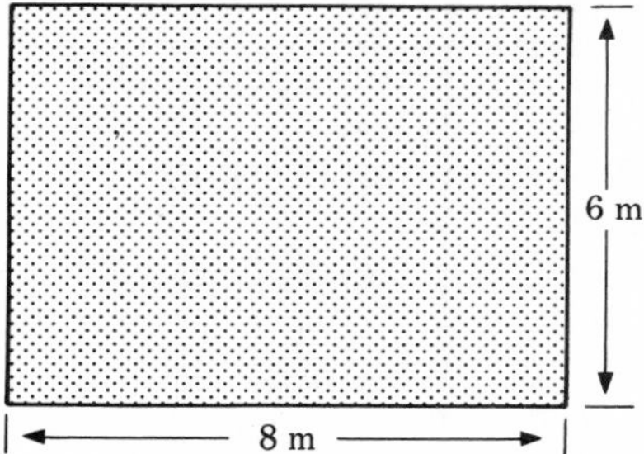

The floor surface can be measured, just as lengths can be measured, by choosing an appropriate unit scale and determining how many of these units can be fitted onto the surface.

Area: 

An appropriate unit scale is one *square meter* since a countable number of square meters can be fitted onto the rectangular surface.

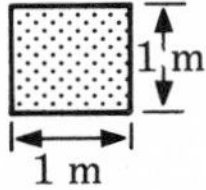

1 *square meter*, written 1 m^2, is the area of a surface that is 1 meter long and 1 meter wide.

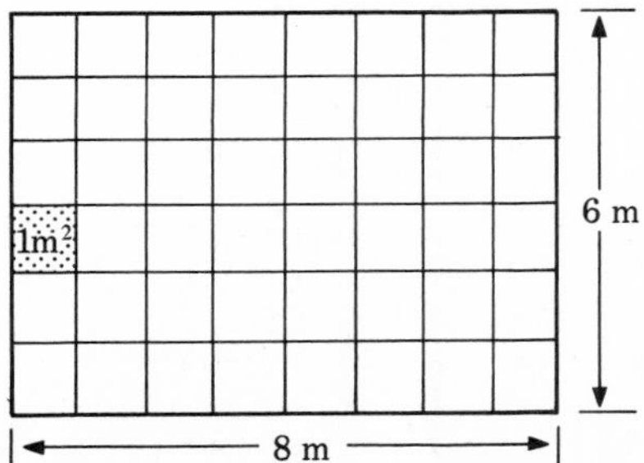

It can be determined by counting that 48 one-square-meter units can be fitted onto the floor surface.

The floor surface is 48 square meters, or 48 m^2.

Measures of surface area are *square units of measure.*

Area: (8)

Fitting and counting a number of square units onto surfaces is not the most efficient way to determine surface areas.

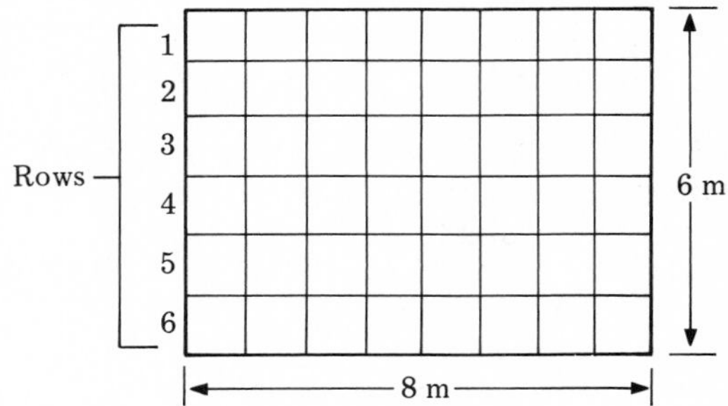

Notice that there are 6 rows of square units and that each row consists of 8 of these units.

$$\underbrace{\text{8 units + 8 units + 8 units + 8 units + 8 units + 8 units}}_{\text{6 rows of square units}}$$

Since repeated additions can be written as multiplications,

6 rows of square units × 8 units per row

or

48 square units

The surface area of the room is 6 m × 8 m or 48 m^2.

Area: (9)

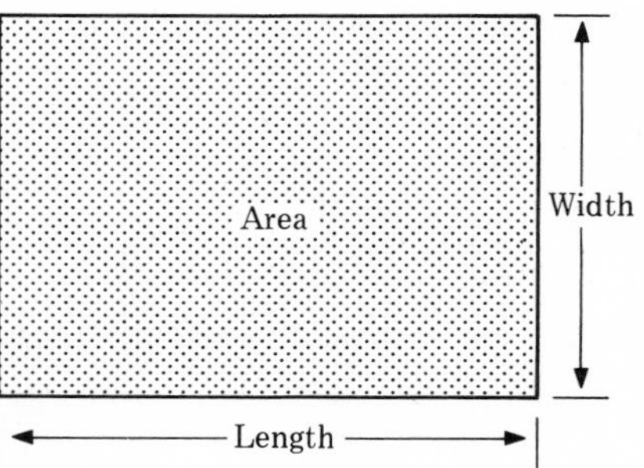

The *area*, *A*, of any rectangular surface can be determined by multiplying the length, *L*, times the width, *W*.

Area = Length times Width

A = L X W

where the measure of length and the measure of width are given in the same units of measure.

It is important to remember that *area* is expressed in *square units* and that *length* and *width* must be expressed in the same measure of length.

Area: 10

The common measures of surface area in the SI system are:

square millimeters	**mm^2**
square centimeters	**cm^2**
square meters	**m^2**
square kilometers	**km^2**

The metric measure of land area, measured in English "acres," is the *hectare*, abbreviated *ha* (no period).

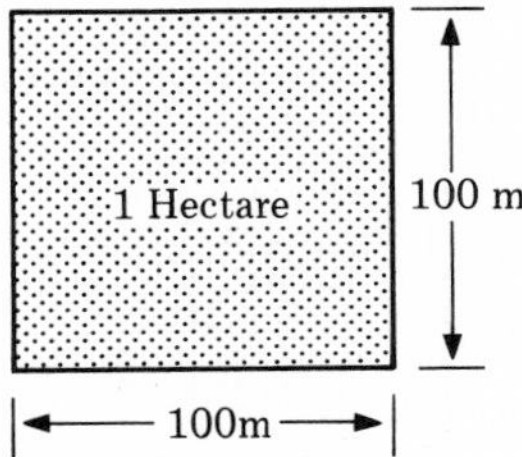

1 hectare = 100 m X 100 m = 10 000 m^2

How many boxes of grass seed are needed to plant a lawn 30 meters long and 20 meters wide 
if 1 box contains enough seed to plant 10 square meters?

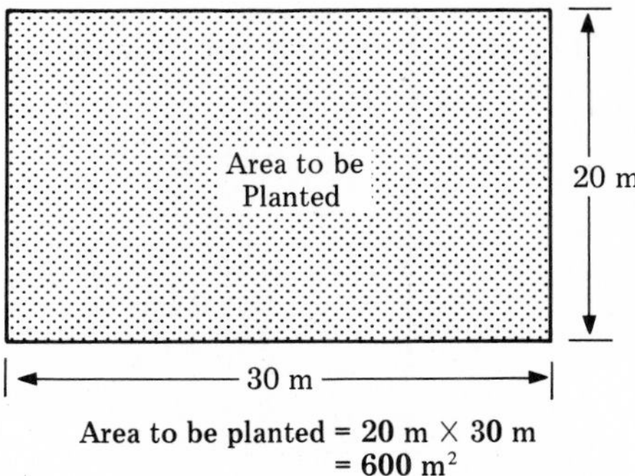

Area to be planted = 20 m X 30 m
= 600 m^2

A proportion can be used to determine how many boxes of seed are necessary to plant the lawn.

Let N = number of boxes of seed.

Number of Boxes	1	N
Area to be Planted	10 m^2	600 m^2

$$\frac{1}{10\text{ m}^2} = \frac{N}{600\text{ m}^2}$$

$$600\text{ m}^2 = 10\text{ m}^2 \times N$$

$$\left(\frac{1}{\overset{}{\underset{1}{\cancel{10\text{ m}^2}}}}\right) \overset{60}{\cancel{600}}\ \cancel{\text{m}^2} = \left(\frac{1}{\underset{1}{\cancel{10\text{ m}^2}}}\right)\left(\overset{1}{\cancel{10}}\ \cancel{\text{m}^2}\right) \times N$$

$$60 = N$$

It will take 60 boxes of seed to plant the lawn.

What is the area of a football field in square meters? (12)

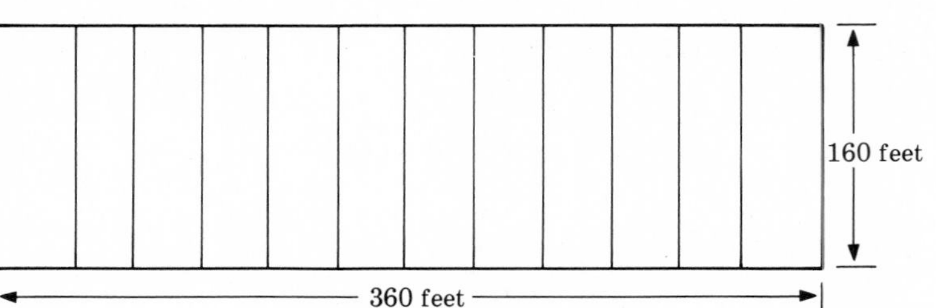

To find the area in square meters, it is necessary to first convert the measures of length and width to meters.

$$160\text{ ft} \approx 160\ \cancel{\text{ft}} \times \frac{0.3\text{ m}}{1\ \cancel{\text{ft}}} \approx 48\text{ m}$$

$$360\text{ ft} \approx 360\ \cancel{\text{ft}} \times \frac{0.3\text{ m}}{1\ \cancel{\text{ft}}} \approx 108\text{ m}$$

Then, the area of the field $\approx 108\text{ m} \times 48\text{ m}$

$\approx 5\ 184\text{ m}^2$

Notice that these are approximate measures in meters since 1 foot is not exactly equal to 0.3 meters.

A postage stamp is approximately 3 centimeters long and 2.5 centimeters wide. What is the area in square centimeters? In square millimeters? 

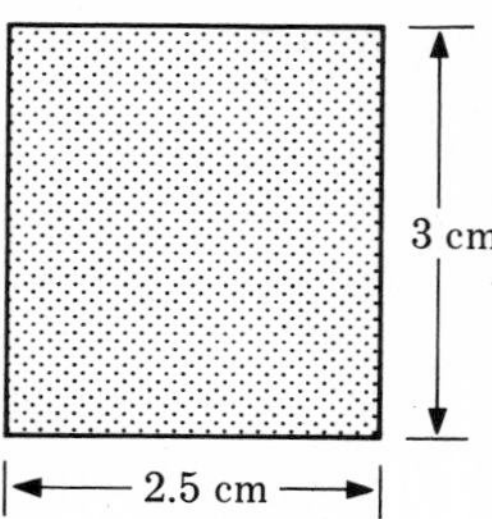

The area in square centimeters:

$$\text{Area} = 3\text{ cm} \times 2.5\text{ cm}$$
$$= 7.5\text{ cm}^2$$

To determine the area in square millimeters, first convert the measures of length and width to millimeters (one "step" down).

$$3 \text{ cm} = 3 \times 10^1 \text{ mm} = 30 \text{ mm}$$
$$2.5 \text{ cm} = 2.5 \times 10^1 \text{ mm} = 25 \text{ mm}$$

Then, the area in square millimeters:

$$\begin{aligned}\text{Area} &= 30 \text{ mm} \times 25 \text{ mm} \\ &= 750 \text{ mm}^2\end{aligned}$$

Notice that 100 times as many square millimeters can be fitted on this surface area as can be fitted in square centimeters.

The area of a school desk top is approximately 2.75 square meters. What is this same area in square centimeters?  14

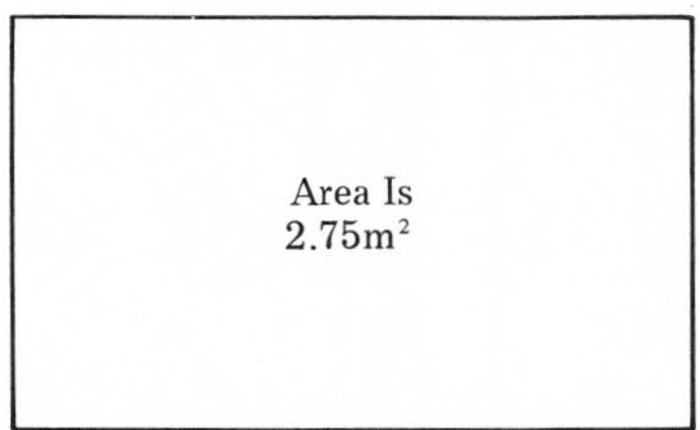

To find the area in square centimeters, think:

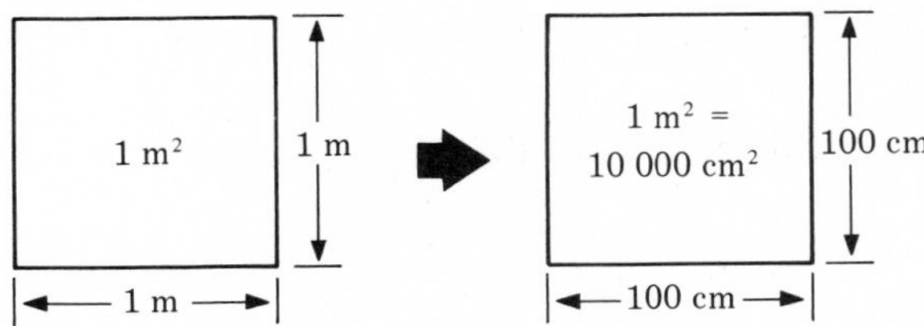

Since 1 meter = 100 centimeters,

1 square meter = 10 000 square centimeters

Using this information for the conversion factor,

$$\begin{aligned}\text{Area} &= 2.75 \text{ m}^2 \\ &= 2.75 \cancel{\text{m}^2} \times \frac{10\,000 \text{ cm}^2}{1 \cancel{\text{m}^2}} \\ &= 2.75 \times 10\,000 \text{ cm}^2 \\ &= 27\,500 \text{ cm}^2\end{aligned}$$

10 000 times as many square centimeters can be fitted onto the area as can be fitted in square meters.

A field is 1 kilometer long and 0.5 kilometer wide. What is the total area in hectares? 

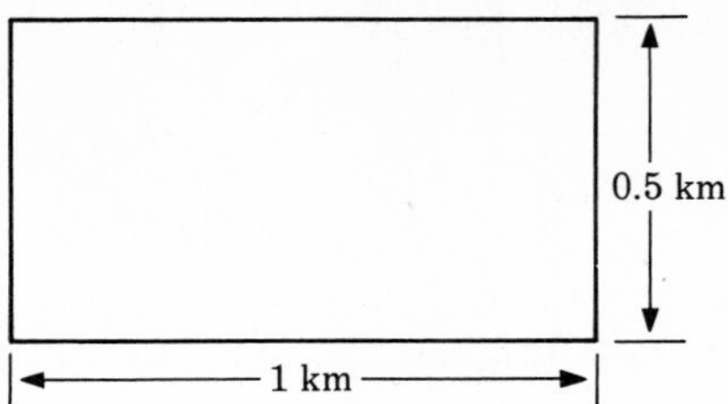

Complete the following:

Since **1 hectare = 10 000 m²**, first convert the measures in kilometers to meters:

$$0.5 \text{ km} = 0.5 \times 10^3 \text{ m} = \triangle \text{ m};$$

$$1 \text{ km} = 1 \times 10^3 \text{ m} = \square \text{ m}$$

$$\text{area} = \square \text{ m} \times \triangle \text{ m}$$

$$= (\square \times \triangle) \text{m}^2$$

$$= \bigcirc \text{m}^2$$

Then, converting to hectares:

$$\text{Area} = \bigcirc \cancel{\text{m}^2} \times \frac{1 \text{ ha}}{10\ 000 \cancel{\text{m}^2}}$$

$$= \frac{\bigcirc}{10\ 000} \text{m}^2$$

$$= 50 \text{ hectares}$$

A picnic table measures 2 meters by 75 centimeters. What is the total area of the table top in square meters? 

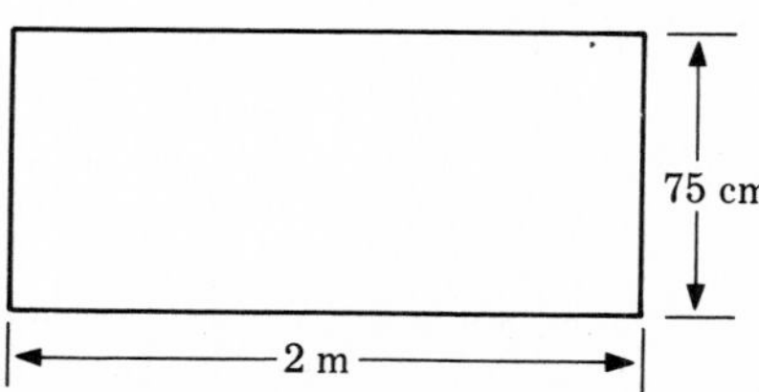

Complete the following:

Converting centimeters to meters:

$$75 \text{ cm} = \frac{75}{10^2} \text{ m} = \triangle \text{ m}$$

Then, the area is:

$$\text{Area} = \square\,\text{m} \times \triangle\,\text{m}$$
$$= (\square \times \triangle)\,\text{m}^2$$
$$= \bigcirc\,\text{m}^2$$

Practice Exercises

1. A standard size baking pan is 13 inches by 9 inches. What is the area of the bottom of the pan in square centimeters? (Round the answer to the nearest square centimeter.)
2. A cement patio slab is 6.6 meters long and 4.5 meters wide. If 1 can of cement paint will cover 10 square meters, how many whole cans of paint will it take to paint the surface of the patio slab? (Round the answer to the next largest whole can of paint.)
3. A brick is approximately 200 millimeters long and 100 millimeters wide. What is the surface area of the face of the brick in square centimeters?
4. Three-fifths of a field 2.25 kilometers long and 1.75 kilometers wide is to be planted in corn. How many hectares will be planted in corn? (Round to the nearest hectare.)

While measures of area provide the answers for many practical problems that cannot be answered by measures of length alone, a further measure is needed to answer questions such as the following:

> An aquarium has a fish tank that is 3 meters long and 2 meters wide. How much water does it take to fill it to a depth of 1 meter?

To answer this question, it is necessary to know something about measures of volume (or capacity).

Volume:

A sketch is useful to help visualize what is to be measured.

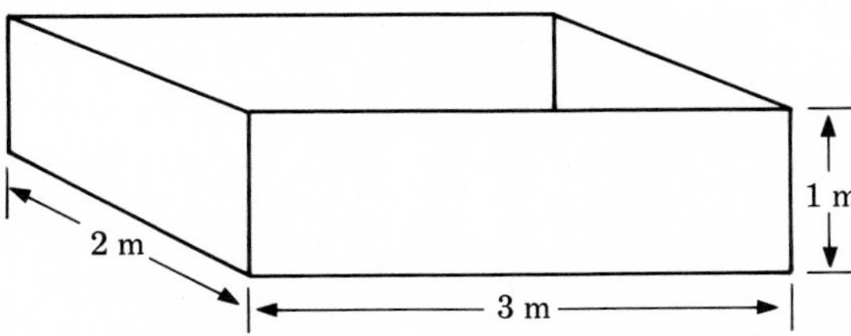

Notice that the measures of *length*, *width*, and *depth* are all measures of length. The *depth* is the distance between the bottom of the tank and the surface of the water.

Volume: 

Even though the dimensions 3 m by 2 m by 1 m are given in measures of length, the question that must be answered is what is the total volume (or capacity) of the tank?

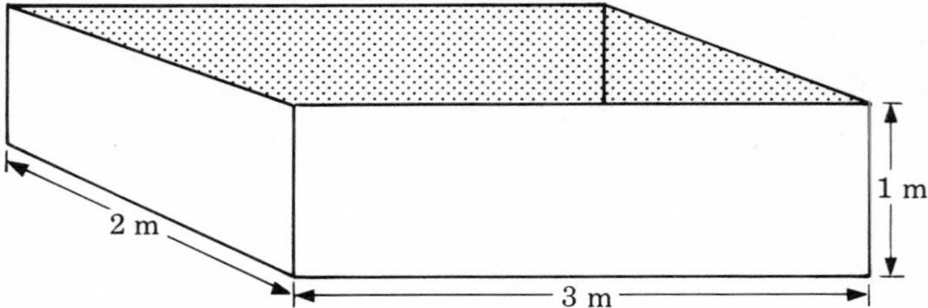

The total capacity can be measured, just as lengths and areas can be measured, by choosing an appropriate unit scale.

Volume: 

An appropriate unit scale is one cubic meter since a countable number of cubic meters can be fitted into the tank.

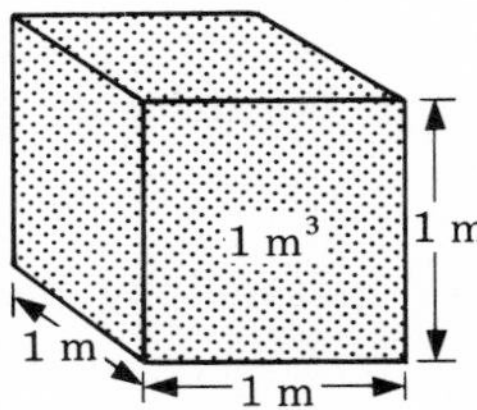

1 *cubic meter*, written 1 m^3, is the measure of the volume (or capacity) of a container that is 1 meter long, 1 meter wide, and 1 meter deep.

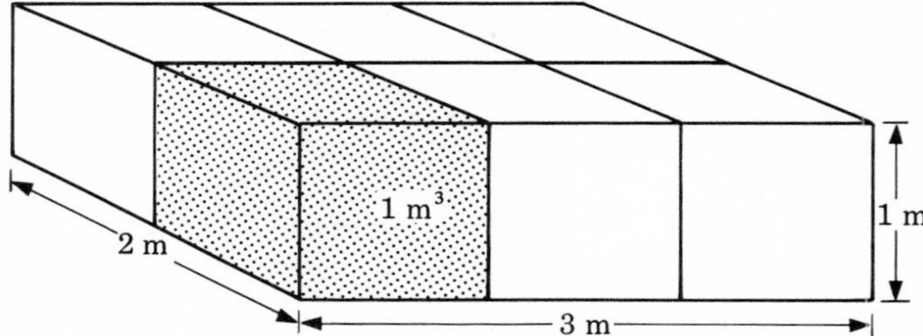

It can be determined by counting that 6 one-cubic-meter units can be fitted into the tank. The volume (or capacity) of the tank is

6 cubic meters or 6 m^3

Measures of volume are *cubic units of measure.*

Volume: 

Fitting and counting a number of cubic units is not the most efficient way to measure the volumes of containers.

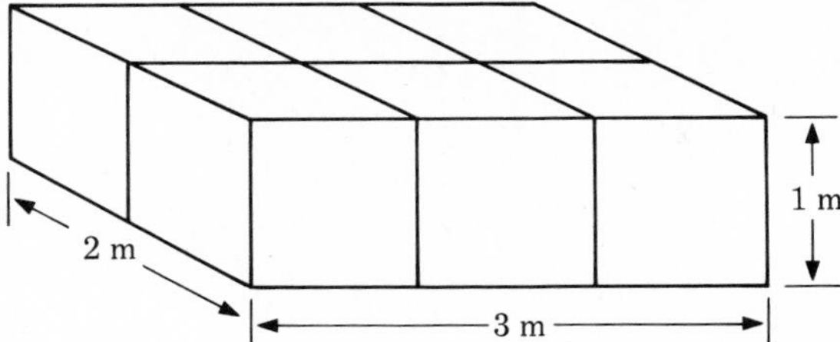

To determine the volume, V, of a rectangular container, find the number of square units of the surface determined by the length, L, and the width, W, and then multiply the surface area by the depth (or height), H, of the container.

Volume = Length × Width × Height

Thus, for the fish tank,

$$V = L \times W \times H$$

$$\begin{aligned} \text{Volume} &= 3\text{ m} \times 2\text{ m} \times 1\text{ m} \\ &= (3 \times 2 \times 1)\text{ m}^3 \\ &= 6\text{ m}^3 \end{aligned}$$

Volume: 

Some common measures of volume in the SI system are:

cubic millimeters	**mm^3**
cubic centimeters	**cm^3 or cc**
cubic meters	**m^3**

Recall from the last unit that the basic metric measure of volume or capacity is the liter. How are these measures related?

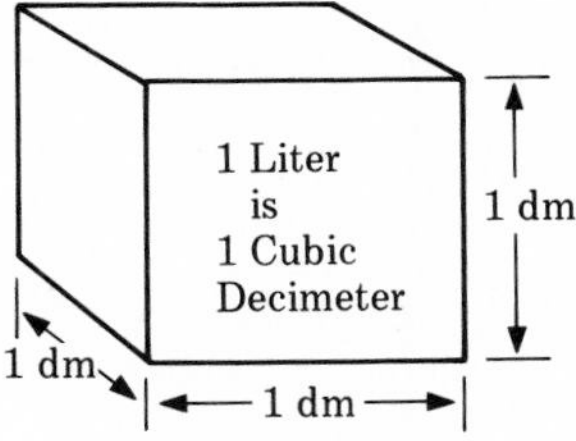

This relationship can also be expressed as follows:

1 liter = 1 000 cm^3 since **1 dm = 10 cm**

1 liter = $\frac{1}{1\,000}$ m^3 since **1 dm = $\frac{1}{10}$ m**

Relationship between metric measures of length, weight, and volume: 

One further measure can be applied to the 1-liter container if it is filled with water and weighed at a specified temperature:

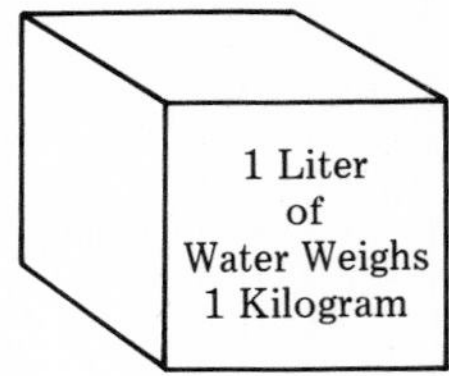

Or, restating this in terms of grams:

1 milliliter or 1 cubic centimeter of water weighs 1 gram at 4° C.

What is the capacity of a 13 inch by 9 inch baking pan, to the nearest cubic centimeter, if it is 2 inches deep? 

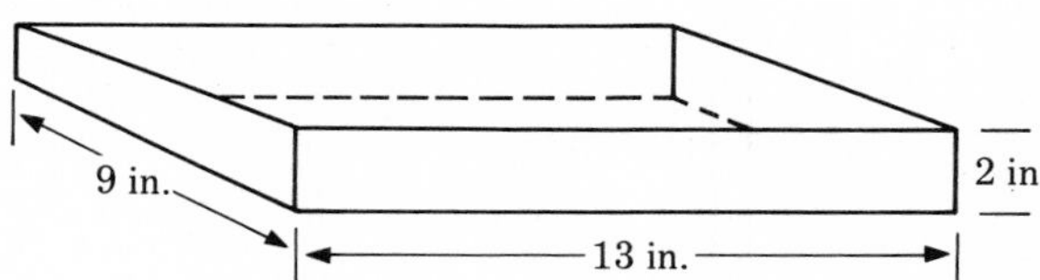

First convert the measures to centimeters:

$$13 \text{ in.} = 13 \not{\text{in.}} \times \frac{2.54 \text{ cm}}{1 \not{\text{in.}}} = 33.02 \text{ cm}$$

$$9 \text{ in.} = 9 \not{\text{in.}} \times \frac{2.54 \text{ cm}}{1 \not{\text{in.}}} = 22.86 \text{ cm}$$

$$2 \text{ in.} = 2 \not{\text{in.}} \times \frac{2.54 \text{ cm}}{1 \not{\text{in.}}} = 5.08 \text{ cm}$$

Then,

$$\text{Volume or capacity} = 33.02 \text{ cm} \times 22.86 \text{ cm} \times 5.08 \text{ cm}$$
$$= 3\,834.572\,976 \text{ cm}^3$$

Rounding to the nearest cubic centimeter:

$$\text{Capacity} \approx 3\,835 \text{ cm}^3$$

What is the volume, in cubic meters, of a cereal box that measures 25 centimeters by 18 centimeters by 6 centimeters? 

Complete the following:

$$25 \text{ centimeters} = \frac{25}{\bigcirc} \text{ meters} = \triangle \text{ meters}$$

$$18 \text{ centimeters} = \frac{18}{\bigcirc} \text{ meters} = \square \text{ meters}$$

$$6 \text{ centimeters} = \frac{6}{\bigcirc} \text{ meters} = ⬡ \text{ meters}$$

Then, $\text{volume in cubic meters} = \triangle \text{ m} \times \square \text{ m} \times ⬡ \text{ m}$

$$= 0.002\,7 \text{ m}^3$$

An Olympic size swimming pool used for competition is 50 meters long and approximately 23 meters wide. If it is filled to a depth of 2 meters, how many cubic meters of water will it hold? What is the weight of this amount of water in kilograms? 27

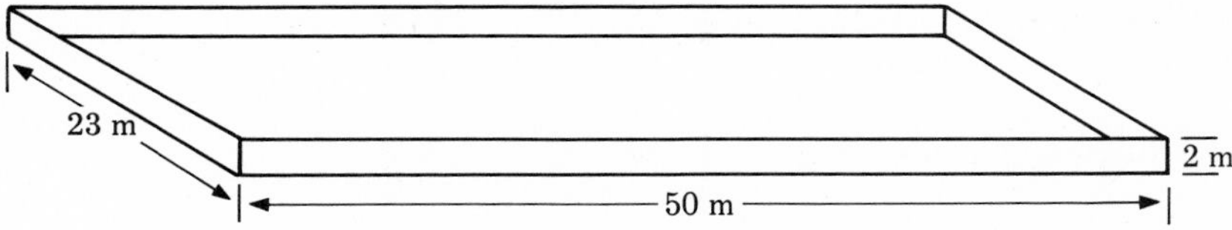

Complete the following:

$$\text{Volume in cubic meters} = 50 \text{ m} \times 23 \text{ m} \times 2 \text{ m}$$
$$= \square \text{ m}^3$$

Since 1 000 liters = 1 cubic meter and 1 liter of water weighs 1 kilogram,

1 cubic meter of water weighs 1 000 kilograms

Thus, 2 300 m^3 of water weighs 1 000 × 2 300 or 2 300 000 kilograms.

Practice Exercises

1. How many cubic meters of soil will it take to fill a rectangular hole that is 17.5 centimeters long, 75 centimeters wide, and 80 centimeters deep?
2. How many cubic meters of storage space is there in a refrigerator if the dimensions of the interior are 4 feet by 2 feet by 2 feet? (Use 1 ft ≈ 0.3 m.)
3. Before planting a vegetable garden, 1 bag of minerals is to be mixed with every 1 cubic yard of soil. If the soil in a garden plot 10 meters long and 8 meters wide is dug out to an overall depth of 0.4 meter, how many bags of minerals should be mixed with this soil? (Round the answer to the next larger whole bag.)
4. A tropical fish tank 60 centimeters long and 30 centimeters wide is filled to a depth of 30 centimeters. What is the weight (in kilograms) of the water in the tank?

One further difference between the metric system and the English system is in the measure used for temperature. The basic SI measure is actually in kelvins, but temperature in metric countries is more commonly measured in degrees *Celsius* (or centigrade). In this country, the common measure of temperature is degrees *Fahrenheit.* (29)

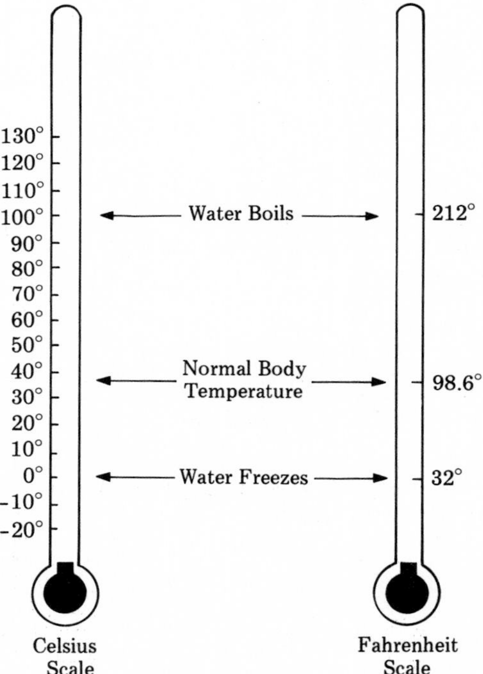

30°C is read "30 degrees Celsius"; -20° means 20 degrees below zero.

Temperature:

The Celsius scale sets 0° as the freezing point of water and 100° as the boiling point of water. Instead of conversion factors, formulas are used to convert temperatures.

To convert degrees Fahrenheit to degrees Celsius:

$$^\circ C = \frac{5}{9}(^\circ F - 32)$$

First subtract 32 from the Fahrenheit temperature; then multiply by $\frac{5}{9}$.

To convert degrees Celsius to degrees Fahrenheit:

$$^\circ F = \frac{9}{5}\ ^\circ C + 32$$

First multiply the Celsius temperature by $\frac{9}{5}$; then add 32.

Temperature conversion: (31)

A. Cakes are usually baked at oven temperatures of 350°F. What is the equivalent oven temperature in degrees Celsius (to the nearest degree)?

$$\begin{aligned} ^\circ C &= \frac{5}{9}(350 - 32) \\ &= \frac{5}{9}(318) \\ &\approx 177 \end{aligned}$$

B. A person's temperature is taken with a thermometer in degrees Celsius. If his temperature is 40°C, is it above or below "normal" body temperature?

$$\begin{aligned} ^\circ F &= \frac{9}{5} \times 40 + 32 \\ &= 72 + 32 \\ &= 104 \end{aligned}$$

which is definitely above the "normal" body temperature of 98.6°F.

Temperature conversion: (32)

Complete the following:

A. A comfortable temperature for a centrally heated office is maintained if the thermostat is set at 68°F. What is the equivalent Celsius temperature?

$$\begin{aligned} ^\circ C &= \frac{5}{9}(68 - 32) \\ &= \frac{5}{9} \times \square \\ &= \bigcirc \end{aligned}$$

B. In the mid-1930's a freak weather condition caused the temperature in a town in Portugal to rise to 70°C for two minutes' duration. What is the equivalent Fahrenheit temperature?

$$°F = \frac{9}{5} \times 70 + 32$$

= ⬡ + 32

= △

Practice Exercises

Answer each of the following, giving a temperature correct to the nearest degree to justify each answer.

1. If the outdoor temperature is 30°C, would it be a better day for swimming or ice skating?
2. Bread is usually baked at an oven temperature of 425°F for 1 hour. If a loaf of bread is baked in an oven set at 270°C, will it be over or under done in the same length of time?
3. If the water temperature in a tub of water is 85°C, would it be comfortable for bathing?

POST-TEST

1. In England, a standard double-bed mattress is 200 centimeters long and 150 centimeters wide. What is the surface area of this mattress in square meters?
2. Wall paneling is sold by the sheet. If each sheet is 8 feet by 4 feet, how many panels will it take to cover a wall 2.4 meters high and 6 meters long? (Use 1 foot $\approx$ 0.3 meters.)
3. What is the capacity in cubic milliliters of a small package that is 6 centimeters long, 6 centimeters wide, and 6 centimeters deep?
4. If water costs $0.135 per cubic meter, what will it cost to fill a pool that is 2.4 meters long, 1.3 meters wide, and 0.5 meters deep? (Round to the nearest cent.)
5. Convert each of the following:
   (a) 100°C to degrees Fahrenheit.
   (b) 32°F to degrees Celsius.

LESSON 28
APPLICATIONS OF MEASUREMENTS

Objectives:

1. Determine the Greatest Possible Error and Relative Error for a measurement. (1–11)
2. Determine the number of significant digits in a measurement. (12–16)
3. Determine the smallest unit of measure for a given measurement. (12–16)
4. Use the concepts of accuracy and precision to round the results of computations with approximate numbers. (17–21)
5. Solve problems dealing with perimeter, surface area, and/or volumes of rectanglar, circular, triangular, spherical, and cylindrical objects. (22–33)

Vocabulary:

Exact Numbers	Circle
Approximate Numbers	Circumference
Precision	Diameter
Greatest Possible Error	Radius
Accuracy	Cylinder
Relative Error (Percent Error)	Perimeter
Significant Digits	Triangle
Square	Sphere

PRE-TEST

1. What is the greatest possible error in a measurement of 9.6 cm?
2. What is the relative error in a measurement of 12.31 centimeters, correct to the nearest ten-thousandth?
3. What is the smallest unit of measure indicated by a measurement of 41 meters?
4. How many significant digits are there in a measurement of 3 200 kilometers?
5. Find the sum or product of the following measurements using precision or accuracy to round.

   (a) $\begin{array}{r} 6.4 \text{ cm} \\ \underline{+3.18} \text{ cm} \end{array}$

   (b) $\begin{array}{r} 8.5 \text{ cm} \\ \underline{\times 0.5} \text{ cm} \end{array}$

6. Find the volume of a box 1.5 meters long, 0.7 meters wide, and 1.2 meters high, rounding the product according to the accuracy of the given dimensions.

In previous lessons, computations with measurements have been carried out as if they were the same as any whole or fractional numbers. Before proceeding with applications, some other important aspects of measurement should be considered. 1

Exact and approximate numbers: 2

In a statement such as, "There are 32 students in the classroom," 32 represents an *exact number* since the number of students can be determined by counting separate, indivisible objects.

In a statement such as, "The book is 32 centimeters long," the 32 represents an *approximate number* in that measurements only approximate "ideal" measures, not subject to human or mechanical error.

Approximate numbers:

All measurements of distance, area, weight, volume, time, and temperature are approximations.

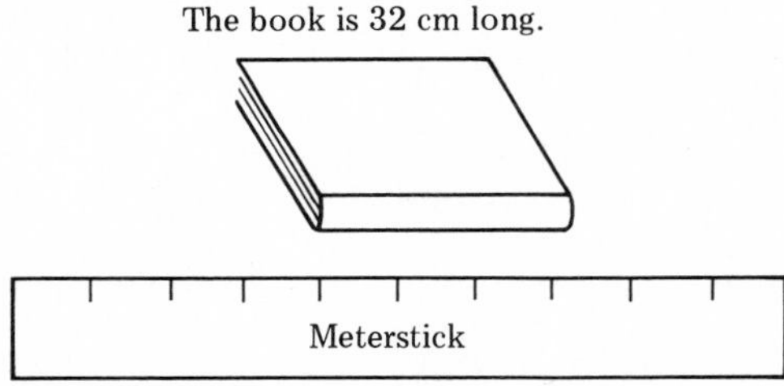
The book is 32 cm long.
Meterstick

This measurement represents a comparison of the length of the book to a replica of a standard of measure such as a meterstick. All measurements represent comparisons between some physical property of an object and a standard scale of measure.

Since all measurements are only approximations, it is necessary to be able to determine how "good" an approximation any measurement is. Two qualities used to describe the "goodness" of a measurement are *precision* and *accuracy*. 

When a book is measured as being 32 centimeters long, how "good" is this measurement as an approximation of the actual length of the book?

Precision: 

The *precision* of a measurement of 32 centimeters depends on the meterstick used to make the measurement.

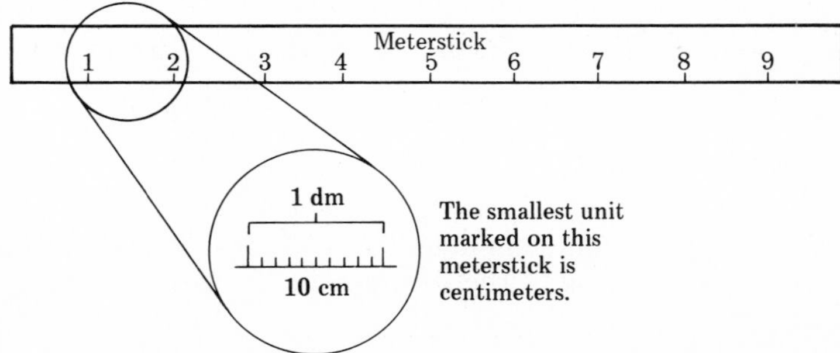
Meterstick
1 2 3 4 5 6 7 8 9
1 dm
10 cm
The smallest unit marked on this meterstick is centimeters.

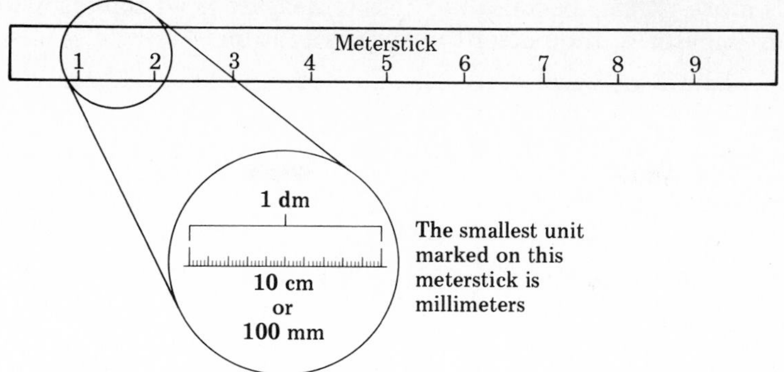

The precision of a measurement is really a reference to the precision of the measuring device used to make the measurement. The precision of the meterstick depends on the smallest unit of measure that is indicated on the meterstick: The more subdivisions of the meter that are indicated, the more closely the measured length approximates the actual length.

Precision: 6

Human and mechanical errors affect the measurement of any object: for example, wooden rulers warp; metal tapes contract and expand; and the eye has trouble distinguishing between 0.01 mm and 0.02 mm.

To account for these sources of error:

> The *Greatest Possible Error*, GPE, of a measurement is defined to be no larger than $\frac{1}{2}$ of the smallest unit used to mark the measuring device. The GPE is used to describe the precision of a measurement.

If the meterstick used to measure the book is marked in centimeters,

GPE is $\frac{1}{2}$ of 1 centimeter or 0.5 cm

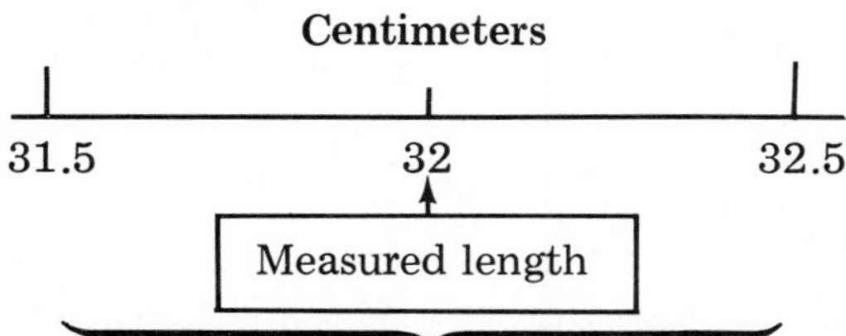

The actual length is somewhere within this range of measures.

If the meterstick is marked in millimeters,

GPE is $\frac{1}{2}$ of 1 millimeter or 0.05 cm

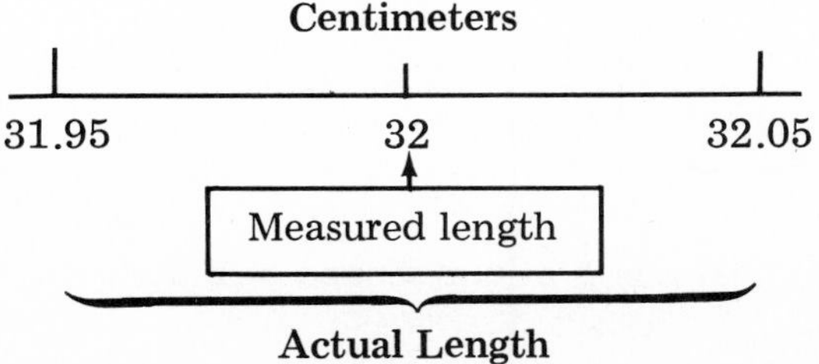

The second measure is more precise because the actual measure is within a smaller range of measures; the smaller the range of measures, the better the approximation.

Precision vs. accuracy: 

The precision of a measurement of 32 centimeters, when measured with a meterstick marked in centimeters, is exactly the same as the precision of a measurement of 32 meters made using the same meterstick since the GPE in both is $\frac{1}{2}$ of 1 centimeter.

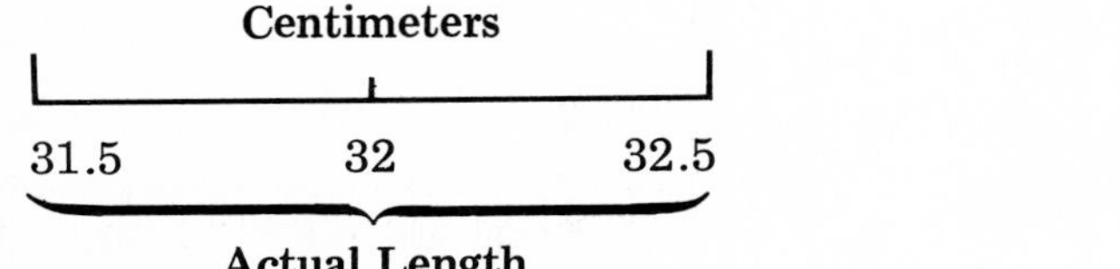

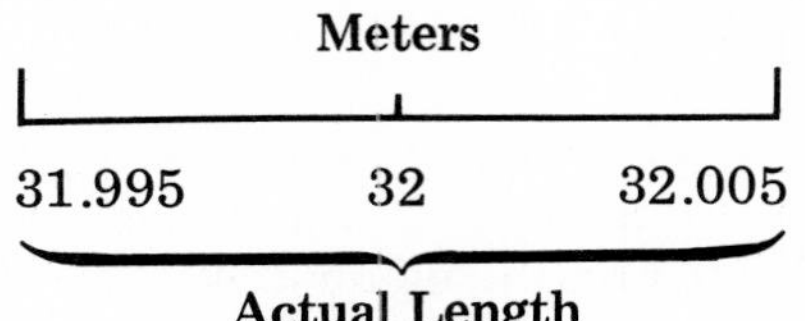

An error of 0.5 cm in a measurement in centimeters could be critical; an error of 0.5 cm in a measurement in meters isn't too significant. Precision alone cannot describe the "goodness" of a measurement.

The *accuracy* of a measurement is described by the *Relative Error* of the measurement, the ratio of the Greatest Possible Error to the measured value.

$$\text{Relative Error} = \frac{\text{GPE}}{\text{Measurement}}$$

Precision vs. accuracy: 

Consider the same measurements of 32 centimeters and 32 meters made with a meterstick marked in centimeters:

Relative Error 32 cm ➔ $\frac{\text{0.5 cm}}{\text{32 cm}}$ or 0.015 625

Relative Error 32 m ➔ $\frac{\text{0.5 cm}}{\text{32 m}}$ ➔ $\frac{\text{0.005 m}}{\text{32 m}}$ or 0.000 156 25

The second measurement, 32 meters, is more accurate since the Relative Error is smaller. Relative Error may be expressed as a percent, called *Percent Error.*

The Percent Error of 32 cm is 1.56% to the nearest hundredth of a percent.

The Percent Error of 32 m is 0.02% to the nearest hundredth of a percent.

Precision and accuracy: 

A time of 2 minutes is measured with a watch that has a second hand.

GPE	Relative Error	Percent Error
0.5 sec	0.004 2	0.42%

GPE is $\frac{1}{2}$ of one second or 0.5 sec

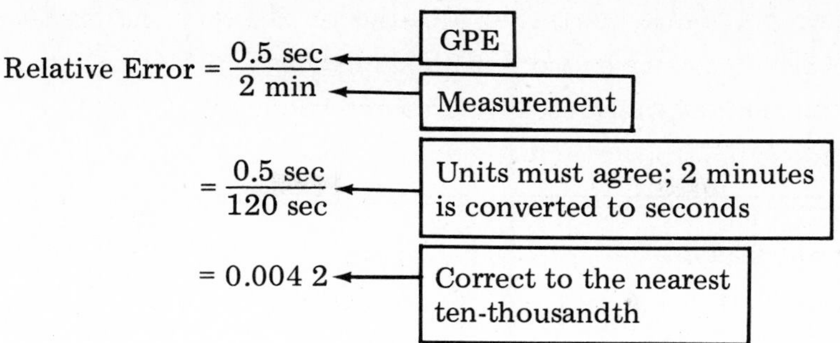

Relative Error = $\frac{0.5 \text{ sec}}{2 \text{ min}}$
GPE
Measurement
= $\frac{0.5 \text{ sec}}{120 \text{ sec}}$
Units must agree; 2 minutes is converted to seconds
= 0.004 2
Correct to the nearest ten-thousandth

The Percent Error is 0.42%, correct to the nearest hundredth of a percent.

Notice that the GPE has units of measure, but the Relative Error does not.

Precision and accuracy: (10)

A man weighs 75 kilograms on a scale marked in $\frac{1}{2}$ kilogram increments.

Complete the following:

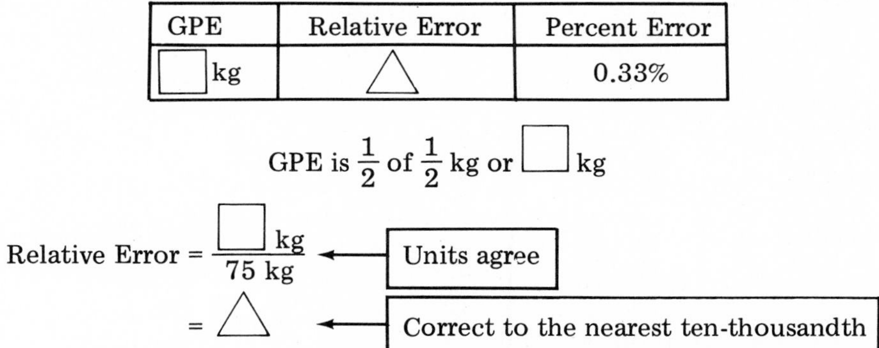

GPE	Relative Error	Percent Error
□ kg	△	0.33%

GPE is $\frac{1}{2}$ of $\frac{1}{2}$ kg or □ kg

Relative Error = $\frac{\square \text{ kg}}{75 \text{ kg}}$
Units agree
= △
Correct to the nearest ten-thousandth

The Percent Error is 0.33%, correct to the nearest hundredth of a percent.

Practice Exercises (11)

Complete the table for the indicated precision for each measurement.

	Measurement	GPE	Relative Error	Percent Error
1.				
2.				
3.				
4.				
5.				
6.				

1. A measurement of 2 cm if the ruler used is marked in centimeters.
2. A measurement of 2 cm if the ruler used is marked in millimeters.

3. A weight of 4 kilograms if the scale used is marked in grams.
4. A weight of 4 grams if the scale used is marked in grams.
5. A time of 5 seconds if the watch used has a second hand.
6. A time of 5 seconds if a stopwatch marked in tenths of a second is used.

Precision and significant digits: (12)

The precision of a measurement always refers to the refinement of the measuring device used to determine the measurement. It describes the maximum amount of variation possible between the measured value and the actual value.

The precision of a measurement can be indicated in the following ways:

1. As a plus or minus correction: 9.78 m $\pm$ 0.005 m

2. Explicitly: The length is 9.78 meters, correct to the nearest centimeter.

Both statements mean that the actual length of the measured distance is between 9.775 m and 9.785 m.

One further way to indicate the precision of a measurement, and the most common, is to determine the smallest unit of measure by the position of a significant digit within the numerical portion of the written measurement.

Significant digits: 

The significant digits in the numerical portion of a measurement tell how many times the smallest unit of measure used to make the measurement is contained within that measurement. For example,

14 kilometers	has 2 significant digits; the smallest unit of measure is 1 kilometer. There are 14 one-kilometer units in 14 kilometers.
14.7 kilometers	has 3 significant digits; the smallest unit of measure is $\frac{1}{10}$ kilometer or 1 hectometer. There are 147 hectometers in 14.7 kilometers.
14.75 kilometers	has 4 significant digits; the smallest unit of measure is $\frac{1}{100}$ km or 1 dekameter. There are 1 475 dekameters in 14.75 km.

The last significant digit on the right indicates the smallest unit of measure. Once the smallest unit of measure is known, the GPE is known and, therefore, the precision is known.

Significant digits: (14)

The number of significant digits in a measurement is determined as follows:

1. All non-zero digits are significant.

   24 kg — 2 significant digits; unit is kilograms.

   43.8 cl — 3 significant digits; unit is $\frac{1}{10}$ of a centiliter or 1 milliliter.

2. All zeros *between* non-zero digits are significant.

   207 km — 3 significant digits; unit is kilometers.

   3.004 g — 4 significant digits; unit is milligrams or $\frac{1}{1,000}$ of a gram.

3. All terminal zeros to the *right* of the decimal point are significant.

   6.0 m — 2 significant digits; unit is decimeters—
   60 decimeters in 6 meters.

   6.00 m — 3 significant digits; unit is centimeters—
   600 centimeters in 6 meters.

   6.000 m — 4 significant digits; unit is millimeters—
   6 000 millimeters in 6 meters.

4. In numbers less than one, any zeros between the decimal point and the first non-zero digit are *not* significant (they are placeholders used only to show the correct value of the smallest unit of measure).

   0.04 l — 1 significant digit; unit is centiliters—
   4 centiliters in 0.04 liters

   0.004 l — 1 significant digit; unit is milliliters—
   4 milliliters in 0.004 liters.

   0.040 l — 2 significant digits; unit is milliliters—
   40 milliliters in 0.040 liters.

5. Terminal zeros in a whole number are significant *only* if they are so indicated. The smallest unit may be indicated by *underlining* if it is a zero.

   13 000 km — 2 significant digits; unit is 1,000 kilometers—
   13 one-thousand kilometer units.

   $13\ \underline{0}00$ km — 3 significant digits; unit is 100 kilometers—
   130 one-hundred kilometer units.

   $13\ 0\underline{0}0$ km — 4 significant digits; unit is 10 kilometers—
   1 300 ten-kilometer units.

   $13\ 00\underline{0}$ km — 5 significant digits; unit is 1 kilometer—
   13 000 one-kilometer units.

These rules apply only to approximate numbers and not to exact numbers.

Precision, accuracy, and significant digits: 

One more fact should be kept in mind about significant digits as they relate to precision and accuracy.

Consider the following measurements:

0.20 m	2.0 m
GPE $= \frac{1}{2}$ of 1 cm or 0.005 m	GPE $= \frac{1}{2}$ of 1 dm or 0.05 m
Relative Error $= \dfrac{0.005\text{ m}}{0.20\text{ m}}$	Relative Error $= \dfrac{0.05\text{ m}}{2.0\text{ m}}$
$= \dfrac{5}{200}$	$= \dfrac{5}{200}$
Percent Error = 2.5%	Percent Error = 2.5%

Greater precision does *not* imply greater accuracy. The *number* of significant digits does relate to the accuracy:

0.20 m — **2 significant digits; unit is 1 cm**
2.0 m — **2 significant digits; unit is 1 dm**

0.20 m is more precise than 2.0 m since it is measured with a smaller unit, but the accuracy of 0.20 m and 2.0 m is *exactly* the same.

But,

0.20 m is *less accurate* than 0.200 m
2.0 m is *less accurate* than 2.00 m

As the total number of significant digits increases, the accuracy increases.

Practice Exercises 

Determine the number of significant digits and name the smallest unit of measure for each of the following:

1. 2 400 kg
2. 2 4$\underline{0}$0 kg
3. 0.03 sec
4. 0.30 sec
5. 10.01 m
6. 10.10 m
7. 73 l
8. 73.0 l

Computations involving measurements: 

Sums and differences of measured amounts are no more precise than the *least* precise measurement used in the computation.

For example, adding 0.06 cm and 0.3 cm:

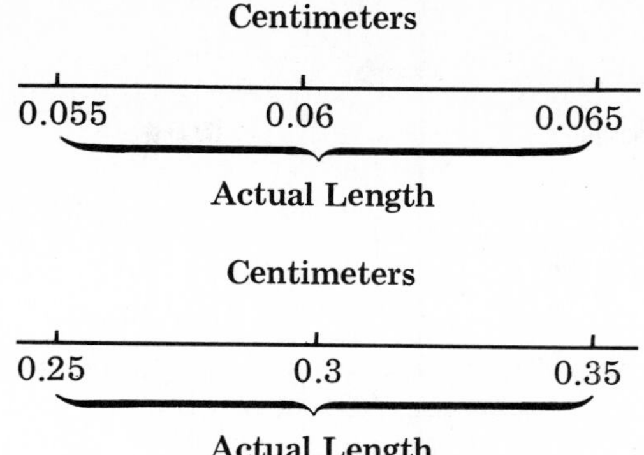

Adding the extreme values and the measured values:

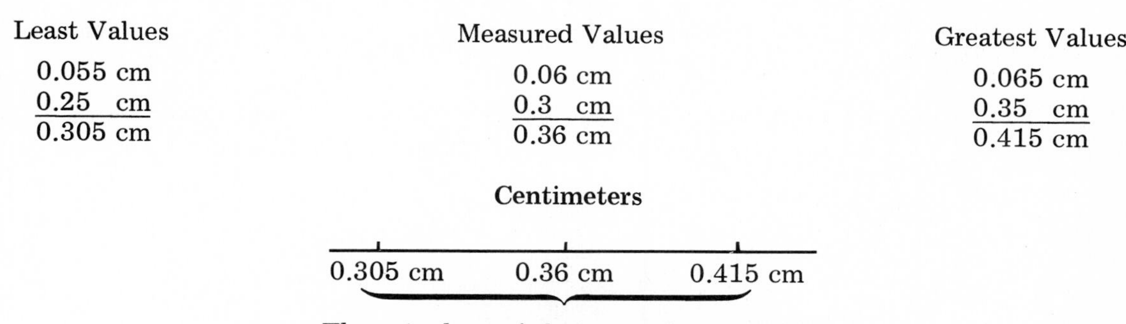

Least Values	Measured Values	Greatest Values
0.055 cm	0.06 cm	0.065 cm
0.25 cm	0.3 cm	0.35 cm
0.305 cm	0.36 cm	0.415 cm

The actual sum is between these values.

The difference between the extreme sums is 0.415 - 0.305 or 0.110 cm.

If 0.06 cm is rounded to tenths of a centimeter (the smallest unit in 0.3 cm, the least precise measure) before adding,

0.06 cm 0.3 cm	→	0.1 cm 0.3 cm 0.4 cm

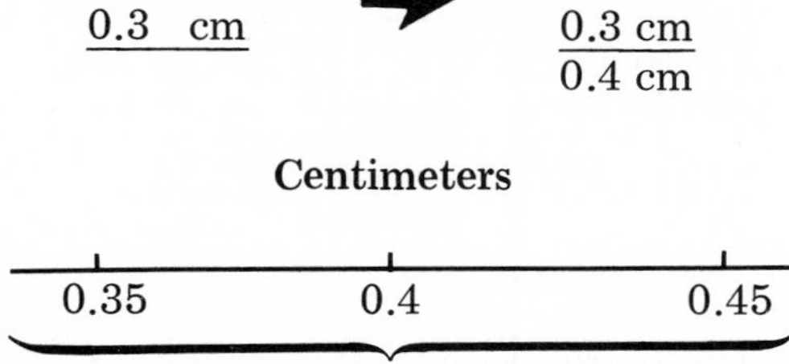

The actual sum is between these values.

The difference between these values is 0.1 cm. 0.4 cm is a better approximation of the actual sum since the difference between the extreme values is smaller.

Computations involving measurements: (18)

The following will be used to determine the number of significant digits to be used in the results of computations involving measurements.

1. When adding or subtracting measurements, round all measurements to the same precision as the *least* precise measurement before doing the computation.

$$\begin{array}{r} 3.14\text{ m} \\ +2.0\ \ \text{m} \\ \hline \end{array} \quad \rightarrow \quad \begin{array}{r} 3.1\text{ m} \\ +2.0\text{ m} \\ \hline 5.1\text{ m} \end{array}$$

$$\begin{array}{r} 3.14\text{ m} \\ -2.0\ \ \text{m} \\ \hline \end{array} \quad \rightarrow \quad \begin{array}{r} 3.1\text{ m} \\ -2.0\text{ m} \\ \hline 1.1\text{ m} \end{array}$$

2. When multiplying or dividing, *first* complete the computation with *all* significant digits in each measure; round the product or quotient to the number of digits as appear in the measurement with the *least* number of significant digits, the least accurate measurement.

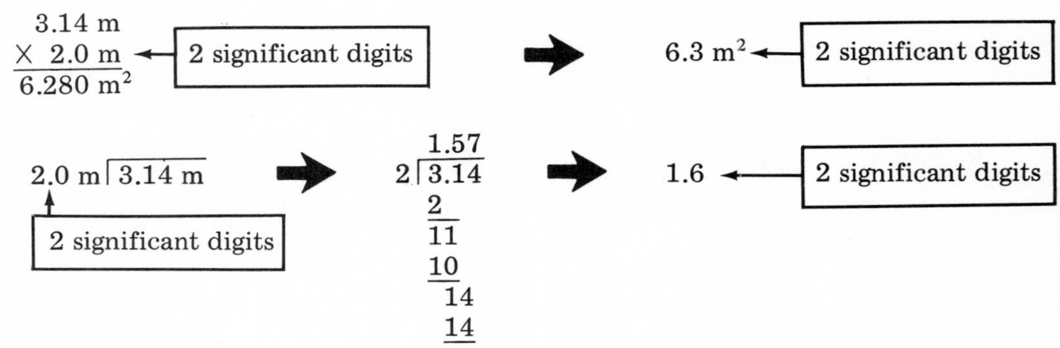

Computations involving measurements: (19)

Multiplying a measurement by an exact number: 3.14 m X 2

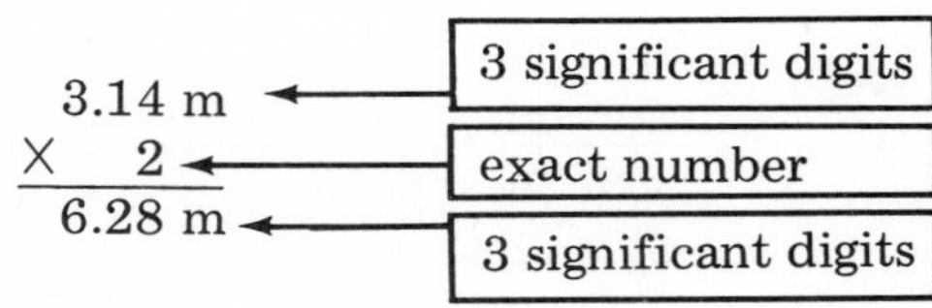

Dividing a measurement by an exact number: 3.14 m ÷ 2

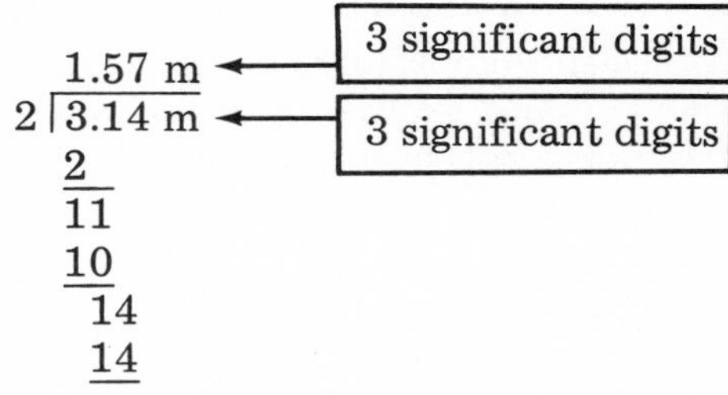

Since the concepts of precision and accuracy apply only to *approximate* numbers, the number of significant digits in the product and the quotient are determined *only* by the number of significant digits in the denominate numbers.

Computations involving measurements:

Complete the following by filling in the blanks:

A. Add 325 kilograms and 140 kilograms.

325 kg has ________ significant digits.

140 kg has ________ significant digits.
The sum should have two significant digits.

B. Multiply 0.003 m by 1.02 m.

0.003 m has ________ significant digits.

1.02 m has ________ significant digits.
The product should be rounded to one significant digit.

C. Divide 701 ml by 3.

701 ml has ________ significant digits.
3 is an ____________ number.
The quotient should be rounded to three significant digits.

Practice Exercises

Do *not* complete the following computations; write only the number of significant digits that should appear in each result if the operation were completed.

1. $7\underline{0}0$ g + 302 g
2. 0.32 l − 0.006 l
3. 101.0 km × $43\underline{0}$ km
4. 52.6 min ÷ 3

Areas and volumes of some rectangular figures were discussed in the last lesson. The areas and volumes of other figures need to be discussed since so many applications of measurement relate to other rectangular and non-rectangular figures. For example, consider the following problem:

Cara wishes to glue a plastic cover onto the seat of a circular kitchen stool. Will a 36-cm square piece of plastic be sufficient to cover the circular stool seat, if the seat measures 115 cm around the edge?

Two different geometric figures must be compared to answer this question: a *square* and a *circle.*

Solving measurement problems:

A *square* is a rectangular figure that has four equal *sides;* that is, the length and the width are equal.

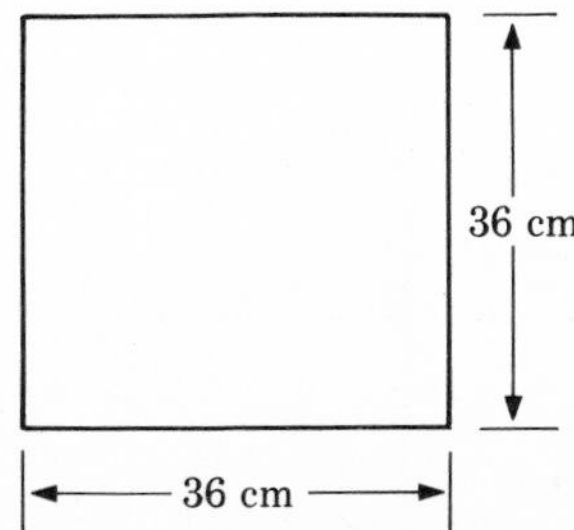

The outer dimensions of a square are measured in units of length.

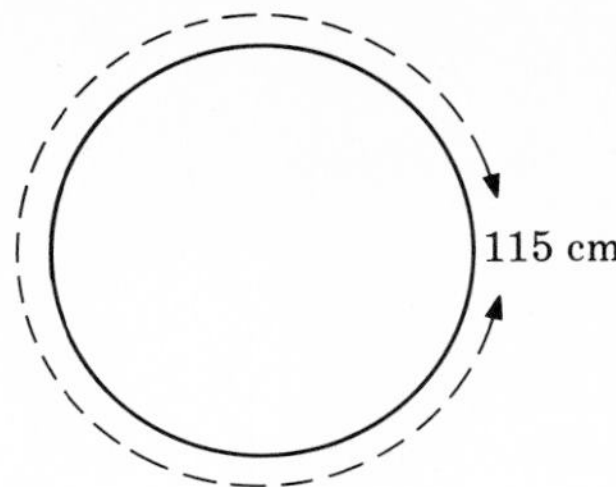

The distance around the edge of a circle, the *circumference*, is also measured in units of length.

Solving measurement problems: 

To solve the problem in Frame 22, it is necessary to determine the largest circular piece that can be cut from the square piece of plastic.

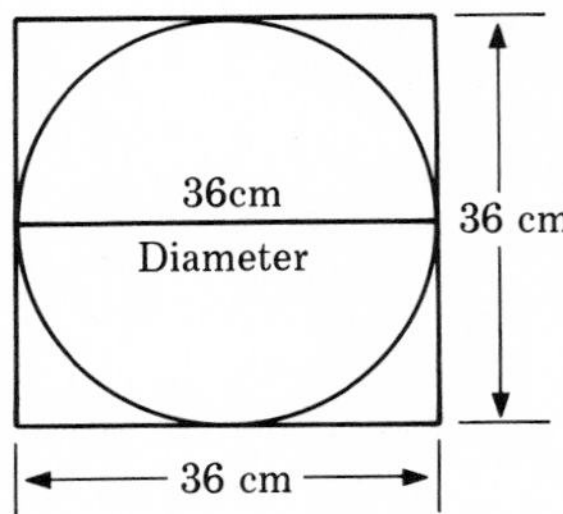

A straight line between the edges of the circle that passes through the center is called the *diameter* of the circle. The diameter is also measured in units of length.

The largest circle that can be cut from the piece of plastic has a diameter of 36 cm. To find the circumference of this circle, the following formula is used:

$$\text{Circumference} = \pi \times \text{Diameter}$$
$$C = \pi D$$

π, the Greek letter *pi*, is the constant ratio of the circumference to the diameter of the circle. A decimal approximation is used for π in computations.

$$\pi \approx 3.141\ 592\ 65$$

Generally, the approximation of π that is used depends on the least accurate measurement in a problem: If the least accurate measurement has 2 significant digits, the 3-digit approximation of π, 3.14, is used; if the least accurate measurement has 3 significant digits, the 4-digit approximation of π, 3.142, is used; and so on. (This and all other formulas are given in the appendix.)

Solving measurement problems: 

To answer the question in Frame 22, the circumferences of the stool top and the circular piece of plastic must be compared.

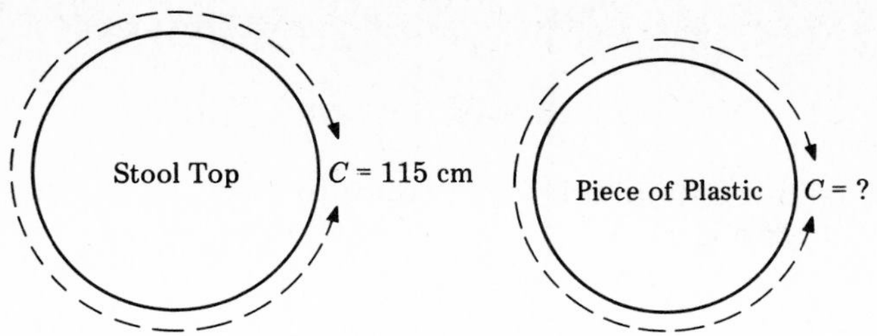

The circumference of the circular piece cut from the plastic is:

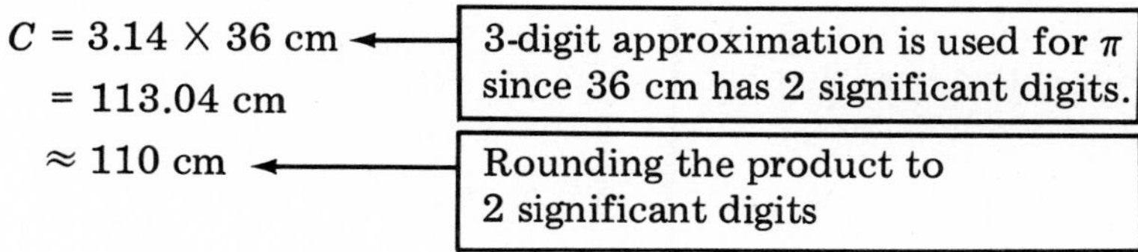

C = 3.14 × 36 cm
= 113.04 cm
≈ 110 cm
3-digit approximation is used for π since 36 cm has 2 significant digits.
Rounding the product to 2 significant digits

Thus, a 36-cm square piece of plastic is not large enough to cover the top of the stool. The dimensions of the smallest square piece of material that can be used to cover the stool must correspond to the diameter of the seat with a circumference of 115 cm.

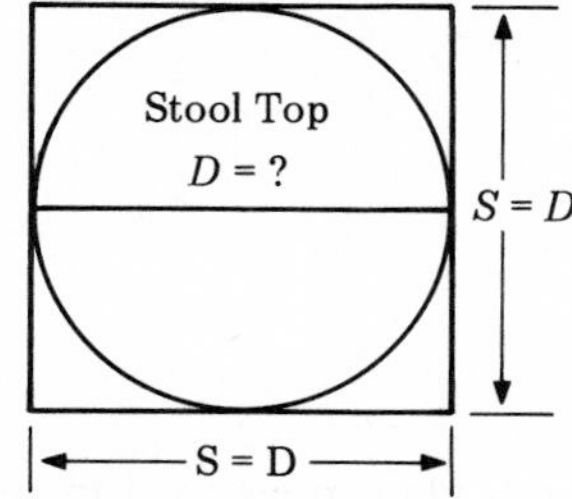

Using $C = \pi D$ and solving for D, we get

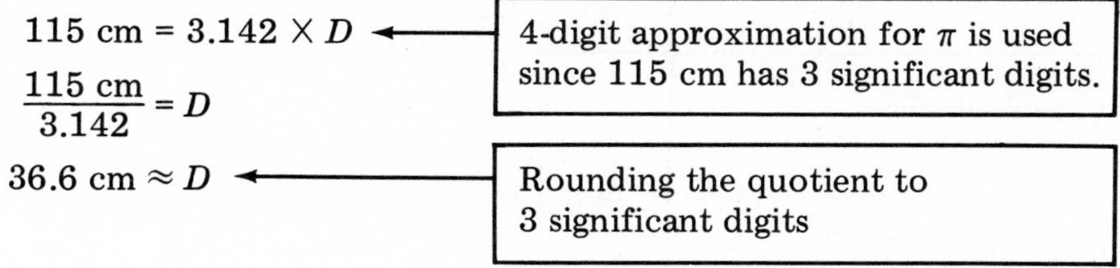

115 cm = 3.142 × D
115 cm / 3.142 = D
36.6 cm ≈ D
4-digit approximation for π is used since 115 cm has 3 significant digits.
Rounding the quotient to 3 significant digits

A circular piece that will cover the top of the stool can be cut from a 36.6-cm square piece of plastic.

Solving measurement problems:

A cyclindical storage tank 15 meters high is to be built with a base that has an inside diameter of 5.0 meters. What is the maximum capacity, in cubic meters, of this tank?

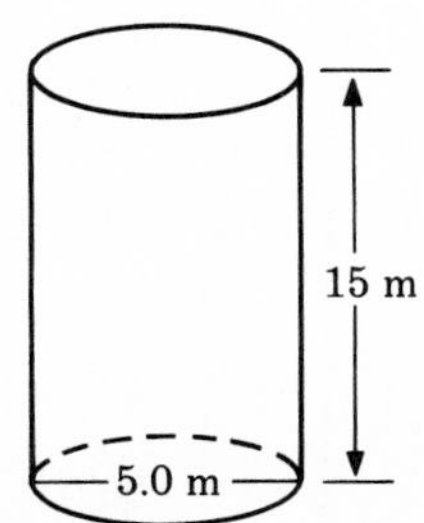

The capacity (or volume) of any container that has sides perpendicular to the bases:

Volume = Area of the base × Height

To determine the volume of this cylinder, it is necessary to first find the area of the circular base. The formula for the area, A, of a circle is

$$A = \pi r^2$$

where r, the *radius*, is $\frac{1}{2}$ the diameter of the circle; $r = \frac{1}{2}D$.

Thus, for this cylinder with a base that has a 5.0-meter diameter:

$$r = \frac{1}{2} \text{ of } 5.0 \text{ m or } 2.5 \text{ m}$$

And

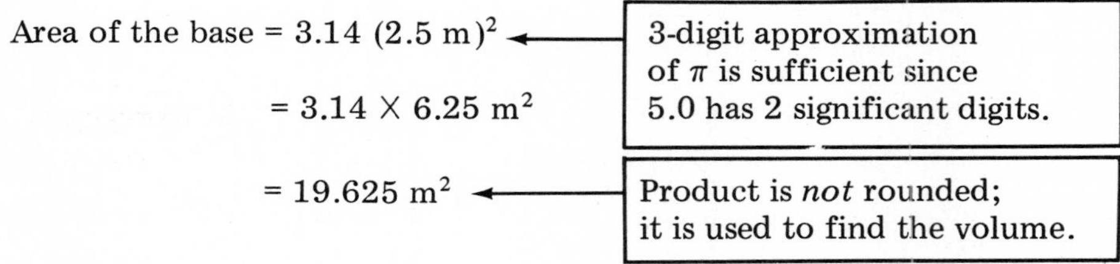

Area of the base = 3.14 (2.5 m)²
= 3.14 × 6.25 m²
= 19.625 m²
3-digit approximation of π is sufficient since 5.0 has 2 significant digits.
Product is not rounded; it is used to find the volume.

Then, the volume, V, is

$$V = 19.625 \text{ m}^2 \times 15 \text{ m}$$
$$= 294.375 \text{ m}^3$$
$$= 290 \text{ m}^3$$

Rounded to 2 significant digits

Solving measurement problems: 

Refer to the storage tank in Frame 26. What is the total inside surface area (in square meters) that will have to be painted once it is built?

The inside walls and the floor and ceiling of the tank will need to be painted. To determine the total surface area (the surface area of the floor and ceiling are known), it is necessary to determine the surface area of the wall and add it to the area of the floor and ceiling.

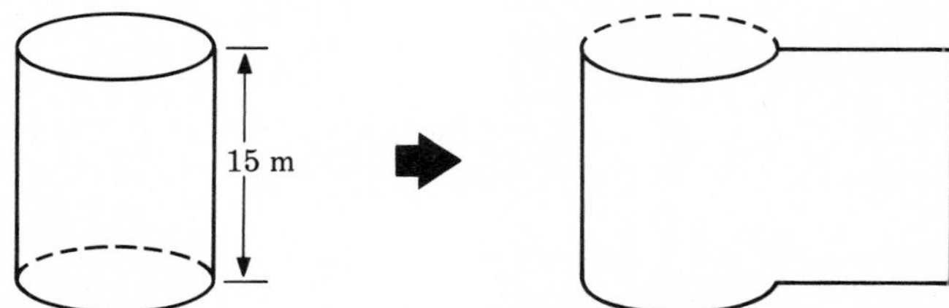

If the wall of the cylinder could be laid out flat, it would be rectangular.

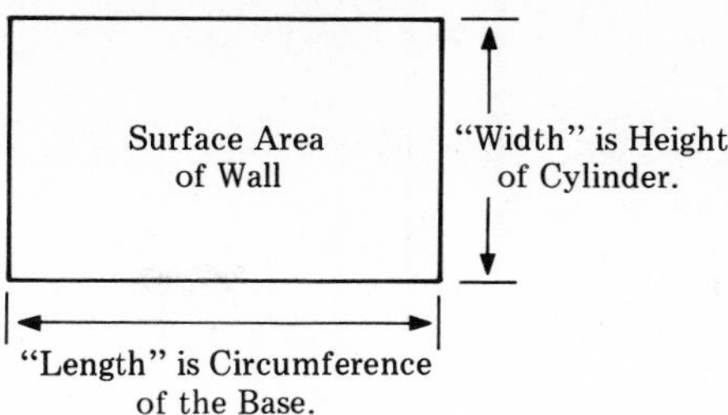

The circumference, C, of the base is

$$C = 3.14 \times 5.0 \text{ m}$$
$$= 15.7 \text{ m}$$

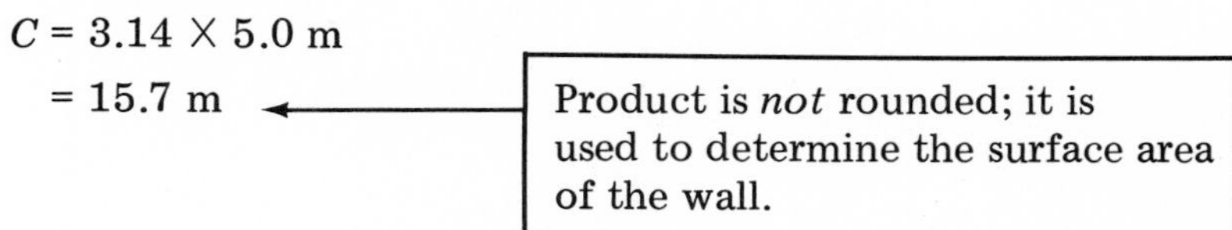

The surface area, A, of the wall is

$$A = 15.7 \text{ m} \times 15 \text{ m}$$
$$= 235.5 \text{ m}^2$$
$$\approx 240 \text{ m}^2$$

Rounded to 2 significant digits

Then, since the surface area of the floor or base is 19 m^2 (from Frame 26) and since the surface area of the ceiling is the same as that of the floor,

$$\text{Total surface area} = 240 \text{ m}^2 + 19 \text{ m}^2 + 19 \text{ m}^2$$
$$\approx 240 \text{ m}^2 + 20 \text{ m}^2 + 20 \text{ m}^2$$
$$\approx 280 \text{ m}^2$$

240 m^2 is precise to 10 square meters

Solving measurement problems:  28

Anne wants to make and frame a needlepoint piece to hang on the wall of her den. If the design she intends to use is a 24-centimeter square, what is the total amount of material (in square centimeters) she will cover with her needlepoint? How long a piece of wood that is 5.0 centimeters wide will she need for the frame?

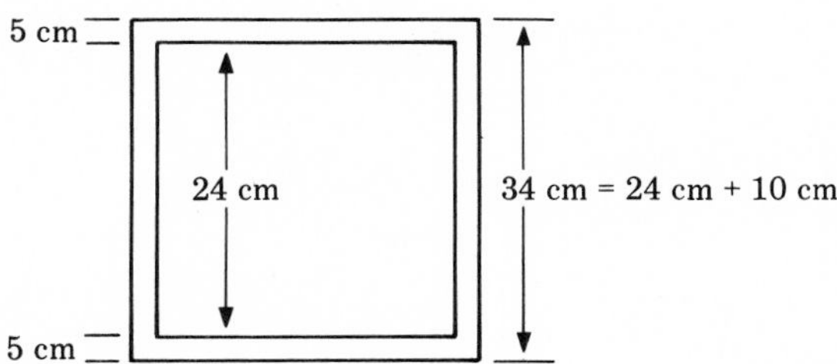

Two separate problems are involved: one involves a surface area; the other involves a *perimeter* or the distance around the outside of a straight-edged figure (the wood used to make the frame must be as long as the perimeter of the finished frame).

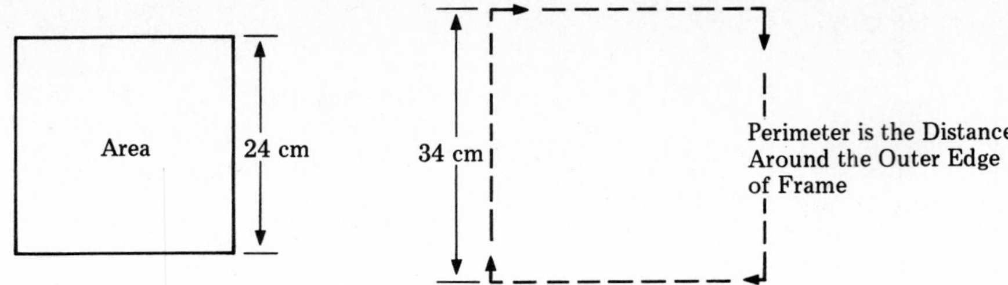

The area, A, of a square is

$$A = S^2 \text{ where } S \text{ is the length of a side.}$$

Thus,

$$\begin{aligned} A &= (24 \text{ cm})^2 \\ &= 24 \text{ cm} \cdot 24 \text{ cm} \\ &= 576 \text{ cm}^2 \\ &= 580 \text{ cm}^2 \end{aligned}$$

Rounded to 2 significant digits

To find the perimeter, P, it is only necessary to add the lengths of all the sides of the frame:

$$\begin{aligned} P &= 34 \text{ cm} + 34 \text{ cm} + 34 \text{ cm} + 34 \text{ cm} \\ &= 136 \text{ cm} \end{aligned}$$

Notice that perimeter, a measure of distance, has units of length and that the precision of the sum is the same as the precision of the given measurements.

Solving measurement problems:

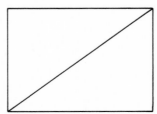

Paul found the following puzzle in a magazine: "Given the following rectangle,

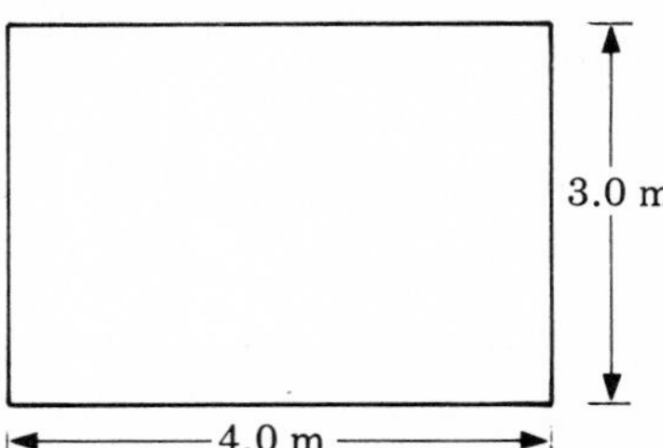

without measuring draw one straight line that divides the area of the rectangle into two equal parts."

After giving the puzzle some thought, Paul came up with the following solution:

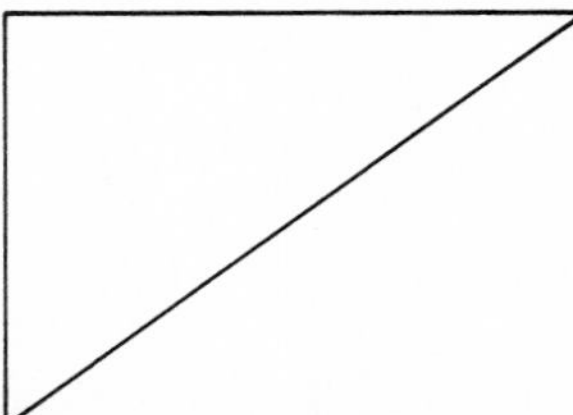

Is his solution correct?

The area of the given rectangle is

$$4.0 \text{ m} \times 3.0 \text{ m} = 12 \text{ m}^2$$

By drawing a diagonal line through the rectangle, two *triangles* (three-sided figures) are formed. The formula for the area of a triangle is

$$\text{Area} = \frac{1}{2} \text{ base} \times \text{height}$$

$$A = \frac{1}{2} BH$$

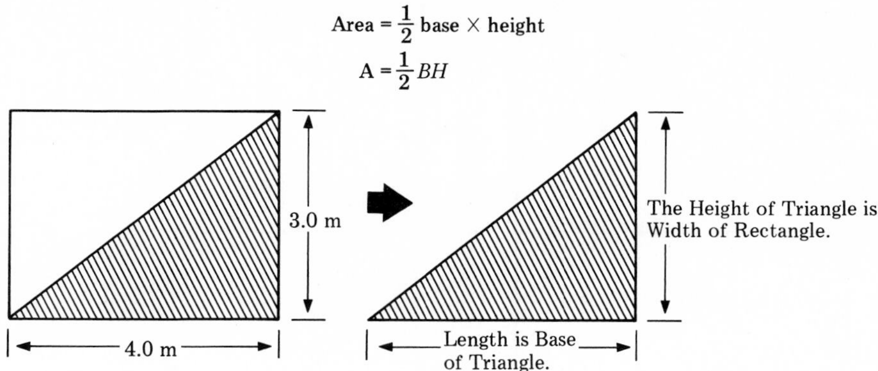

Thus,

$$\text{Area of one triangle} = \frac{1}{2}(4.0)(3.0)$$

$$= 6.0 \text{ m}^2$$

And, the area of both triangles is

$$6.0 \text{ m}^2 + 6.0 \text{ m}^2 = 12.0 \text{ m}^2$$

Paul's solution is correct. The diagonal line divides the area of the rectangle into two equal parts. The area of either triangle is exactly $\frac{1}{2}$ the area of the rectangle.

Solving measurement problems: (30)

Refer to the problem in Frame 29. What is the perimeter of the triangles formed by drawing a diagonal line through the rectangle?

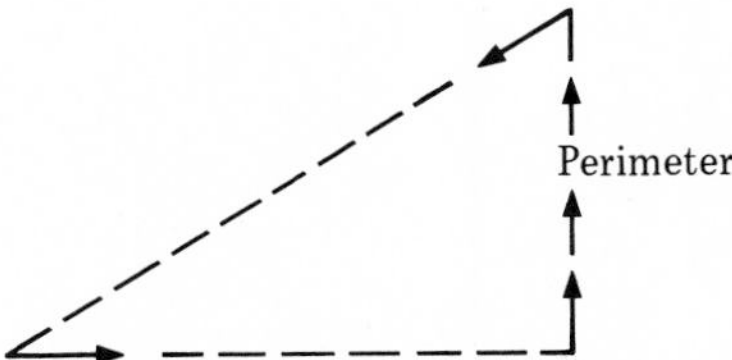

To find the perimeter, it is only necessary to add the lengths of the three sides. The lengths of two sides are known, but what is the length of the diagonal?

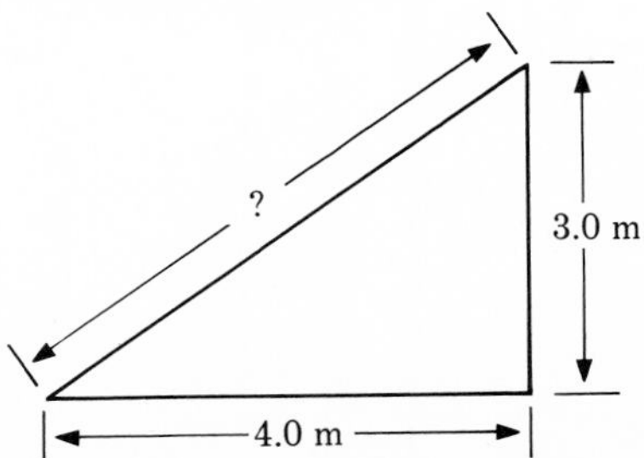

The Distance Formula (first given in Unit 4) is used to find the length of the diagonal:

$$d = \sqrt{(L^2 + W^2)}$$

$$\begin{aligned}\text{Length of the diagonal} &= \sqrt{([4.0\text{ m}]^2 + [3.0\text{ m}]^2)}\\ &= \sqrt{(16\text{ m}^2 + 9.0\text{ m}^2)}\\ &= \sqrt{25\text{ m}^2}\\ &= 5.0\text{ m}\end{aligned}$$

Notice that the square root of "m²" is "m."

Then, the perimeter of the triangle is 4.0 m + 3.0 m + 5.0 m = 12.0 m.

Solving measurement problems: (31)

What is the total surface area and the volume of the box shown below?

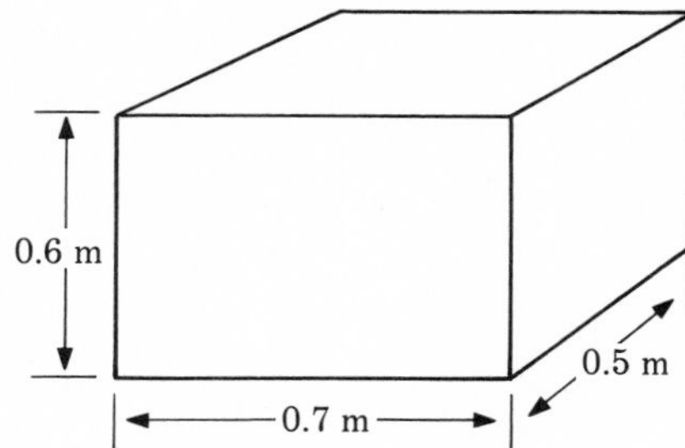

Complete the following:

To find the surface area, "open up" the box and "lay it flat."

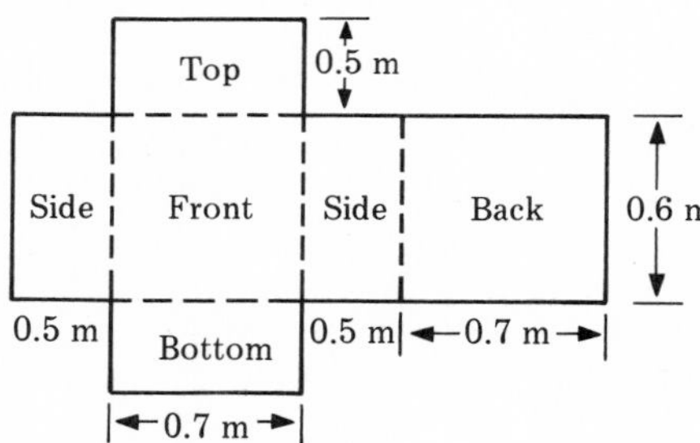

Since the top and the bottom are the same size, the total surface area can be determined by finding the areas of the following component parts:

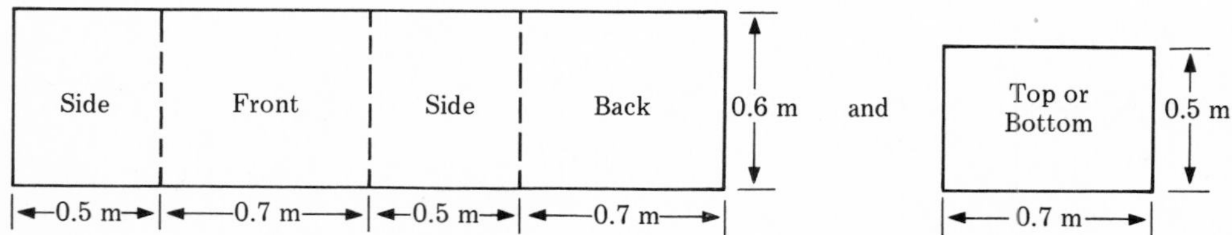

Thus, the area of the front, back, and sides is

$$2.4 \text{ m} \times 0.6 \text{ m} = 1.44 \text{ m}^2$$

$$\approx \bigcirc \text{ m}^2$$ ← Rounded to 1 significant digit

And the area of the top is

$$0.5 \text{ m} \times 0.7 \text{ m} = 0.35 \text{ m}^2$$

$$\approx \triangle \text{ m}^2$$ ← Rounded to 1 significant digit

The area of the top and bottom is

$$\triangle \text{ m}^2 + \triangle \text{ m}^2$$

Finally, the total surface area is

$$\bigcirc \text{ m}^2 + \triangle \text{ m}^2 + \triangle \text{ m}^2 = 1.8 \text{ m}^2$$

The total surface area is approximately 2 m^2.

To find the volume, the formula is $V = L \times W \times H$:

$$V = 0.7 \text{ m} \times 0.5 \text{ m} \times 0.6 \text{ m}$$

$$= 0.21 \ \square$$ ← Fill in correct unit of measure.

The volume is approximately 0.2 cubic meters.

Solving measurement problems: 

How much air does it take (in milliliters) to blow up a balloon until it has a diameter of 20 centimeters? (Assume that the shape of the balloon is spherical—the balloon is perfectly round.)

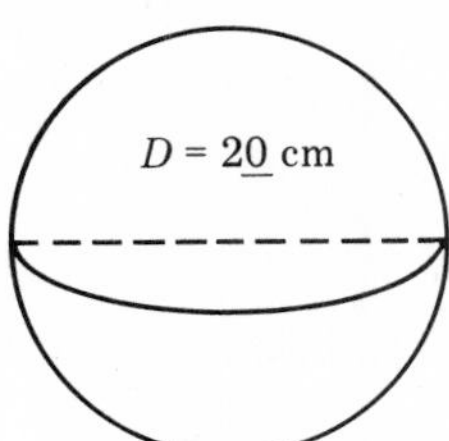

Complete the following:

The formula for the volume, V, of a sphere is

$$V = \frac{4}{3}\pi r^3$$

The radius, r, is $\frac{1}{2}$ the diameter:

$$r = \frac{1}{2}(2\underline{0}\text{ cm})$$

And $$= \bigcirc\text{ cm}$$

$$V = \frac{4}{3} \cdot 3.14 \cdot (\bigcirc\text{ cm})^3$$

$$= \frac{4}{3} \cdot 3.14 \cdot \square\text{ cm}^3$$

$$= \frac{4}{3}(3\ 140\text{ cm}^3) \longleftarrow \text{Using decimal-shift multiplication}$$

$$\approx ⬡\text{ cm}^3 \longleftarrow \text{Rounding to 2 significant digits}$$

Then, since 1 cm^3 = 1 ml, the volume in milliliters is approximately 4 200 ml.

Practice Exercises 

1. The outside wall of a house is approximately 12 meters high. How long must a ladder be (in meters) so that the top will reach exactly to the roof line when the base of the ladder is placed 5.0 meters from the wall and rested against the house?

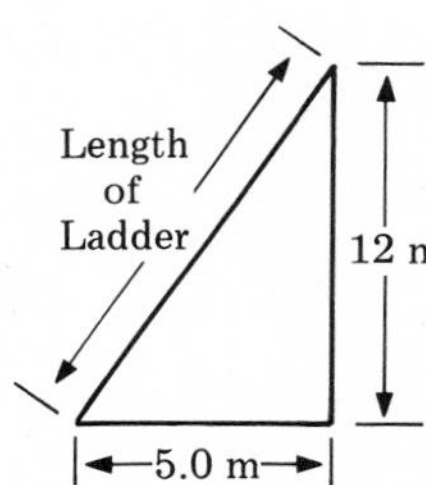

2. A circular roller-skating rink is to be built that has a diameter of 6$\underline{0}$ meters.
   (a) If the cost of the flooring material is \$6.50 per square meter, what will the total cost of the floor be to the nearest thousandth dollar?
   (b) If a railing is built around the entire edge of the rink, how many linear meters of material will be needed for a strip around the top of the railing?

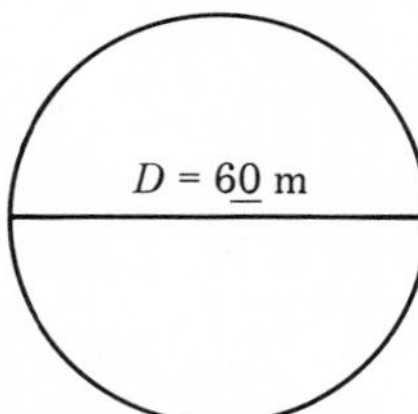

3. A pool that is 7.0 meters long, 3.0 meters wide, and 2.0 meters deep is to be drained and resurfaced.
   (a) How many square meters of area must be resurfaced?
   (b) How many cubic meters of water will it take to refill the pool?

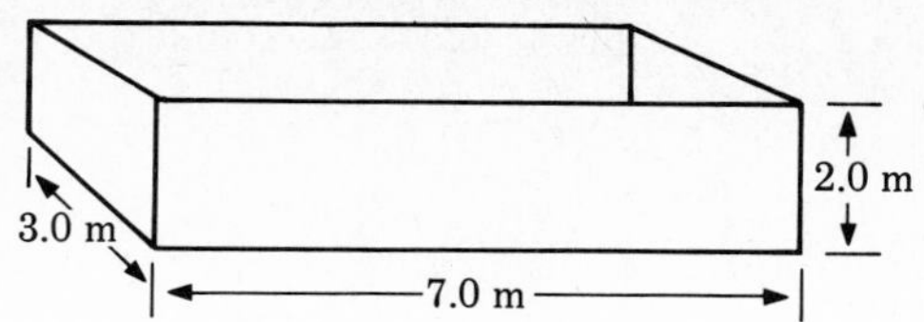

4. An irregular-shaped garden plot has the following dimensions:

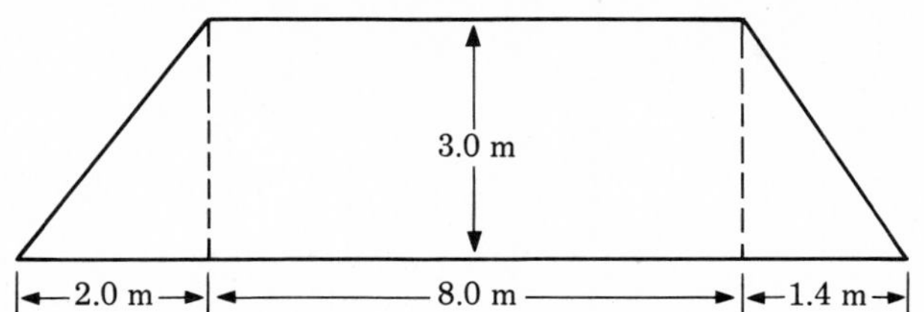

What is the total (in square meters) of the garden?

POST-TEST

1. What is the Greatest Possible Error in a measurement of 23 meters?
2. What is the Relative Error in a measurement of 2.3 centimeters, correct to the nearest hundredth.
3. What is the smallest unit of measure indicated by a measurement of 1.3 meters?
4. How many significant digits are there in a measurement of 0.00342 grams?
5. Find the sum or product of the following measurements, using precision or accuracy to round.

   (a) 4.53 cm
+5.1 cm

   (b) 7.5 m
×2.1 m

6. Find the circumference of a circle whose diameter is 2.03 meters, rounding the product according to the accuracy of the given dimension.

DRILL EXERCISES: UNIT 7

Convert the following by using dimension analysis:

1. 15 ft to inches
2. 12 ft to inches
3. 6 yd to inches
4. $\frac{2}{3}$ yd to inches
5. $\frac{2}{3}$ ft to inches
6. 7 pt to fluid ounces
7. $4\frac{1}{2}$ pt to fluid ounces
8. 3.5 pt to fluid ounces
9. 8 pt to fluid ounces
10. 1 gal to fluid ounces
11. 12 pt to quarts
12. $2\frac{1}{4}$ gal to quarts
13. $\frac{3}{4}$ gal to quarts
14. 4 yds to feet
15. $5\frac{2}{3}$ yd to feet
16. $\frac{3}{4}$ mi to feet
17. 1.5 mi to feet
18. $\frac{1}{3}$ yd to feet
19. 5 qt to pints
20. 3 gal to pints
21. 0.7 gal to pints
22. 5 lb to ounces
23. $3\frac{1}{2}$ lb to ounces
24. $6\frac{1}{2}$ yr to months
25. $\frac{1}{3}$ yr to months
26. 48 oz to pounds
27. 1.7 tons to pounds
28. 9 months to years
29. $2\frac{1}{2}$ yr to weeks
30. $2\frac{1}{5}$ yr to days
31. $3\frac{1}{2}$ weeks to days
32. $2\frac{1}{3}$ days to hours
33. 1,460 days to years
34. 30 months to years

Find each sum or difference and simplify.

35.
```
  4 ft 8 in.
 +3 ft 1 in.
```

36.
```
  3 lb 10 oz
  2 lb  5 oz
 +1 lb 15 oz
```

37.
```
   5 yd 1 ft 10 in.
  11 yd 2 ft  6 in.
 + 5 yd 1 ft  3 in.
```

38.
```
  8 hr 11 min 8 sec
 +4 hr 55 min 2 sec
```

39.
```
  8 ft 10 in.
 -4 ft  8 in.
```

40.
```
  5 mi 2,000 ft
 -     3,500 ft
```

41. 7 yds 1 ft 2 in.
−6 yds 2 ft 7 in.

42. 18 hr 35 min 5 sec
−11 hr 34 min 12 sec

Find each product or quotient and simplify.

43. (5 ft 4 in.) × 3

44. (6 lb 3 oz) × 5

45. (2 yds 2 ft 5 in.) × 2

46. (3 hr 18 min 8 sec) × 5

47. (15 ft 10 in.) ÷ 5

48. (30 mi 200 ft) ÷ 10

49. (6 gal 4 qt 1 pt) ÷ 3

50. (7 hr 11 min 20 sec) ÷ 2

51. Abbreviate the following units:
(a) deciliter (b) milliliter (c) hectogram (d) centimeter
(e) dekameter (f) centiliter (g) meter (h) kilogram

Convert the following:

52. 16 cm to millimeters

53. 3.5 dm to centimeters

54. 45 cm to decimeters

55. 8 m to centimeters

56. 4.07 m to millimeters

57. 8 dam to meters

58. 1.8 dam to kilometer

59. 11 hm to kilometer

60. 28 dg to centigrams

61. 865 mg to grams

62. 1,472 mg to grams

63. 8.74 hg to dekagrams

64. 52 l to centiliters

65. 6,827 ml to liters

66. 97 l to kiloliters

67. 5 l to dekaliters

Convert the following. (Round answers to the nearest hundredth of a unit wherever appropriate.)

68. 4 liquid quarts to liters

69. 10.1 liquid quarts to liters

70. 15 liters to liquid quarts

71. 6.3 liters to liquid quarts

72. 4 in. to centimeters

73. 4.2 in. to centimeters

74. 40 cm to inches

75. 3.7 cm to inches

76. 2 kg to pounds

77. 30.2 kg to pounds

78. 2.2 lbs to kilograms

79. $1\frac{1}{2}$ lbs to kilogram

80. 5.4 mi to kilometers

81. $1\frac{1}{2}$ mi to kilometers

82. 40 km to miles

83. 1,000 km to miles

84. 8 m to yards

85. 1,321 m to yards

86. 7 ft to meters

87. 3,000 yards to meters

88. 5 m to feet

89. 76.3 m to feet

90. 11.5 ft to meters

91. 100 ft to meters

92. 98.6°F to °C

93. 102°F to °C

94. 40°C to °F

95. 41°C to °F

In exercises 96–106, round answers to the nearest hundredth of a unit whenever necessary.

96. Find the height in meters of a man who is 6 feet tall.

97. A popular event in the Olympic games is the 100-meter dash. Find the distance in yards.

98. If a road sign indicated 193 km to San Diego, how far is this in miles?

99. If an artillery shell weighs 80 kg, what is its weight in pounds?

100. If an automobile weighs 4,800 pounds, determine its weight in kilograms.

101. If a female model has measurements of 36 in., 24 in., and 36 in., determine her measurements in centimeters.

102. How much would 2 kilograms of bacon cost at $1.50 a pound?

103. How many liters is a $\frac{1}{2}$ gallon of milk?

104. If milk costs $1.59 a gallon, what is the price of 1 liter of milk?

105. Determine the weight of a can of corn in ounces if it weighs 230 grams.

106. If a European auto averages 10 km per liter of gasoline and an American model averages 16 miles per gallon, which auto gets the better "mileage"?

In exercises 107–115, state the number of significant figures in each measurement.

107. 0.002 in. 108. 0.020 in. 109. 12.1 in.

110. 20.03 in. 111. 28.5 lbs 112. 187,000 m

113. 30.0 ft 114. 4 57$\underline{0}$ m 115. 32,$\underline{0}$00 ft

In exercises 116–119, for each of the pairs, which measurement has the greater precision? Which measurement has the greater accuracy?

116. 14.8 pounds; 14.80 pounds 117. 43,25$\underline{0}$ ft; 0.008 in.

118. 0.0304 grams; 750 grams 119. 8.0 mm; 0.080 mm

In exercises 120–131, perform the indicated operations on the given numbers and measurements, according to the rules specified in Lesson 28.

120. $$\begin{array}{rl} 4.7 & \text{m} \\ 3.83 & \text{m} \\ \underline{+21.000} & \text{m} \end{array}$$

121. $$\begin{array}{rl} 20.83 & \text{in.} \\ 6.783 & \text{in.} \\ \underline{+14.85} & \text{in.} \end{array}$$

122. $$\begin{array}{rl} 6.049 & \text{cm} \\ \underline{-3.5} & \text{cm} \end{array}$$

123. $$\begin{array}{rl} 5.08 & \text{ft} \\ \underline{-3.082} & \text{ft} \end{array}$$

124. $$\begin{array}{r} 16.3 \text{ ft} \\ \underline{\times\ 2.4 \text{ ft}} \end{array}$$

125. $$\begin{array}{r} 4.6 \text{ cm} \\ \underline{\times 4.1} \text{ cm} \end{array}$$

126. $$\begin{array}{l} 5.1 \text{ m} \\ \underline{\times\pi} \end{array}$$

127. $$\begin{array}{l} 20.0 \text{ in.} \\ \underline{\times\ \pi\qquad} \end{array}$$

128. 7 in. $\div\ \pi$ 129. 23 mm $\div\ \pi$

130. 448 g $\div$ 3.8 131. 252 miles $\div$ 2

In the following exercises, perform the computations with measurements according to the rules specified in Lesson 28.

132. Find the area of a rectangle whose length is 10.4 meters and whose width is 4.7 meters.

133. Find the volume in cubic meters of a room 6 meters long, 5 meters wide, and 3 meters high.

134. If a man weighs 153.0 lbs and picks up an orange which weighs 0.27 lbs, what is the total weight of the man and the orange?

135. If the man in exercise 134 weighs 153.00 lbs, what is the combined weight?

136. Find the circumference of each of the following circles:
(a) diameter = 2.5 ft
(b) diameter = 25.1 cm
(c) diameter = 100 m
(d) radius = 4.78 dm

137. Find the area of a circle whose radius is 7.8 meters.

138. Find the diameter of a circle whose circumference is 16.8 meters.

139. Find the volume of a box which measures 2.4 meters long, 1.3 meters wide, and 1.6 meters deep.

140. Find the surface area of the box in exercise 139.

141. The volume of a sphere is given by the following formula:

$$V = \frac{4}{3} \pi r^3$$

Find the volume of a sphere that has a radius of 2.0 meters.

142. A can of corn measures 12 cm high and 7 cm in diameter. How many square centimeters of paper are required for the label to cover the surface area of the wall of the can? (Refer to Frame 27 in Lesson 28.)

143. A cylindrical storage tank has a diameter of 8 meters and a height of 12 meters. What is the capacity in cubic meters? (Refer to Frame 26 in Lesson 28.)

144. If Bill is standing $4\underline{0}$ meters from a building which is $3\underline{0}$ meters high and if Bill is holding a rope attached to the top of the building, how long is the rope?

1. Jackie weighs 63 kilograms. Compute her weight in pounds correct to the nearest pound.
   (a) 138 pounds
   (b) 128 pounds
   (c) 139 pounds
   (d) 129 pounds
   (e) None of the above
2. Convert 0.72 meters to millimeters.
   (a) 0.00072 millimeters
   (b) 0.000072 millimeters
   (c) 720 millimeters
   (d) 7 200 millimeters
   (e) None of the above
3. Convert 27.83 milliliters to liters.
   (a) 2,783 liters
   (b) 0.2783 liter
   (c) 0.02783 liter
   (d) 0.27830 liter
   (e) None of the above
4. Convert 62.708 centigrams to hectograms.
   (a) 0.062708 hectograms
   (b) 62 708 hectograms
   (c) 627 080 hectograms
   (d) 0.62708 hectograms
   (e) None of the above
5. Find the surface area of a square box 3.0 meters long, 3.0 meters wide, and 3.0 meters high.
   (a) 36 square meters
   (b) 9 square meters
   (c) 54 square meters
   (d) 27 square meters
   (e) None of the above
6. Determine the circumference of a circle whose diameter is 8.21 centimeters, rounding the result according to the accuracy of the given measurement.
   (a) 25.8 centimeters
   (b) 25.787 centimeters
   (c) 25.80 centimeters
   (d) 24 centimeters
   (e) None of the above
7. A chemist mixes 0.38 grams of water, 1.85 grams of alcohol, and 0.367 grams of iodine together. How much does the resulting mixture weigh? (The precision of the given measurements will determine the precision of the results.)
   (a) 2.597 grams
   (b) 2.6 grams
   (c) 260 grams
   (d) 2.60 grams
   (e) None of the above

8. A measurement of 4 180.35 meters has how many significant digits?
   (a) 6
   (b) 2
   (c) 3
   (d) 0
   (e) None of the above

9. Find the Greatest Possible Error given the following measurement: 52.36 liters.
   (a) 0.06 liter
   (b) 0.01 liter
   (c) 0.001 liter
   (d) 0.005 liter
   (e) None of the above

10. Determine the Percent Error of a measurement of 0.03 grams.
   (a) 3%
   (b) ≈16.7%
   (c) ≈0.005%
   (d) 0.05
   (e) None of the above

11. Which of the following measurements has the greatest precision?
   (a) 3.0 meters
   (b) 30 meters
   (c) 0.003 meter
   (d) 30.03 meters
   (e) $30\underline{0}$ meters

12. Find the product of the measures 8.3 dm and 72.5 dm, correct to the nearest 10 dm^2.
   (a) 601.8 dm^2
   (b) 700 dm^2
   (c) 601.75 dm
   (d) $6\underline{0}0$ dm^2
   (e) None of the above

13. What is the area in square meters of a rectangular plot of ground measuring 62.1 feet by 20 feet, correct to the nearest square meter.
   (a) 1 242 square meters
   (b) 164 square meters
   (c) 1240 square meters
   (d) 82 square meters
   (e) None of the above

14. A classroom with the dimensions 25 m by $3\underline{0}$ m by $1\underline{0}$ m contains how many cubic meters of air? (Use the accuracy of the given measurements to correctly round the product.)
   (a) 65 m^3
   (b) 7 500 m^3
   (c) 1 300 m^3
   (d) 260 m^3
   (e) None of the above

15. A measurement of 43.03 meters is precise to the nearest
   (a) Meter
   (b) Centimeter
   (c) Deciliter
   (d) Kilometer
   (e) None of the above

16. Convert 15 yards to inches.

(a) $\frac{5}{12}$ inches
(b) 180 inches
(c) $\frac{5}{12}$ inches
(d) 540 inches
(e) None of the above

17. Given the formulas: $F = \frac{9}{5}C + 32$ and $C = \frac{5}{9}(F - 32)$. Convert 90°F to °C.

(a) 194°C
(b) 18°C
(c) 90°C
(d) $32\frac{2}{9}$°C
(e) None of the above

18. Find the sum and simplify the following:

	3 days	14 hours	47 minutes
+	2 days	18 hours	50 minutes

(a)	5 days	33 hours	37 minutes
(b)	6 days	8 hours	97 minutes
(c)	5 days	9 hours	37 minutes
(d)	6 days	9 hours	37 minutes
(e)	None of the above		

19. Find the following difference:

	5 quarts	3 pints	2 cups
−	2 quarts	3 pints	3 cups

(a)	2 quarts	0 pints	0 cups
(b)	2 quarts	0 pints	9 cups
(c)	2 quarts	1 pint	0 cups
(d)	2 quarts	1 pint	1 cup
(e)	None of the above		

20. Convert 8,734.2017 centigrams to dekagrams.

(a) 8.7342017 dekagrams
(b) 8 734 201.7 dekagrams
(c) 87.342017 dekagrams
(d) 87 342 017 dekagrams
(e) None of the above

LESSON 25:

Pre-test

1. 10,800 seconds

2. (a) 8 pints 1 cup 3 ounces
   (b) 2 pints 1 cup 5 ounces
   (c) $2\frac{3}{16}$ feet
   (d) $14\frac{1}{2}$ ounces

Practice Exercises

Frame 17

1. 48 pints
2. 5,800 pounds
3. $2\frac{5}{8}$ feet or 2.625 feet
4. $2\frac{7}{8}$ pounds or 2.875 pounds

Frame 25

1. 9 feet 3 inches
2. 12 weeks 5 days 16 hours
3. 1 gallon 2 quarts
4. 5 minutes 48 seconds

Frame 32

1. 168 gallons 3 quarts
2. 31 yards 1 foot 4 inches
3. 3 quarts
4. 4 hours 37 minutes 22 seconds

Post-test

1. 10,080 minutes

2. (a) 13 yards 2 feet 5 inches
   (b) 1 yard 2 feet 11 inches
   (c) 1.86 inches
   (d) 1 day 17 hours

LESSON 26:

Pre-test

1. (a) 1,000 meters (b) cg
2. 0.278 l
3. 75 miles
4. 6.65 l

Practice Exercises

Frame 18

1. 100
2. ml
3. kilometers
4. kilogram
5. 1
6. liters
7. hg
8. dekameters

Frame 31

1. 34.2 dm
2. 13 700 g
3. 0.53 l
4. 62 000 mm
5. 0.082 hg
6. 1 230 dal

Frame 34

1. 77.47 cm
2. 0.2 qt
3. 1 174.57 km
4. 0.8 oz
5. 0.285 l
6. 11 yd

Post-test

1. (a) 1 (b) km
2. 24 000 ml
3. 8 lbs
4. 14 m

LESSON 27:

Pre-test

1. 1 535 cm^2
2. 3 cans
3. 0.003 m^3
4. $150
5. (a) 22.2°C (b) 98.6°F

Practice Exercises

Frame 17

1. 755 cm^2
2. 3 cans
3. 200 cm^2
4. 394 ha

Frame 28

1. 0.105 m^3
2. 0.432 m^3
3. 42 bags
4. 54 kg

Frame 33

1. Swimming, 86°F
2. Overdone, 518°F
3. No, 185°F

Post-test

1. 3 m^2
2. 5 panels
3. 216 ml
4. $0.21
5. (a) 212°F (b) 0°C

LESSON 28:

Pre-test

1. 0.05 cm
2. 0.0004
3. 1 meter
4. Two
5. (a) 9.6 cm (b) 4 cm^2
6. 1 m^3

Practice Exercises

Frame 11

1. GPE = $\frac{1}{2}$ cm, Relative Error = 0.25, Percent Error = 25%
2. GPE = $\frac{1}{20}$ cm, Relative Error = 0.025, Percent Error = 2.5%
3. GPE = $\frac{1}{2}$ g, Relative Error = 0.000 125, Percent Error = 0.0125%
4. GPE = $\frac{1}{2}$ g, Relative Error = 0.125, Percent Error = 12.5%
5. GPE = $\frac{1}{2}$ sec, Relative Error = 0.1, Percent Error = 10%
6. GPE = $\frac{1}{20}$ sec, Relative Error = 0.01, Percent Error = 1%

Frame 16

1. 2; 100 kg 2. 3; 10 kg 3. 1; 0.01 sec 4. 2; 0.01 sec
5. 4; 0.01 m 6. 4; 0.01 m 7. 2; 1 l 8. 3; 0.1 l

Frame 21

1. 2 2. 2 3. 3 4. 3

Frame 33

1. 13 m 2. (a) $18,200 (b) 190 m
3. (a) 61 m^2 (b) 42 m^3 4. 29 m^2

Post-test

1. 0.5 meter 2. 0.02 3. 0.1 meter or 1 decimeter
4. 3 5. (a) 9.6 cm (b) 16 m^2 6. 6.38 m

Answers to Drill Exercises: Unit 7

1. 180 in.
2. 144 in.
3. 216 in.
4. 24 in.
5. 8 in.
6. 112 fl oz
7. 72 fl oz
8. 56 fl oz
9. 128 fl oz
10. 128 fl oz
11. 6 qt
12. 9 qt
13. 3 qt
14. 12 ft
15. 17 ft
16. 3,960 ft
17. 7,920 ft
18. 1 ft
19. 10 pt
20. 24 pt
21. 5.6 pt
22. 80 oz
23. 56 oz
24. 78 mo
25. 4 months
26. 3 lb
27. 3,400 lb
28. $\frac{3}{4}$ yr
29. 130 weeks
30. 803 days
31. $24\frac{1}{2}$ days
32. 56 hr
33. 4 yr
34. $2\frac{1}{2}$ yr
35. 2 yds 1 ft 9 in.
36. 7 lb 14 oz
37. 22 yd 2 ft 7 in.
38. 13 hr 6 min 10 sec
39. 4 ft 2 in.
40. 4 mi 3,780 ft
41. 1 ft 7 in.
42. 7 hr 53 sec
43. 16 ft
44. 30 lb 15 oz
45. 5 yd 1 ft 10 in.
46. 16 hr 30 min 40 sec
47. 3 ft 2 in.
48. 3 mi 20 ft
49. 2 gal 1 qt 1 pt
50. 3 hr 35 min 40 sec
51. (a) dl (b) ml (c) hg (d) cm
    (e) dam (f) cl (g) m (h) kg
52. 160 mm
53. 35 cm
54. 4.5 dm
55. 800 cm
56. 4 070 mm
57. 80 m
58. 0.018 km
59. 1.1 km
60. 280 cm
61. 0.865 g
62. 1.472 g
63. 87.4 dag
64. 5 200 cl
65. 6.827 l
66. 0.097 kl
67. 0.5 dal
68. 3.8 l
69. 9.6 l
70. 15.79 qt
71. 6.63 qt
72. 10.16 cm
73. 10.67 cm
74. 15.75 in.
75. 1.46 in.
76. 4.41 lb
77. 66.58 lb
78. 1 kg
79. 0.68 kg
80. 8.69 km
81. 2.41 km
82. 24.86 mi
83. 621.50 mi
84. 8.75 yd
85. 1,444.66 yd
86. 2.13 m
87. 2 743.20 m
88. 16.4 ft
89. 250.33 ft
90. 3.51 m
91. 30.48 m
92. 37°C
93. 38.89°C
94. 104°F
95. 105.8°F
96. 1.83 m
97. 109.36 yd
98. 119.95 mi
99. 176.37 lb
100. 2 177.28 kg
101. 91.44 cm, 60.96 cm, 91.44 cm
102. $6.61
103. 1.9 l
104. $0.42
105. 8.11 oz
106. European
107. One
108. Two
109. Three
110. Four
111. Three
112. Three
113. Three
114. Four
115. Three
116. Greater precision, 14.80 lb; greater accuracy, 14.80
117. Greater precision, 0.008 in.; greater accuracy, 43,25$\underline{0}$ ft
118. Greater precision, 0.0304 g; greater accuracy, 0.0304 g
119. Greater precision, 0.080 mm; equally accurate.
120. 29.5 m
121. 42.46 in.
122. 2.5 cm
123. 2.00 ft
124. 39 ft^2
125. 19 cm^2
126. 16 m
127. 62.8 in.
128. 2 in.
129. 7.3 mm
130. 118 g
131. 126 mi
132. 49 m^2
133. 90 m^3
134. 153.3 lb
135. 153.27 lb

136. (a) 7.9 ft (b) 78.9 cm (c) 300 m (d) 30.0 dm
137. 190 m^2 138. 5.35 m 139. 5.0 m^3 140. 18 m^2
141. 33 m^3 142. 300 cm^2 143. 600 m^3 144. $5\underline{0}$ m

Answers to Self Test: Unit 7

1.	(c)	2.	(c)	3.	(c)	4.	(e)	5.	(c)	6.	(a)	7.	(d)
8.	(a)	9.	(d)	10.	(b)	11.	(c)	12.	(d)	13.	(e)	14.	(b)
15.	(b)	16.	(d)	17.	(d)	18.	(d)	19.	(d)	20.	(a)		

UNIT EIGHT

Using Numerical Information

Lesson 29 DESCRIPTIVE STATISTICS

Lesson 30 GRAPHS AND TABLES

LESSON 29
DESCRIPTIVE STATISTICS

Objectives:

1. Determine the range, mode (or modes, if any), mean, median, and standard deviation for a distribution of numerical values. (1-12)
2. Graph the frequency polygon for a distribution of numerical values. (10-19)
3. Graph the histogram for a distribution of numerical values. (20-25)

Vocabulary:

Statistics
Descriptive Statistics
Tally
Distribution
Frequency Distribution
Frequency
Mode
Range
Cumulative Frequency
Median
Mean
Average
Frequency Polygon
Standard Deviation
Deviation
Histogram

PRE-TEST

1. The number of people living in each of the 20 houses on a particular street are as follows: 2, 3, 1, 3, 5, 2, 2, 4, 4, 6, 5, 2, 4, 1, 7, 3, 3, 2, 4, 5. Determine each of the following correct to the nearest whole number:
   (a) Range
   (b) Mode (or modes, if any)
   (c) Median
   (d) Mean
   (e) Standard deviation

Number of People	*Tally*	*Frequency*	*Cumulative Frequency*	*Frequency × Number*	*Deviation*	*Deviation Squared*	*Frequency × Deviation Squared*
7							
6							
5							
4							
3							
2							
1							

2. Graph the frequency polygon and histogram for the distribution given in Problem 1.

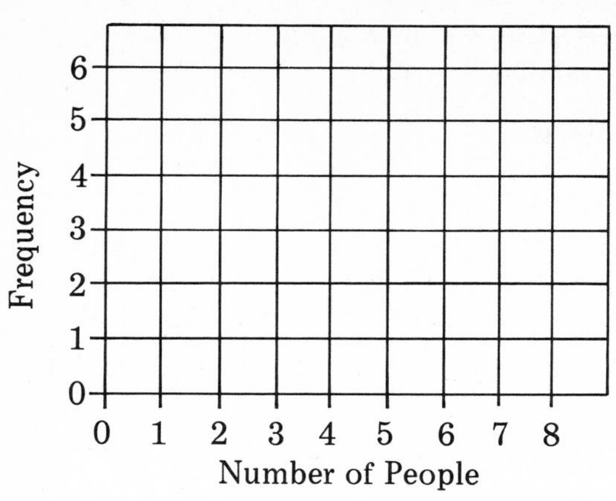

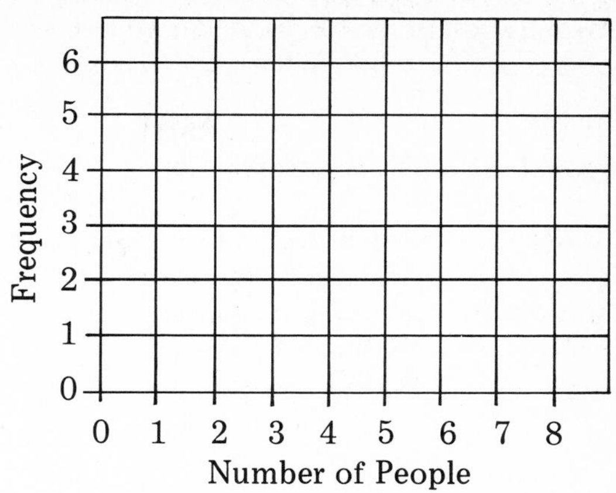

Many of the "real life" problems that a person encounters have no exact answers easily determined by arithmetic computations alone. Most "real" problems have no ready answers; decisions must be made on the basis of the best available information. Some questions have no exact answers: How many people will be alive in the year 2000? Who will be the next president? What will the consumer want in the way of home entertainment 10 years from now? (1)

These are only a few examples of questions that have no exact answers; they can only be answered with an opinion or by an educated guess. But all these questions can be dealt with knowledgeably; they can all be answered reasonably by using the methods of the branch of mathematics called *statistics.*

For most people, the word "statistics" calls to mind endless lists of numbers. These people are certain that they could never hope to, nor would they care to, learn anything about statistics because of this notion. But statistics, instead of adding to the confusion, is the branch of mathematics that can bring order and meaning to the endless masses of numerical data generated in any complex society.

Why study statistics? Even if the average person will never have a reason to do a statistical analysis in order to solve a problem, it is impossible for that person to spend one day of his life without encountering statistics or being affected by the results of statistical analysis. Descriptive statistics are used in such diverse areas as baseball batting averages, public opinion polls, and advertising. Inferential statistics, statistics used to make decisions about the best alternative available, are used in medical research, in determining the effect of population growth on the ecology, to analyze the results of interest and aptitude tests, and in determining where and when a community will need a new school, for example. In order to be an intelligent voter, homeowner, consumer, parent, sports fan, or informed reader, everyone should know something about statistics and statistical methods.

This lesson deals with *descriptive statistics.* The purpose of descriptive statistics is to provide a way for extracting meaningful information from "endless lists" of numbers or countable, related facts. 

For example, suppose that the following is a list of the test scores from a midterm exam given in a history class:

90, 100, 75, 85, 95, 40, 95, 80, 90, 45, 35, 85, 80, 100, 85
95, 60, 90, 90, 50, 85, 95, 45, 55, 50, 95, 85, 30, 65, 50

If Jack has a score of 75 as his test result, is it "good" or "bad"? That is, how does his score compare with all the other test scores for the students in his class? Is he above or below the class "average"? Before these questions can be answered, before something can be said about the "goodness" of Jack's score, it will be necessary to somehow organize and summarize the list of test scores.

Organizing numerical information: (3)

The first thing that is done to organize the list of test scores is to do a *tally*, a simple counting of the number of times that each score appears in the list. The information from the tally is then entered in a table called a *frequency distribution.* In statistics, a *distribution* is any list of related numerical information. A *frequency distribution* shows the number of times that each of the values, in this case test scores, appears in the original list; that is, it shows the *frequency* of each different score.

Test Scores	*Tally*
100	||
95	~~||||~~
90	||||
85	~~||||~~
80	||
75	|
70	
65	|
60	|
55	|
50	|||
45	||
40	|
35	|
30	|

→

Test Scores	*Frequency*
100	2
95	5
90	4
85	5
80	2
75	1
70	0
65	1
60	1
55	1
50	3
45	2
40	1
35	1
30	1
	30

Notice that the tally and the frequency distribution both have 15 categories, the scores from 30 to 100, in numerical order entered in the first columns. The frequency of each score, the entry in the second column of the frequency distribution, is obtained by counting the total number of tally marks for each of the categories.

Analyzing numerical information: 

With only this much organization completed, it is possible to begin making "statistical statements" about the test scores. The *modes*, the scores that appear with the greatest frequency in the distribution, are 95, which occurs 5 times, and 85, which also occurs 5 times. The *range* of the distribution, the difference between the highest and lowest test score, is 100 minus 30, or 70.

Analyzing numerical information: 

Still more meaningful numerical measures of this distribution must be determined before saying whether or not Jack's score of 75 is "good" or "bad" for this exam. By adding another column to the frequency distribution for the *cumulative frequency*, it is possible to determine how many scores were below 75, or below any other score in this distribution. This column is filled in by adding the frequencies row by row, starting with the frequency of the lowest score, 30. The entry in each row shows the sum of all the frequencies up to and including that row.

Test Scores	*Frequency*	*Cumulative Frequency*
100	2	30
95	5	28
90	4	23
85	5	19
80	2	14
75	1	12
70	0	11
65	1	11
60	1	10
55	1	9
50	3	8
45	2	5
40	1	3
35	1	2
30	1	1
	30	

Notice that the cumulative frequency column shows the total number of scores in the distribution in the row opposite a score of 100, the highest score. Using the fact that there are 30 scores in all, we can now say that Jack's score is in the bottom half of the class results. Multiplying 30, the total number of test scores, by $\frac{1}{2}$, we get the product, 15, the number of values that are in the bottom 50% of the distribution. Fifteen is also the number of values in the top 50% of the distribution.

Analyzing the information: 6

The cutoff point that separates the top 50% from the bottom 50% of the distribution occurs between the 15th and 16th scores that appear in the frequency distribution. By reading up the cumulative frequency column, the 15th and 16th scores can be identified:

Test Scores	*Cumulative Frequency*
90	23
85	19
80	14

Looking at this excerpt from the frequency distribution, we see that the entry of 14 opposite the score of 80 means that the 1st through the 14th scores are scores of 80 or less. An entry of 19 opposite the score of 85 means that the 15th, 16th, 17th, 18th, and 19th scores are all scores of 85.

This cutoff point is the midpoint of this distribution. A numerical value, called the *median*, is assigned to the midpoint of a distribution. Since this distribution has an even number of scores, 30, the median is determined by adding the 15th and 16th scores and then dividing the sum by 2:

$$\text{Median} = \frac{85 + 85}{2} \text{ or } 85$$

If a distribution has an odd number of values, 29, for example, the median can be read directly from the cumulative frequency column. It will be the 15th score in a distribution with 29 values since there will be 14 scores below and 14 above the median.

Analyzing the information: (7)

The median may indicate an "average" or typical score for this (or any) distribution in that it may be indicative of how well the class did as a whole. More precisely, the median is one measure of where the values "cluster," that is, how they are grouped, within the distribution. In this example, since at least $\frac{1}{2}$ of the class scored 85 or more on the test, and since there are but a few extremely low scores, the median is probably a "good" indication of how well the class did on this particular test. If the results of an analysis of a distribution are given without showing a frequency table or the individual values, a relatively large median and a large range will usually indicate that most of the values are in the upper categories with a few extremely low values.

Analyzing the information:

Although the median of a distribution may be used to indicate a typical value for a distribution, the "average" usually refers to another measure of the clustering of the scores: the *mean* or arithmetic average of the distribution. The mean is one of the most important measures used in statistics in that most other measures used in statistical analysis are determined by or related to the mean. The mean (the arithmetic average) is calculated by adding all the values in a distribution and dividing the sum by the total number of values in the distribution. Since the values, the test scores, in this distribution have been tabulated, the frequency distribution can be used to provide the sum of all the scores rather than having to add all the scores in the original list.

To calculate the mean from the frequency distribution, a new column is added to the table: frequency times scores. This column is filled in with the product of the score and the frequency of its occurrence that appear in the same row as the new entry. After all the products have been entered, the column entries are added to obtain the sum of all the values. (Multiplying a score such as 85 by its frequency, 5, is the same as adding five 85's since multiplication can be used for repeated additions of the same number.)

Test Scores	*Frequency*	*Frequency* × *Scores*
100	2	200
95	5	475
90	4	360
85	5	425
80	2	160
75	1	75
70	0	0
65	1	65
60	1	60
55	1	55
50	3	150
45	2	90
40	1	40
35	1	35
30	1	30
	30	2,220

Calculating the mean:

$$\text{Mean} = \frac{\text{Sum of } \textit{frequency} \times \textit{score} \text{ column}}{\text{Sum of } \textit{frequency} \text{ column}}$$

$$\text{Mean} = \frac{2{,}220}{30} \text{ or } 74$$

Because the mean reflects all of the scores or values in a distribution, it is the most reliable "average" for most statistical purposes. Notice that the mean value, 74, is *not* one of the test scores in the original list. For any distribution, neither the mean nor the median is necessarily numerically equivalent to a value included in the distribution. The mode, if there is one, will always be one of the values that appears in the distribution.

Summarizing the information: 9

The information that has been drawn from the distribution of test scores could simply be stated at this point: The mean is 74; the median is 85; the modes are 85 and 95; and the range is 70. This still doesn't "say" all it could; perhaps a better way to summarize and convey this information is to display it graphically. One type of graph used for frequency distributions is called a *frequency polygon.*

The first thing that must be done is to set up a horizontal scale and a vertical scale by using the information from the frequency distribution.

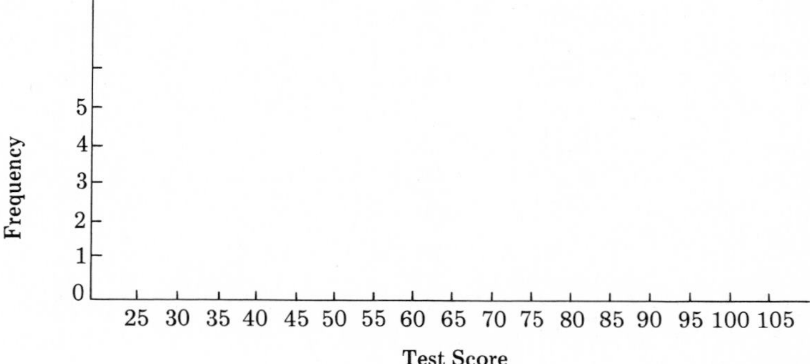

Since all frequency polygons must begin and end at some point on the horizontal line, the first and last scores shown on this scale are scores that do not actually appear in the frequency distribution. On the vertical scale, the frequencies are always numbered from zero.

Graphing a frequency polygon: 10

Once the vertical and horizontal scales are drawn, the information from the frequency distribution can be transferred to the graph.

The first point that is entered on the graph is shown on the horizontal line above the score of 25. This point is also opposite the frequency of 0. This means that there are no scores of 25 in this distribution.

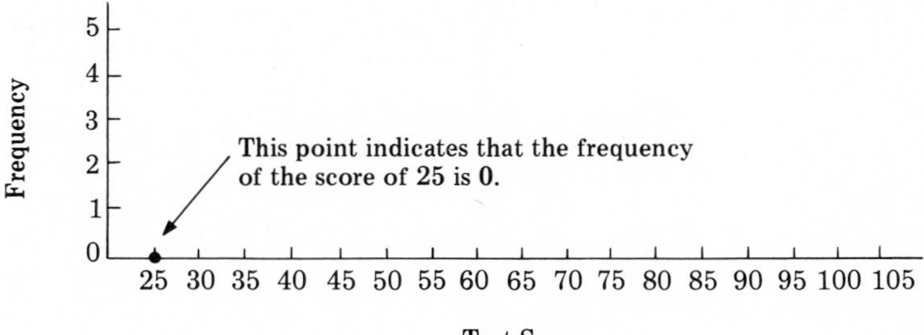

Every point shown on a graph represents two values: in this case, a test score and its frequency. To determine the other points on the frequency polygon, locate the score on the horizontal scale. Move directly up from this mark, in a straight line, to the correct frequency for the score (a ruler or other straightedge is useful).

Test Scores	Frequency
100	2
95	5
90	4
85	5
80	2
75	1
70	0
65	1
60	1
55	1
50	3
45	2
40	1
35	1
30	1
	30

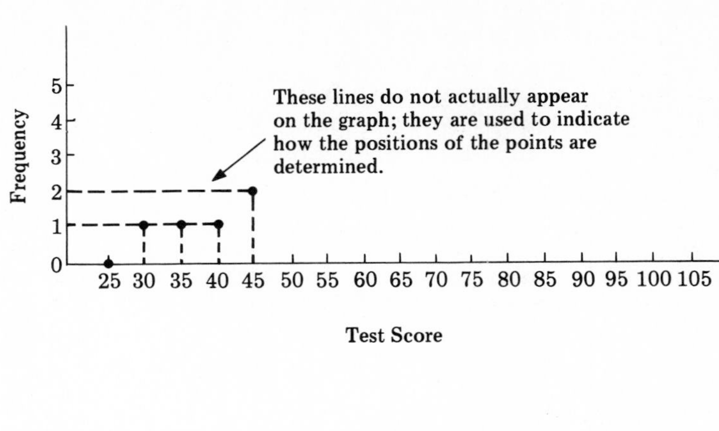

Graphing a frequency polygon: (11)

Finally, after all the points have been indicated, they are connected with straight lines drawn between each adjacent pair of points. The mean and the median can also be indicated on the graph (shown here as broken lines).

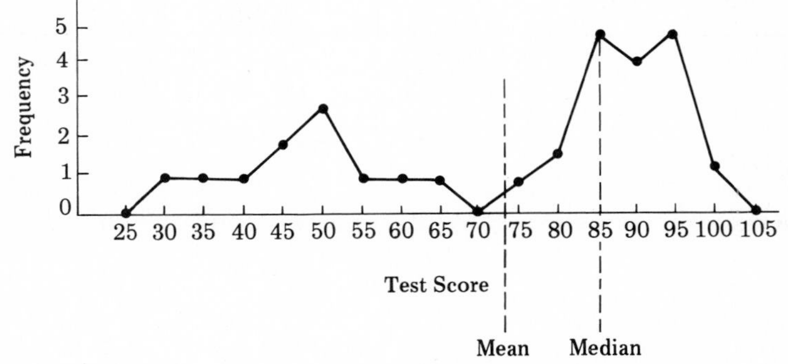

This visual representation of the distribution helps to show that the median is perhaps the better indication of the average or typical score for this class. The mean, calculated from all of the scores, is perhaps not as good an indication since the few extreme low scores affected the mean value, but not the median.

The Frequency polygon shows how the scores are grouped within the distribution, but there is another measure that gives additional information about the "shape" of the distribution: the *standard deviation.* The standard deviation is a measure of the distances between the mean value and all other values in a distribution. If this measure is relatively small, the values are clustered around the mean (the mean is a good representative score); if the standard deviation is relatively large, the values are more scattered (the mean may or may not be the most representative score). This statement will be more meaningful if the standard deviations of two comparable distributions are calculated. (12)

Calculating the standard deviation: 

In order to use the tabulated scores to calculate the standard deviation, three additional columns are added to the frequency distribution table. The first additional column is labeled "Deviation"; the *deviations* are the differences between the mean and the values (in the first column). The next column is filled in by squaring these deviations. Finally, the last column's entries are the products of the numbers in the "Deviation Squared" column and the frequencies appearing in the same row. When the last column is completed, all of the entries are added.

Test Scores	*Frequency*	*Deviation*	*Deviation Squared*	*Frequency × Deviation Squared*
100	2	26	676	1,352
95	5	21	441	2,205
90	4	16	256	1,024
85	5	11	121	605
80	2	6	36	72
75	1	1	1	1
70	0	4	16	0
65	1	9	81	81
60	1	14	196	196
55	1	19	361	361
50	3	24	576	1,728
45	2	29	841	1,682
40	1	34	1,156	1,156
35	1	39	1,521	1,521
30	1	44	1,936	1,936
	30			13,920

Once the sum of the "*Frequency × Deviation Squared*" column is known, the standard deviation is calculated by dividing this sum by the total number of scores and then taking the square root of this quotient.

$$\text{Standard Deviation} = \sqrt{\frac{\textit{Frequency} \times \textit{deviation squared}}{\text{Sum of } \textit{frequency} \text{ column}}}$$

$$\text{Standard Deviation} = \sqrt{\left(\frac{13{,}920}{30}\right)}$$

$$= \sqrt{464}$$

The square root of 464 is between 21 and 22. (See Unit 1 if review is needed on approximating square roots.) Since 464 is closer to $22^2 = 484$ than it is to $21^2 = 441$, $\sqrt{464}$ is approximately equal to 22. Whole number approximations will be used for all standard deviations in this unit.

The same test was also given to a second history class. The scores in the second class were as follows:

80, 70, 80, 50, 60, 80, 95, 70, 90, 60, 85, 14
80, 40, 75, 80, 50, 40, 85, 95, 75, and 50

Tallying the scores and filling in the frequency distribution for the second class:

Test Scores	Tally
95	ǀǀ
90	ǀ
85	ǀǀ
80	𝍸
75	ǀǀ
70	ǀǀ
65	
60	ǀǀ
55	
50	ǀǀǀ
45	
40	ǀǀ

Test Scores	Frequency	Cumulative Frequency
95	2	21
90	1	19
85	2	18
80	5	16
75	2	11
70	2	9
65	0	7
60	2	7
55	0	5
50	3	5
45	0	2
40	2	2
	21	

It is now possible to determine the mode, range, and median for this distribution.

The greatest frequency shown in the *Frequency* column indicates the mode:
The mode is 80 with a frequency of 5.

The range is the difference between the highest and lowest values:
The highest value is 95; the lowest is 40.
The range for this distribution is 95 - 40 or 55.

Since there are an odd number of scores, 21, the median can be determined without calculations by locating the 11th score in the distribution. The median will have the same numerical value as the 11th score since there will be 10 values in the distribution both above and below this value.

Reading up the *Cumulative Frequency* column, the 11th score occurs opposite a test score of 75; the median is 75.

Calculating the mean and standard deviation for the scores of the second history class:

Test Scores	Frequency	Frequency × Score	Deviation	Deviation Squared	Frequency × Deviation Squared
95	2	190	24	576	1,152
90	1	90	19	361	361
85	2	170	14	196	392
80	5	400	9	81	405
75	2	150	4	16	32
70	2	140	1	1	2
65	0	0	6	36	0
60	2	120	11	121	242
55	0	0	16	256	0
50	3	150	21	441	1,323
45	0	0	26	676	0
40	2	80	31	961	1,922
	21	1,490			5,831

$$\text{Mean} = \frac{1{,}490}{21}$$

≈ 70.9

≈ 71, rounded to 2 significant digits

The approximate mean value of 71 is used to calculate the deviations.

$$\text{Standard Deviation} = \sqrt{\left(\frac{5\,831}{21}\right)}$$

$\approx \sqrt{278}$

≈ 17

The square root of 278 is between 16 and 17, but it is closer to 17 than to 16 since 278 is closer to $17^2 = 289$ than it is to $16^2 = 256$.

Comparing the results of the two history classes: (16)

Looking at the frequency polygons of both classes' scores and comparing the means, medians, and standard deviations of both distributions:

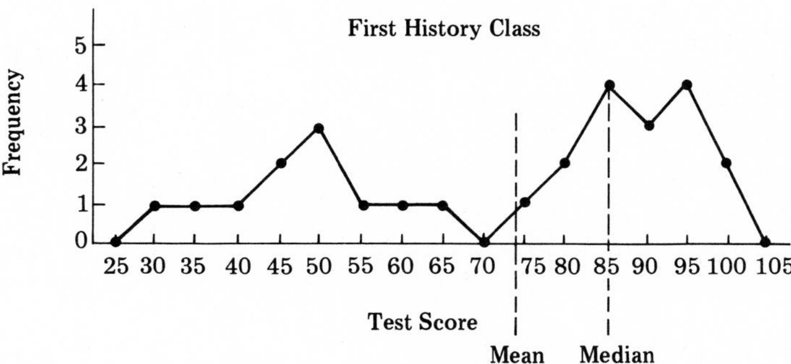

First Class

Modes: 85 and 95
Range: 70
Mean: 74
Median: 85
Standard deviation: 22

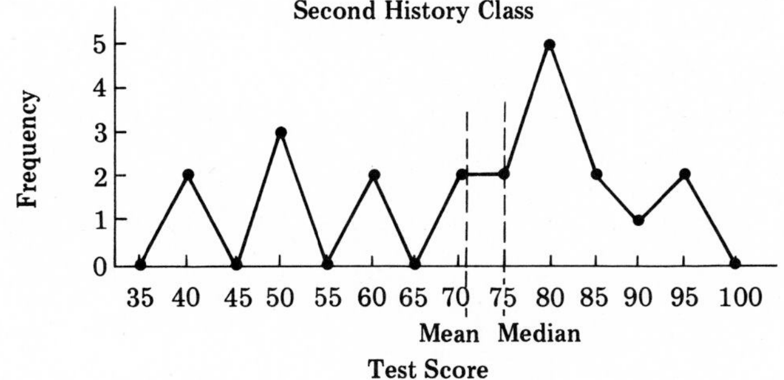

Second Class

Mode: 80
Range: 55
Median: 75
Mean: 71
Standard deviation: 17

Notice that there is a greater amount of spread—the range is greater—in the first class and a greater standard deviation. In the first example, the median is a better indication of the representative score than is the mean; in the second example, either the mean or the median is a good indication of the representative score. The greater the standard deviation, the farther the distances between the mean and the extreme scores (either lower or higher scores). It is not necessarily true that the mean will never be the best representative score just because the standard deviation is relatively large, but it does mean that a closer examination should be made of all the measures for the distribution before stating that the mean is the best "average," or typical score.

The following distribution represents the number of hours of overtime worked in one month by the employees of a particular company. (17)

Hours of Overtime	*Frequency*	*Cumulative Frequency*	*Frequency × Hours*	*Deviation*	*Deviation Squared*	*Frequency × Deviation Squared*
11	9	100	99	3	9	81
10	20	91	200	2	4	80
9	17	71	153	1	1	17
8	12	54	96	0	0	0
7	12	42	84	1	1	12
6	11	30	66	2	4	44
5	7	19	35	3	9	63
4	6	12	24	4	16	96
3	1	6	3	5	25	25
2	2	5	4	6	36	72
1	1	3	1	7	49	49
0	2	2	0	8	64	128
	100		765			667

From the frequency distribution:

The range is 11 – 0 or 11.
The mode is 10 hours with a frequency of 20.
The median is between the values of the 50th and 51st hour:

$$\frac{8+8}{2} \text{ or } 8$$

Calculating the mean:

$$\text{Mean} = \frac{765}{100}$$
$$= 7.65$$
$$\approx 8$$

The approximate mean value of 8 is used to calculate the standard deviation.

$$\text{Standard Deviation} = \sqrt{\frac{667}{100}}$$
$$= \sqrt{6.67}$$
$$\approx 3$$

The square root of 6.67 is between 2 and 3, but closer to 3 than to 2 since 6.67 is closer to $3^2 = 9$ than it is to $2^2 = 4$. (9.00 – 6.67 is only 2.33 but 6.67 – 4.00 is 2.67.)

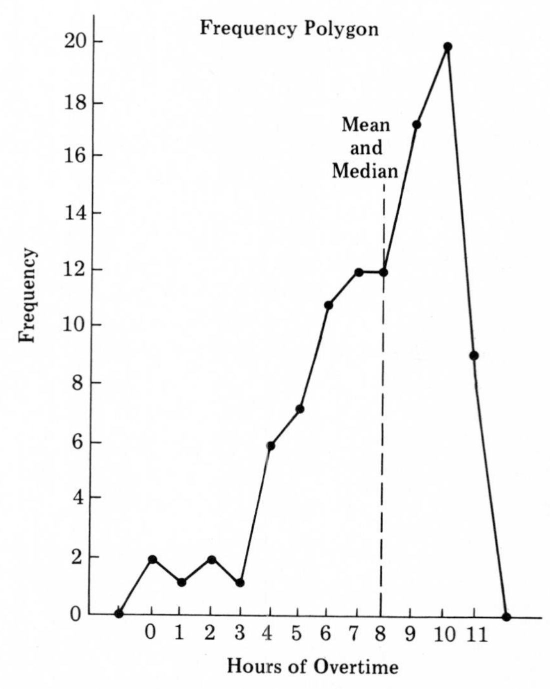

A "good" average for this distribution is 8 hours.

A study is conducted for a manufacturer of radios to determine the average number of defective radios that are produced in the plant per day. The number of radios per day that are found to be defective during the first month of the study are (18)

4, 1, 3, 6, 5, 6, 4, 0, 1, 3, 2,
6, 3, 1, 4, 2, 0, 5, 2, and 2

Complete the following table:

Number of Defective Radios	*Tally*
0	||
1	|||
2	||||
3	|||
4	|||
5	||
6	|||

Number of Defective Radios	*Frequency*	*Cumulative Frequency*	*Frequency × Number of Radios*	*Deviation*	*Deviation Squared*	*Frequency × Deviation Squared*
0		20	0	3		18
1		18	3	2		12
2		15	8	1		4
3		11	9	0		0
4		8	12	1		3
5		5	10	2		8
6		3	18	3		27
	20					72

From the frequency distribution: the range is 6 – 0 or 6; the mode is 2; the median is between the 10th and 11th scores, $\frac{3 + 3}{2} = 3$.

Complete the following calculations:

Mean = $\frac{\bigcirc}{\square} = \triangle$

Standard Deviation = $\sqrt{\left(\frac{⬡}{\square}\right)} \approx 2$

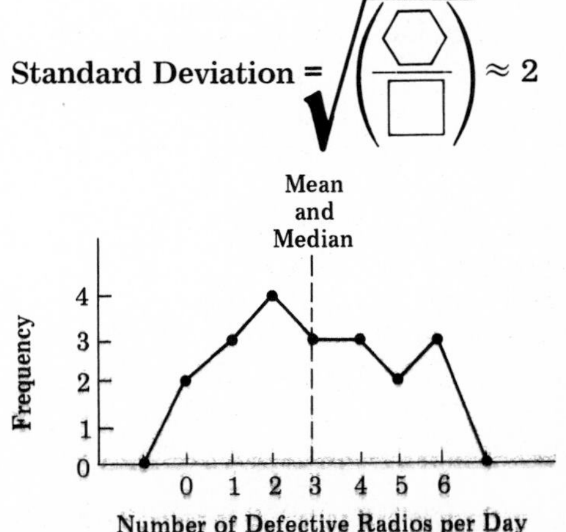

A "good" average for this distribution is 3 defective radios per day.

For each of the following, determine the range, the mode (or modes), if any, the median, the mean, and the standard deviation, all correct to the nearest whole number. (Estimate the approximate value of the standard deviation by determining which whole number it is nearest in value.)

1. A consumer analyst has recorded the selling prices of a particular brand of cereal at twelve supermarkets in one community. The selling prices, in cents, were as follows:

89, 85, 89, 88, 83, 86, 88, 89, 91, 87, 84, and 85

Price	*Tally*	*Frequency*	*Cumulative Frequency*	*Frequency × Price*	*Deviation*	*Deviation Squared*	*Frequency × Deviation Squared*
91							
90							
89							
88							
87							
86							
85							
84							
83							

2. The following distribution represents the scores of the students in a particular algebra class for one 20-point quiz.

Score	*Frequency*	*Cumulative Frequency*	*Frequency × Score*	*Deviation*	*Deviation Squared*	*Frequency × Deviation Squared*
20	3					
19	4					
18	5					
17	2					
16	5					
15	3					
14	2					
13	1					
12	2					
11	1					
10	2					

3. The weight loss, in kilograms, per person for a group of people on the same diet for a 1-month period is as follows:

10, 6, 11, 7, 5, 5, 12, 4, 6, 10, 8, 7, 8, 10,
8, 4, 9, 11, 10, 6, 11, 5, 8, 6, 8

Weight	Tally	Frequency	Cumulative Frequency	Frequency × Weight	Deviation	Deviation Squared	Frequency × Deviation Squared
12							
11							
10							
9							
8							
7							
6							
5							
4							

A second type of graph commonly used to represent a frequency distribution is the *histogram*. A histogram is a bar graph. Refer to the distribution in Frame 14. The information needed to draw the graphs and the frequency polygon for the distribution are shown below: (20)

Test Score	Frequency
95	2
90	1
85	2
80	5
75	2
70	2
65	0
60	2
55	0
50	3
45	0
40	2

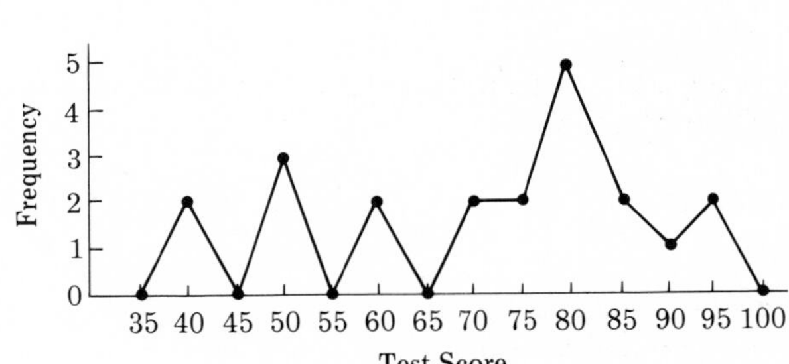

The corresponding histogram is shown below. Notice that many of the bars in this graph share the same edge. So that there will be no unnecessary "gaps" or spaces in the graph, the width to be used for each bar must be carefully chosen. The width chosen for the bars in this graph is 5 units; all bars must be equally wide. In order for the bars to share "common" sides, the intervals used on the horizontal scale must begin and end exactly halfway *between* the test score values used to graph the frequency polygon. For example, the number that is halfway between 35 and 40 is 37.5. To find this number, add 35 and 40 and then divide the sum, 75, by 2.

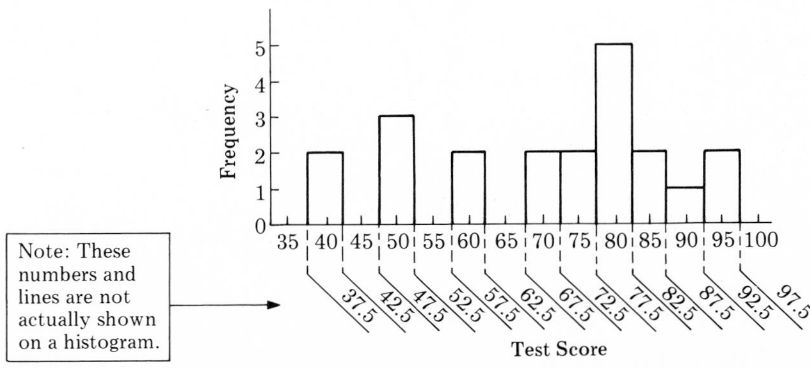

Graphing from a frequency distribution:

So that the corresponding graphs may be compared for the distribution in Frame 14, both the histogram and the frequency polygon are shown below on the same graph.

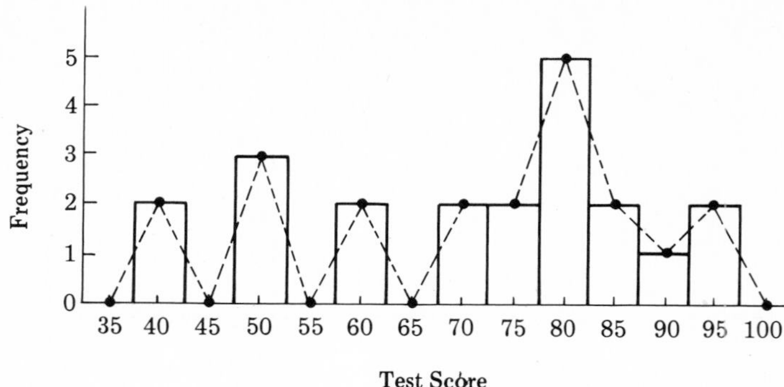

Notice that the points used to draw the frequency polygon are exactly in the middle of the tops of the bars in the histogram since the tops of the bars are drawn opposite the frequency of the test score for each value.

Graphing a frequency distribution:

The frequency distribution shown below refers to the distribution given in Frame 17. To graph the histogram, the vertical scale is drawn so that each bar will be 1 unit wide. Notice that the first bar will start at -0.5, one-half unit to the left of 0.

Hours of Overtime	*Frequency*
11	9
10	20
9	17
8	12
7	12
6	11
5	7
4	6
3	1
2	2
1	1
0	2

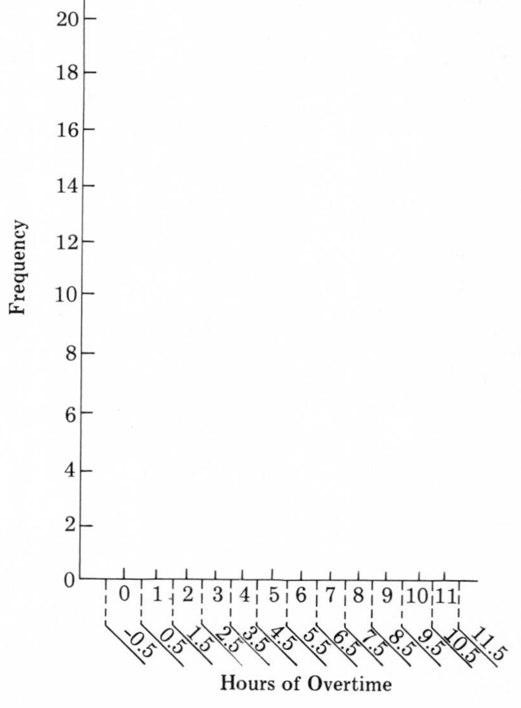

To check the completed histogram (shown on the left below), construct the frequency polygon by connecting the midpoints of the tops of the bars of the histogram. Compare this frequency polygon (on the right below) with the frequency polygon shown in Frame 17.

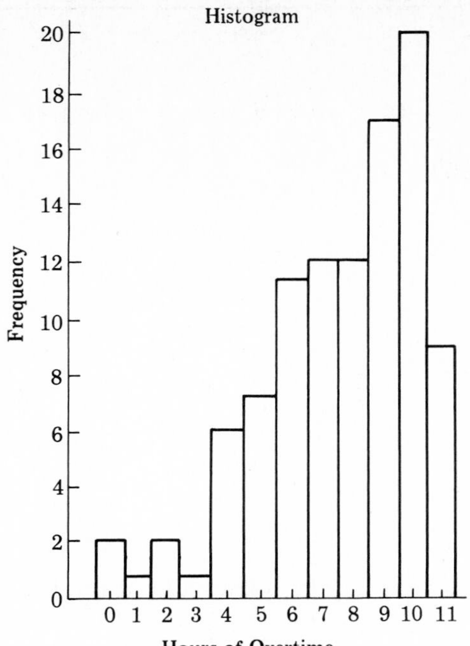

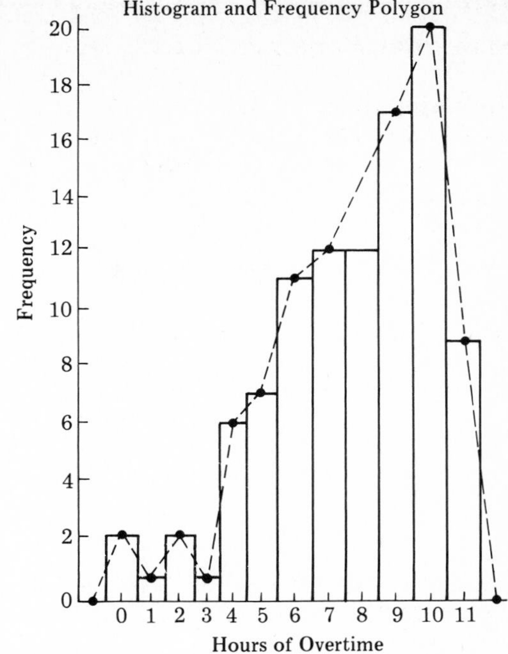

Graphing from a frequency distribution: ㉓

Refer to the distribution in the example in Frame 18.

Number of Defective Radios	*Frequency*
0	2
1	3
2	4
3	3
4	3
5	2
6	3

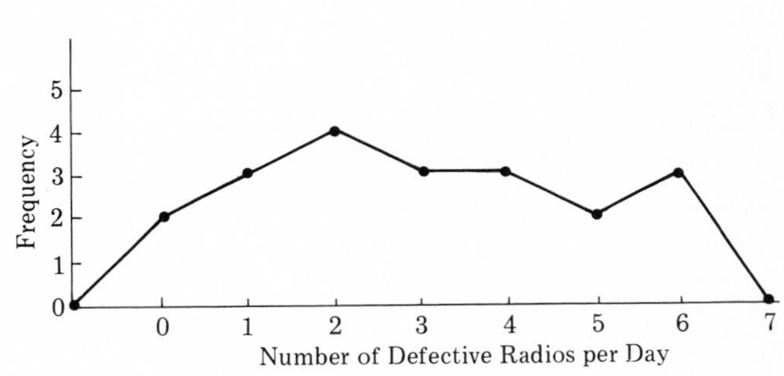

Complete the histogram by drawing in the bars.

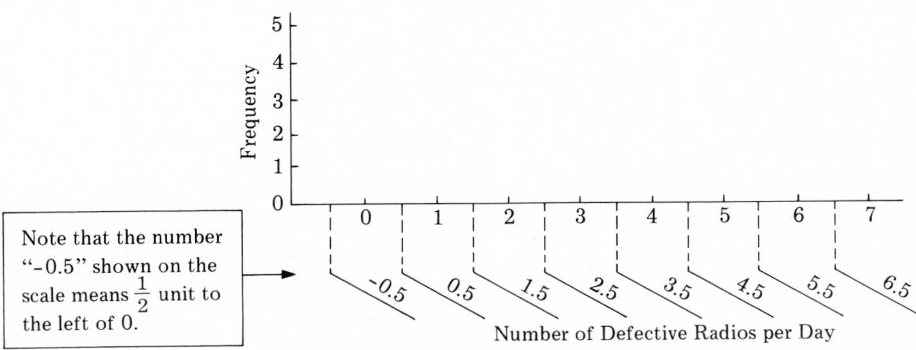

To check the drawing of the histogram, draw the frequency polygon on the same graph by connecting the middles of the tops of the bars of the histogram with a dotted line. Does this frequency polygon match the one shown at the top of this frame?

Graphing from a frequency distribution:

The following frequency distribution represents the IQ's (Intelligence Quotients) for a group of students in the same eighth-grade class.

Complete the histogram and check it by drawing the frequency polygon on the same graph for comparison.

IQ Score	*Frequency*
137	1
128	2
119	5
110	7
101	8
92	6
83	5
74	4
65	2
	40

Complete the histogram and check it by drawing the frequency polygon on the same graph (as was done in preceding frame).

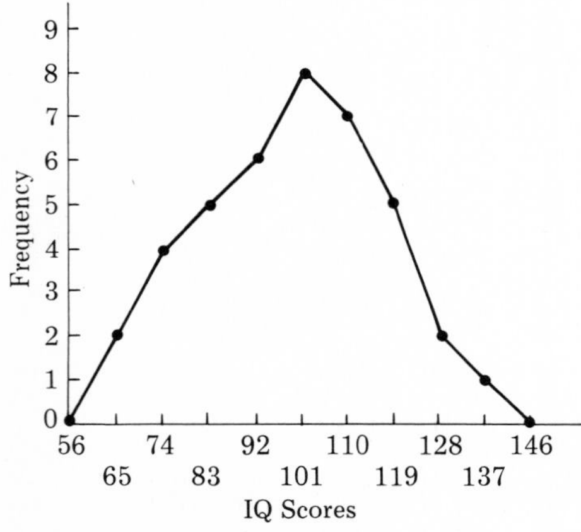

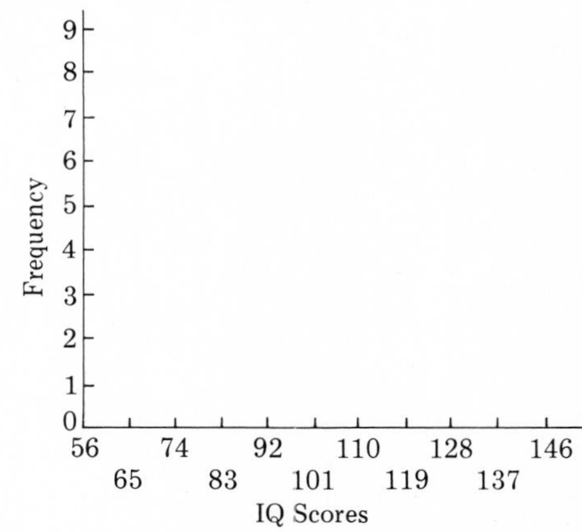

Without doing the actual calculations, estimate the mean of this distribution. What is the mode? The range?

Practice Exercises

Refer to the distributions given in the Practice Exercises in Frame 19 to graph a frequency polygon and a histogram for each distribution.

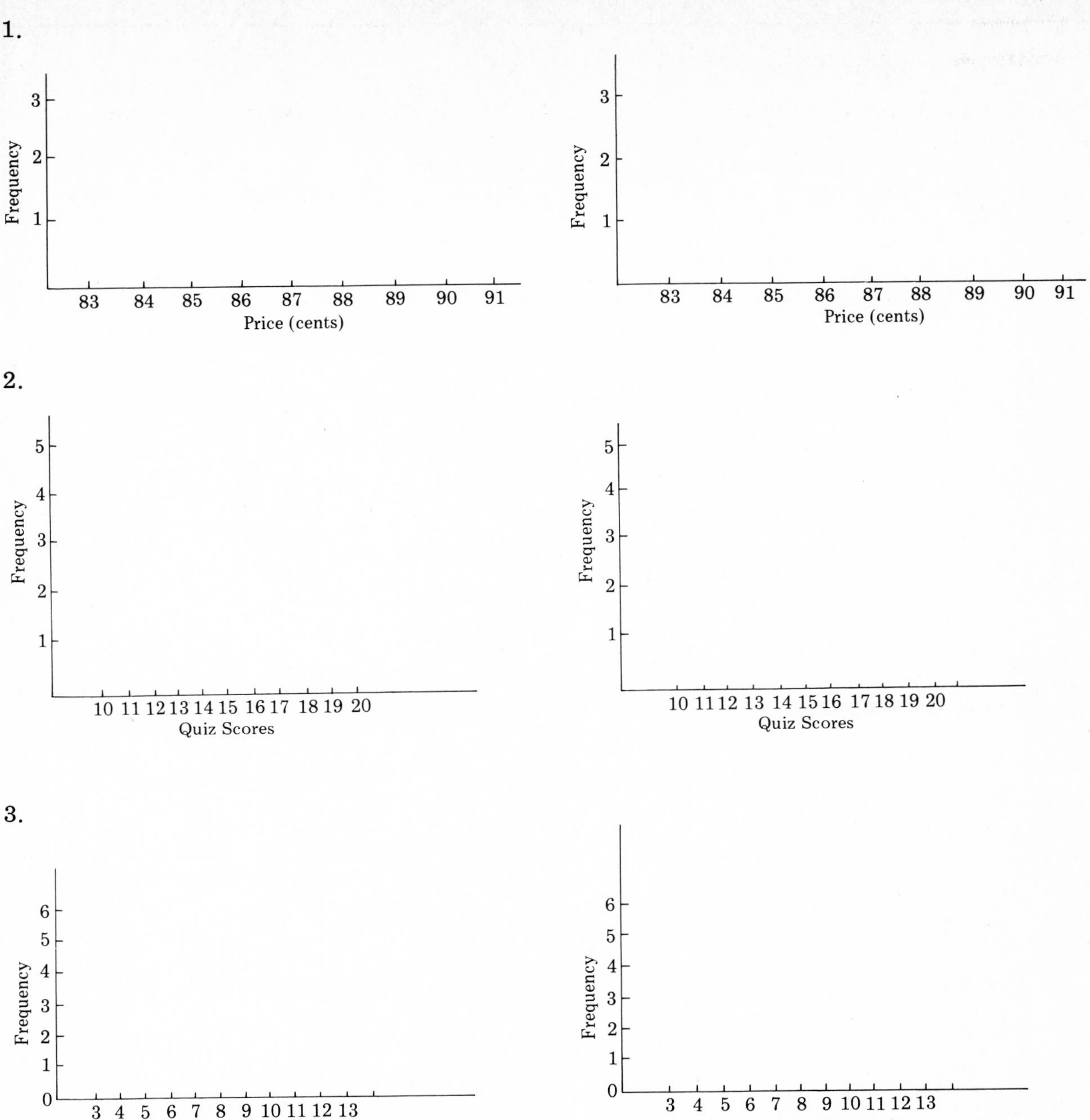
1.
Frequency
1 2 3
83 84 85 86 87 88 89 90 91
Price (cents)
Frequency
1 2 3
83 84 85 86 87 88 89 90 91
Price (cents)
2.
Frequency
1 2 3 4 5
10 11 12 13 14 15 16 17 18 19 20
Quiz Scores
Frequency
1 2 3 4 5
10 11 12 13 14 15 16 17 18 19 20
Quiz Scores
3.
Frequency
0 1 2 3 4 5 6
3 4 5 6 7 8 9 10 11 12 13
Weight Loss (kilograms)
Frequency
0 1 2 3 4 5 6
3 4 5 6 7 8 9 10 11 12 13
Weight Loss (kilograms)

POST-TEST

1. The number of minutes it took Sandra to drive to work for each of 15 days were as follows: 17, 16, 18, 21, 23, 20, 19, 20, 19, 17, 20, 19, 20, 17, 21. Determine each of the following correct to the nearest whole number:
   (a) Range
   (b) Mode (or modes, if any)
   (c) Median
   (d) Mean
   (e) Standard deviation

Number of Minutes	*Tally*	*Frequency*	*Cumulative Frequency*	*Frequency × Number*	*Deviation*	*Deviation Squared*	*Frequency × Deviation Squared*
23							
22							
21							
20							
19							
18							
17							
16							

2. Use the distribution from the above problem, graph the frequency polygon and histogram.

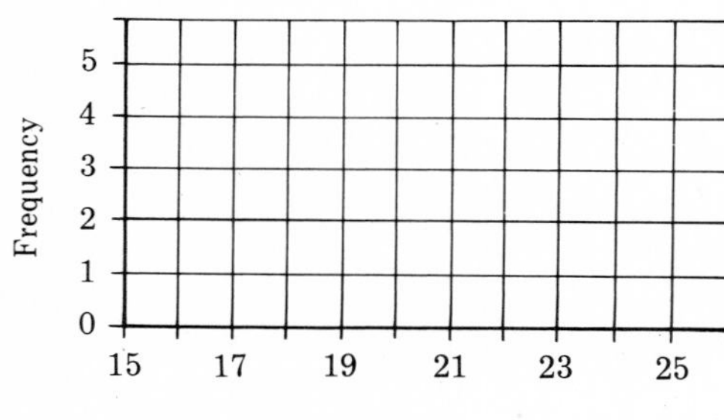

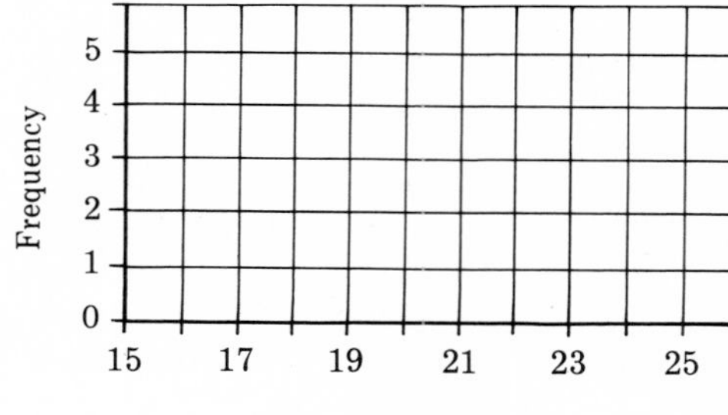

LESSON 30
GRAPHS AND TABLES

Objectives:

1. Name the type of graph shown in a given representation. (1–12)
2. State the purpose for which a line graph, a circle graph, a bar graph, or a pictograph is best suited. (1–12)
3. Solve a problem by using information that can be determined by correctly reading a given graph. (13–21)
4. Solve a problem by using information that can be determined by correctly reading a specified table. (22–29)

Vocabulary:

Vertical Bar Graph
Line Graph
Trend
Key
Horizontal Bar Graph
Pictograph
Circle Graph
Sales Tax Table
Federal Tax Rate Schedule
Compound Interest
Simple Interest
Compound Interest Table

PRE-TEST

1. Use the following graph to answer questions (a) through (e):

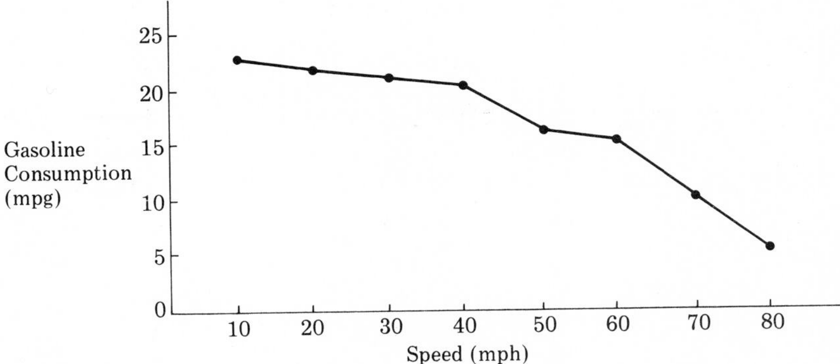

(a) What is the name of this type of graph?
(b) What purpose is best served by this type of graph?
(c) What is the gasoline consumption, mpg, at 70 mph?
(d) What is the difference in gasoline consumption at 40 mph and 70 mph?
(e) What percent decrease in gasoline consumption exists when a car is driven at 80 mph instead of 40 mph?

2. Use the following table to answer questions (a) and (b):

Amount of $1.00 Compounded Annually

Number of Years	*1%*	*2%*	*3%*	*4%*	*5%*	*6%*
1	1.01	1.02	1.03	1.04	1.05	1.06
2	1.02	1.04	1.06	1.08	1.10	1.12
3	1.03	1.06	1.09	1.12	1.16	1.19

(a) Mr. Arlen invested $800 at 5% compounded annually. What is the amount of his investment after 3 years?

(b) Ms. Power's $1,775 earned how much interest when it was invested for 2 years at 4% compounded annually?

In the last lesson, two types of graphs were introduced: the frequency polygon and the histogram. Both types of graphs provide an effective way for visualizing information about a distribution. Graphs provide an effective way to summarize and convey numerical information. 1

For example, the following graphs both represent the number of televisions sold by an independent dealer for a 6-month period:

Monthly Television Sales

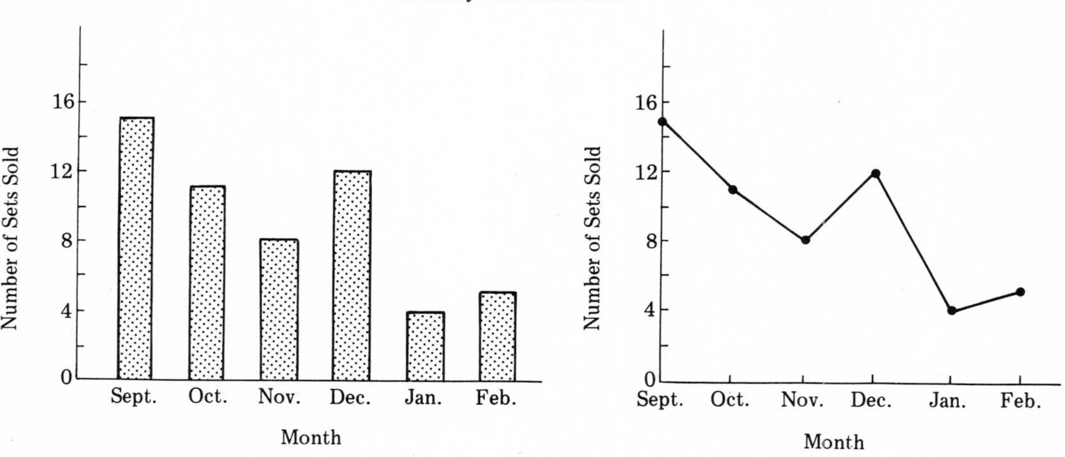

The graph on the left (that looks like a histogram) is called a *vertical bar graph;* the graph on the right (that looks like a frequency polygon) is called a *line graph.* Notice that both graphs have a title and that the horizontal and vertical scales are carefully labeled. Each unit on the vertical scale represents two television sets.

Take a closer look at the graphs of the dealer's monthly television sales. 2

Notice that the bar graph provides a better visualization for comparing the monthly sales than does the line graph. By simply looking at the bar graph, we see that the sales in November, for example, were about twice the number of sales made in January. It is also easy to see that the months with the greatest number of sets sold were September, October, and December.

The line graph shows more about the change in the number of sets sold through the winter months; it shows the *trend,* the general direction or tendency, of the sales. The line graph shows that the number of sales was generally declining throughout this 6-month period, in spite of an "upswing" in December and February.

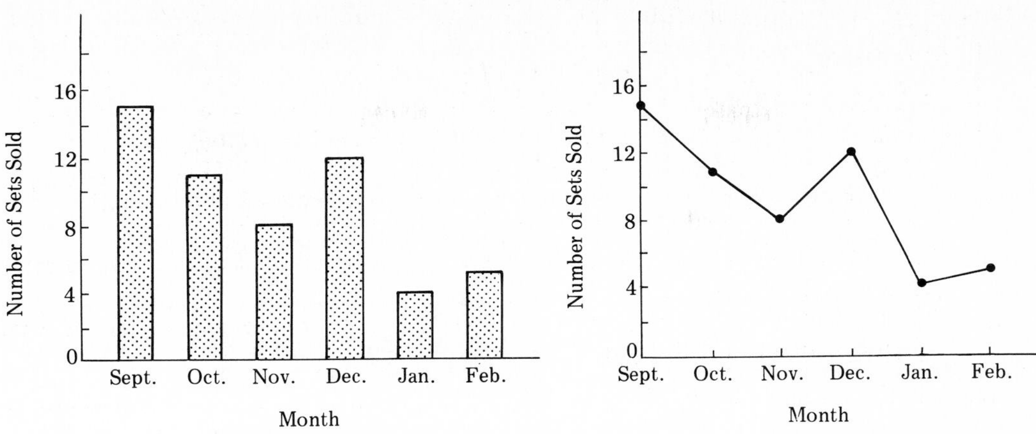

In the example used in Frames 1 and 2, either graph could be used to visualize and summarize the information about the monthly television sales. The choice of which type of graph to use would depend on whether the graph is to be used to compare the monthly sales or to show the trend of the monthly sales. (3)

Now consider comparing the independent dealer's television sales with those of the television department of a major chain store during the same 6-month period.

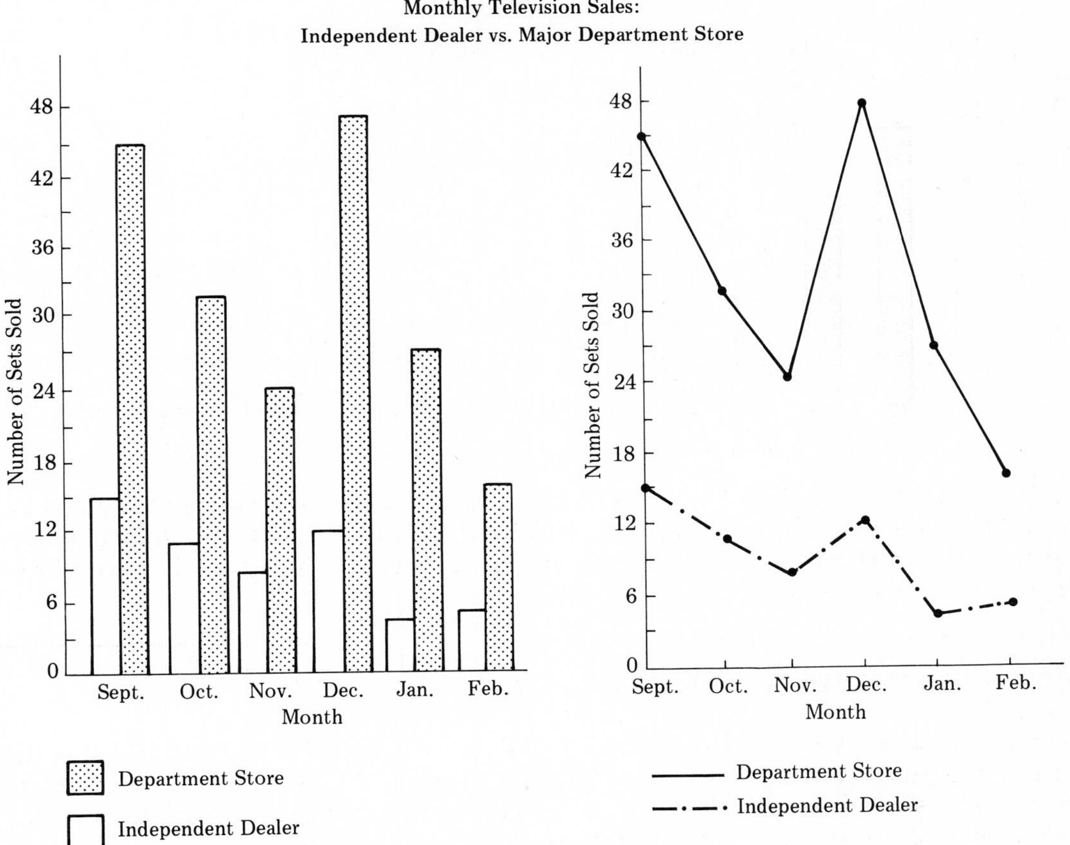

In this example the double line graph (two line graphs drawn on the same horizontal and vertical scale) conveys little information except the fact that the chain store outsold the independent dealer every month. The double bar graph, however, provides a good monthly comparison between the sales of the chain store and the sales of the independent dealer. Clearly, a comparison like this should *not* be made by using a double line graph since the numbers of sets sold by the chain store and the independent dealer are not within the same range of numbers: The number of sets sold per month by the chain store varies between a high of 47 and a low of 16; the number of sets sold by the independent dealer varies between a high of 15 and a low of 4.

Notice that a *key* must be included with a double line graph or a double bar graph:

The minimum monthly salaries of the personnel at Community Hospital are to be compared with the minimum monthly salaries of the personnel at General Hospital. A double bar graph is drawn as follows:

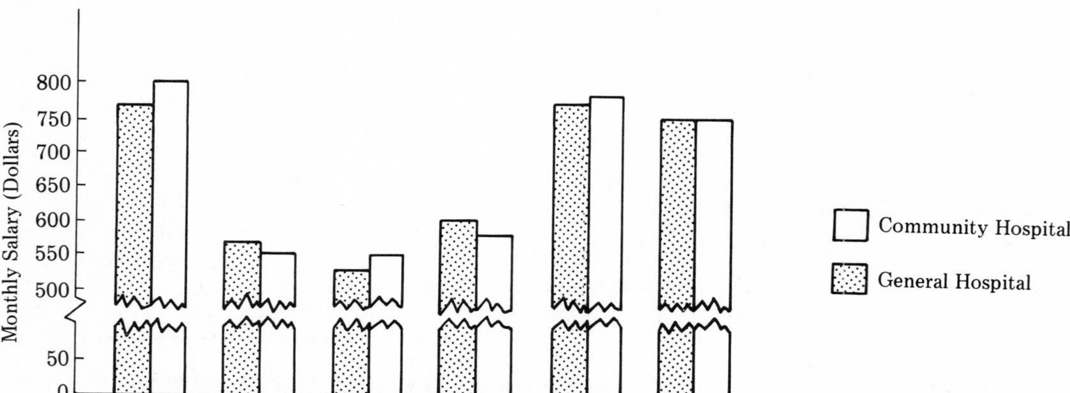

Notice the break shown in the vertical scale between 50 dollars and 500 dollars. A break such as this should not be used to force the fit of large numbers into a double bar or bar graph. Since a bar graph compares the lengths of bars within the graph to convey information, nothing should be done that will distort these lengths. If the vertical scale is changed to correct the distortion, we get the following:

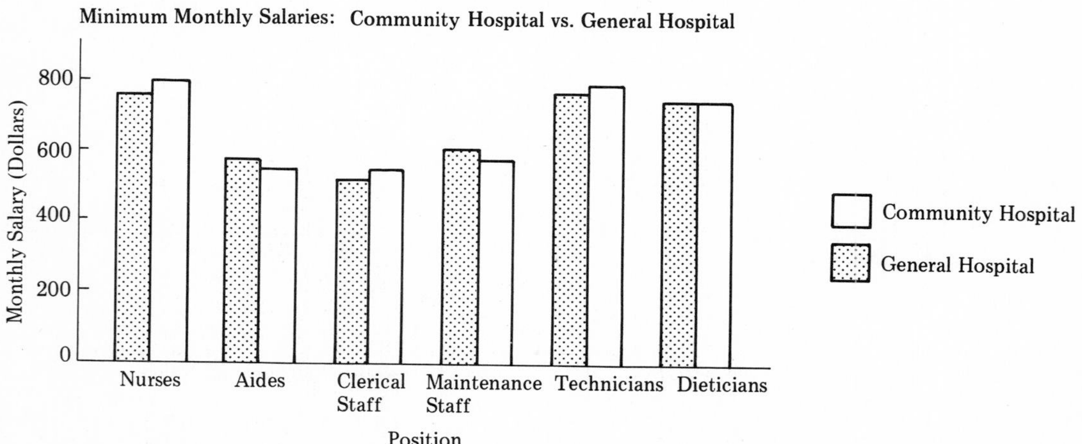

The change in the vertical scale units makes it very difficult to estimate what the salaries are and to differentiate between the comparable salaries, for technicians, for example.

A double line graph can be used to compare the salaries at the two hospitals. (5)

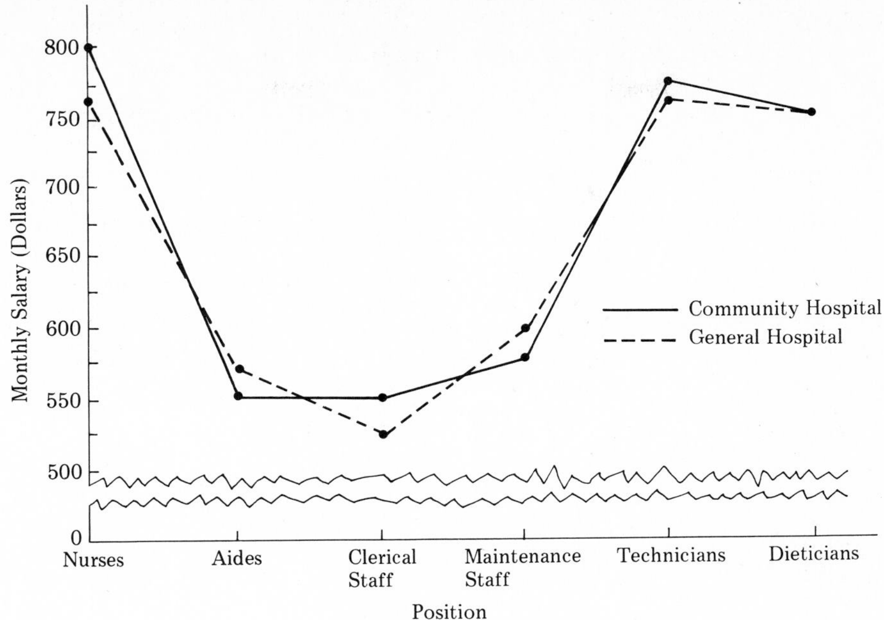

Since this graph is not comparing the lengths of bars from the horizontal scale, a break in the vertical scale is acceptable. This allows for a scale to be chosen so that it is possible to make a good estimate of the comparable salaries. Notice also that the salary ranges for the two hospitals are comparable.

To summarize, a bar graph is used to compare similar categories of numerical values that can be represented by the lengths of the bars used in the graph. A scale should be chosen that does not distort the size of the bars and yet makes it possible to differentiate between the lengths of the bars. (6)

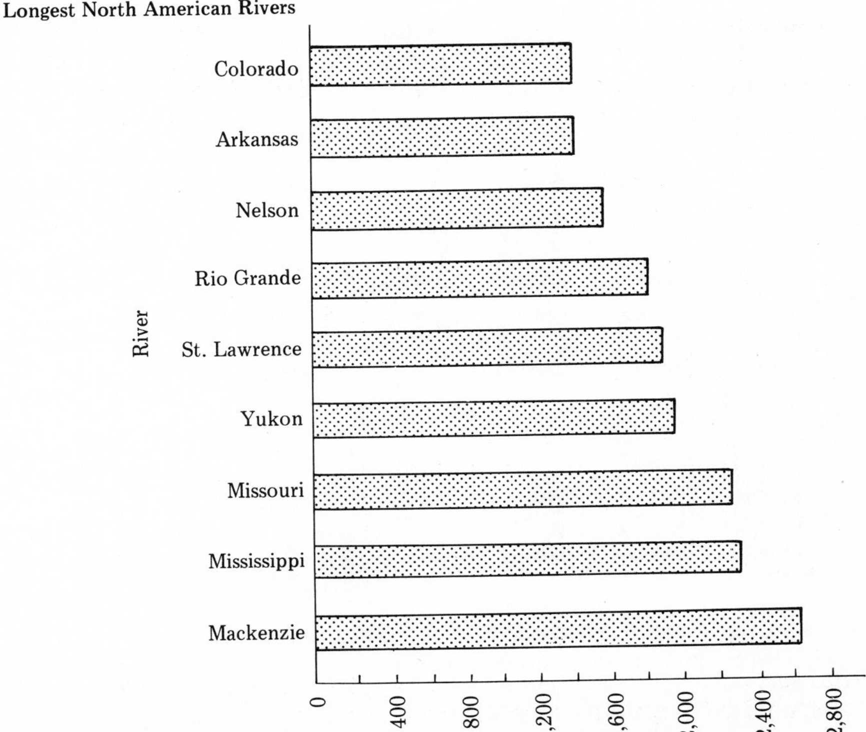

This type of graph is called a *horizontal bar graph*. It is used in this example, instead of a vertical bar graph, so that the names of the rivers can be easily read.

A line graph is best used to show trends over a considerable length of time.  7

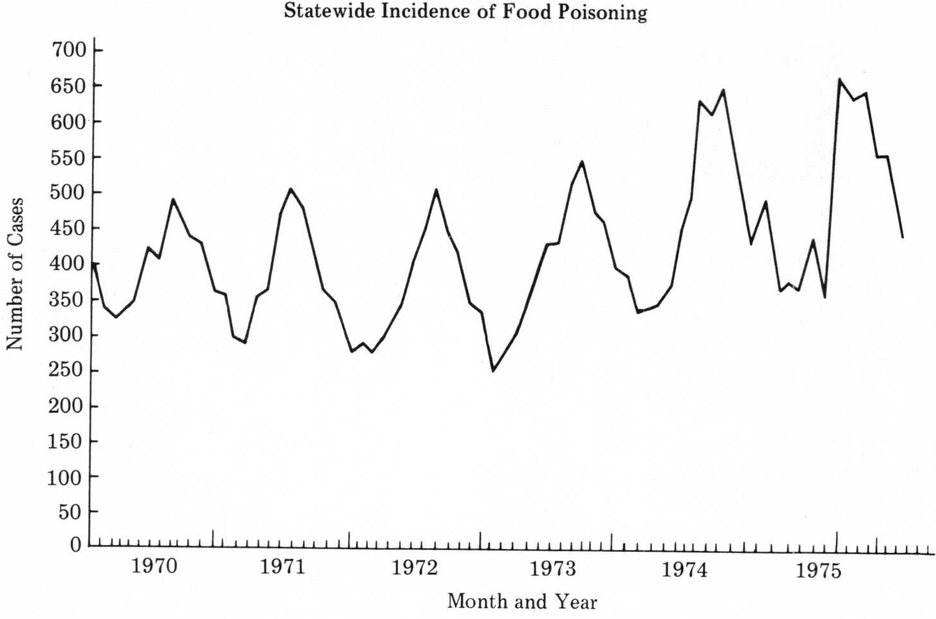

Another type of graph that could be used to visualize the number of televisions sold by the independent dealer in the example in Frame 1 is a *pictograph*. A pictograph, commonly used in magazines and other publications, uses graphic symbols to represent numbers. 8

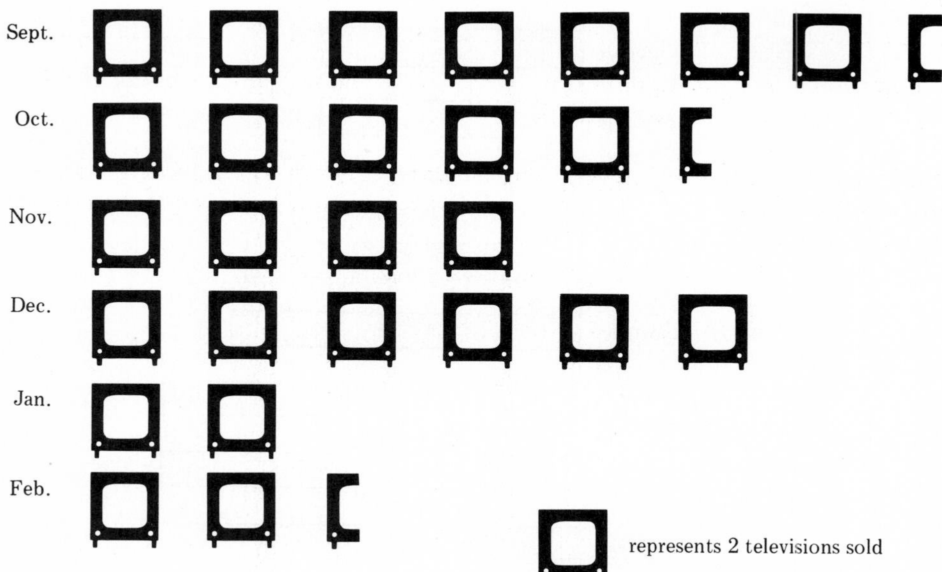

A pictograph is best suited to comparisons, especially if the numbers are not too important. If the numerical values are important, a bar graph is preferable.

Suppose that the following pictograph is used to compare the number of television sets sold for the 3-month period from November to January: 

Monthly Television Set Sales

From the bar graph in Frame 1 and the pictograph in Frame 8, it can be determined that the number of sets sold in November is 8, the number sold in December is 12, and the number in January is 4. Does this pictograph accurately represent the fact that twice as many sets were sold in November as were sold in January and that three times as many sets were sold in December as were sold in January?

Does doubling the length of the side of the square symbol double the area? Does tripling the length of the side triple the area?

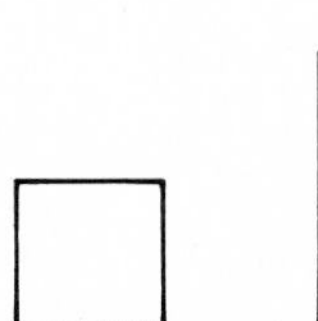

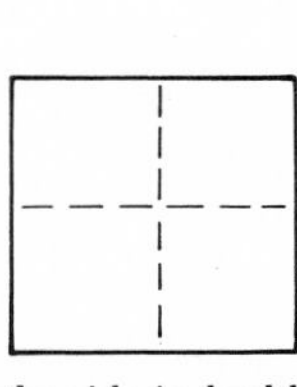

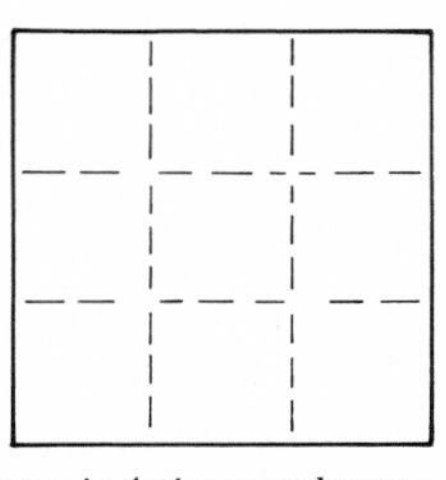

If the length of the side is doubled, the area is 4 times as large; if the length of the side is tripled, the area is 9 times as large.

All the symbols used in a pictograph should be the same size if the correct information is to be accurately conveyed.

The last type of graph to be considered in this lesson is the *circle graph*. A *circle graph* is used when the categories represent the parts of a whole. 

For example, returning to the number of television sets sold by the independent dealer, if the total number of sets sold in the 6-month period, 55 sets, is considered to be one whole, the following graph can be drawn:

Monthly Television Set Sales

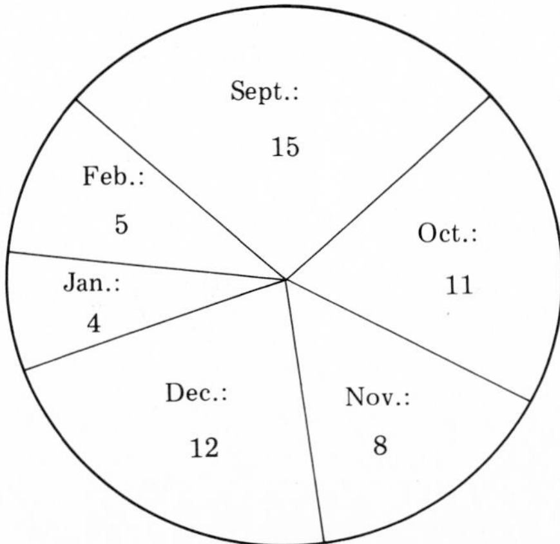

Notice that this circle graph is designed to be read in a clockwise direction. Also, the information within the sections of a circle graph must be readable or not put inside the graph at all.

The Allen family budgets its monthly income as follows: 11

Allen Family's Budget

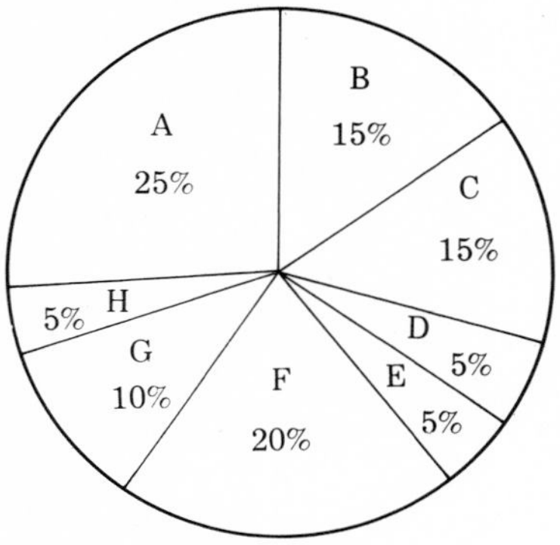

A 25%
B 15%
C 15%
D 5%
E 5%
F 20%
G 10%
H 5%

Key

A: Housing (mortgage payment and utility bills)
B: Food
C: Insurance and savings
D: Medical expenses
E: Clothing
F: Automobiles (including installment payment and maintenance)
G: Installment payments
H: Miscellaneous

In this example there is more information than can be fitted into the sections of the graph. Therefore, a key is used. Also, notice that the fractional parts of the budget are expressed as percents. The percents shown on the circle graph must add up to 100%.

Practice Exercises 12

Complete the following by filling in the blanks:

1. A bar graph is best used to show __________________ .
2. If a trend is to be shown graphically, the best graph to use is a ________ graph.
3. A pictograph can be used for the same purpose as a ________ graph.
4. The vertical scale can be broken on a ______ graph, but it can never be broken on a ________ graph.
5. A circle graph is best used to show the relationships of the ______ of a ________ .

Just as few people can escape statistics in the course of daily living, it is almost impossible to escape graphical information. (Pick up a current weekly news magazine and count the number of graphs that appear in just one issue!) Everyone needs to know something about graphs and how they are constructed to be able to read graphs correctly. Just being aware of the best uses of the various types of graphs discussed in this lesson is the first step in learning how to interpret graphs. A bar graph compares; a line graph shows trends; and a circle graph relates parts of a whole. A ruler or other straightedge will be extremely useful for correctly reading other information from a graph. 13

Using graphical information: 14

Before attempting to solve problems by using graphical information, it will be worthwhile to carefully look at a few things on a graph.

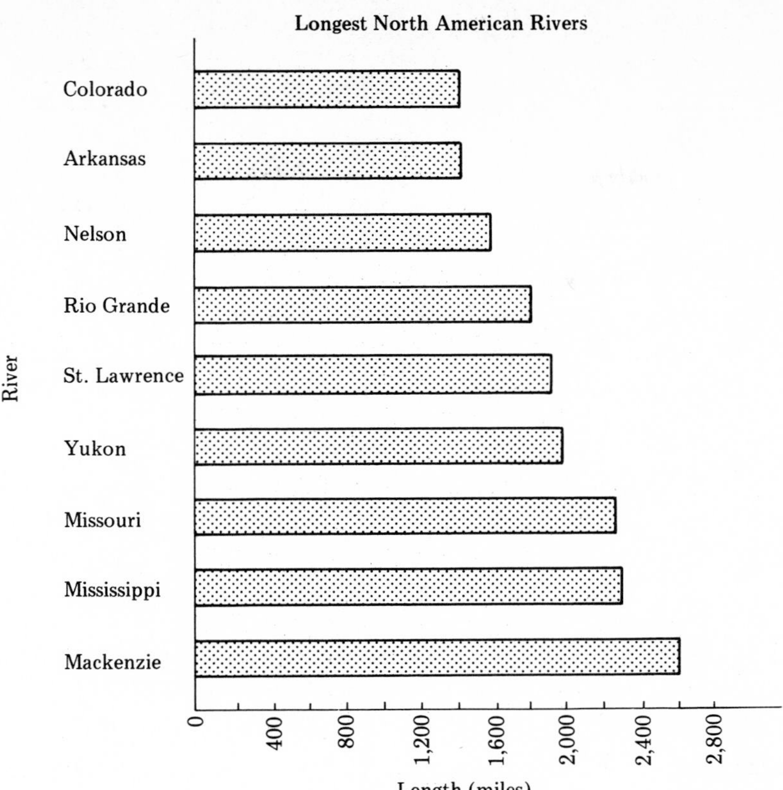

The following questions should be answered before using any information from this or any graph: What is the purpose of the graph? What is being represented? What are the horizontal and vertical scale units? Is it possible that the information is questionable because of the type of graph used or because of distortions in the construction?

Using graphical information: (15)

The purpose of the graph below is to compare the lengths of the rivers of North America. The horizontal scale is in units of 200 miles. The vertical scale names the rivers. Since a bar graph is best suited to comparisons, the information obtained from the graph is adequate.

Suppose now that this graph is to be used to answer the following questions:

A. **Which North American rivers are approximately the same length?**

B. **How much longer is the Mackenzie River than the Nelson River?**

To answer A, slide a ruler or the edge of a piece of paper across the graph. Keep the edge parallel to the vertical scale. The rivers that are approximately the same length are the Colorado and the Arkansas, the St. Lawrence and the Yukon, and the Missouri and the Mississippi.

To answer B, place the ruler parallel to the vertical scale, at the end of the bars representing the lengths of the Mackenzie and Nelson Rivers and read the number of miles from the horizontal scale (broken lines on the graph indicate where these lengths were read):

The Mackenzie River $\approx$ 2,600 miles long; the Nelson $\approx$ 1,600 miles long.
The Mackenzie River is approximately 2,600 - 1,600 or 1,000 miles longer than the Nelson River.

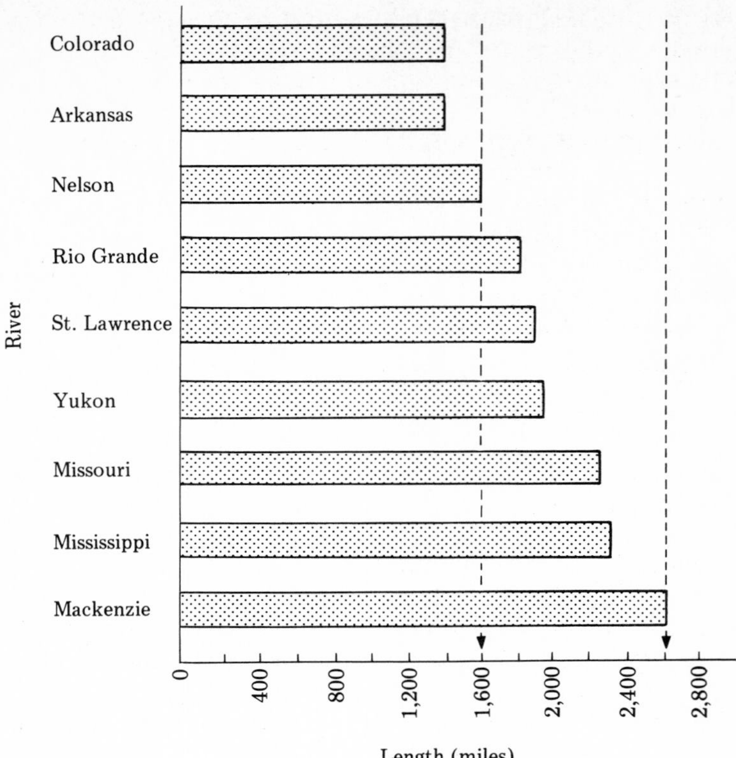

Using graphical information: ⑯

The graph in this frame shows the quarterly profits of a small manufacturing firm for the five-year period from 1928 through 1932. The horizontal scale is in units of one quarter; the vertical scale is in units of 5,000 dollars.

Consider the following questions:

A. **In which quarter of which year was the decline in profits the greatest?**

B. **In which quarter of which year was the increase in profits the greatest?**

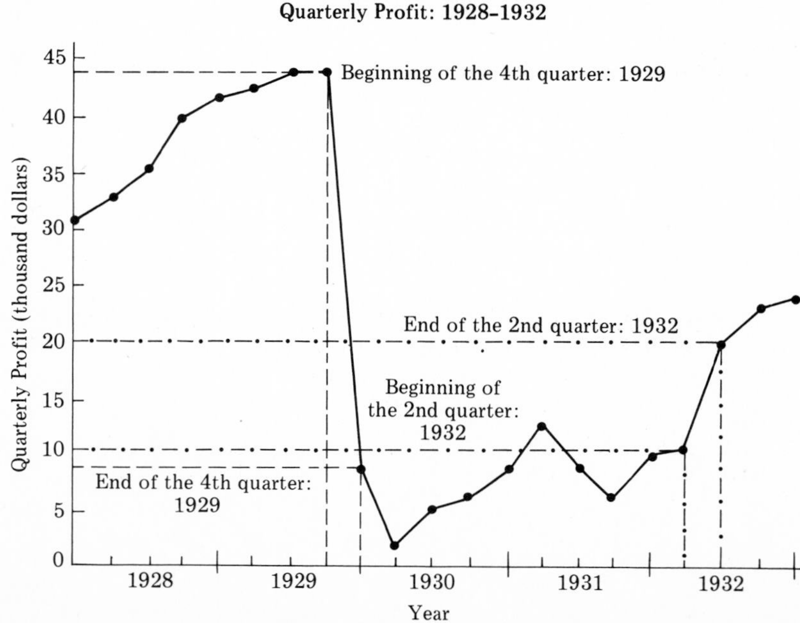

By inspection, it appears that the line graph drops the greatest amount in the fourth (or last) quarter of 1929. Using a straightedge (indicated by the broken lines), it is possible to determine the amount of decrease during this quarter. At the beginning of the fourth quarter of 1929, the reported profit was about 44,000 dollars; at the end of the quarter, the profit had dropped to about 9,000 dollars. The decrease in profits during this quarter amounted to \$44,000 - \$9,000 or \$35,000.

Although the change during the quarter with the greatest increase in profits is not as drastic as that during the quarter with the greatest decline, it is apparent that the greatest increase occurred in the second quarter of 1932. Using a straightedge to determine the amount of increase: at the beginning of the quarter the reported profit was about 10,000 dollars; at the end, about 20,000 dollars. The amount of increase was \$20,000 - \$10,000 or \$10,000.

Using graphical information: 

The purpose of the graph below is to show the relationships between the fractional parts that make up the total major sources of air pollution.

Consider the following questions:

A. **How many times greater is the amount of pollution that is attributable to transportation than the amount attributable to fuel-consuming stationary sources?**

B. **What is the percent of the total sources of pollution that are primarily man-made?**

Major Sources of Air Pollution

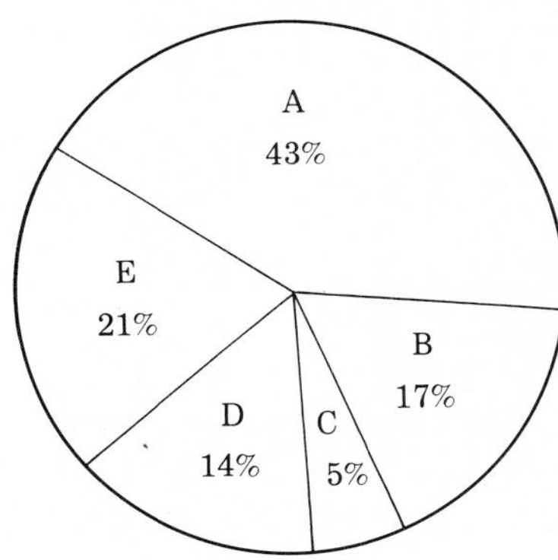

Key

A: Transportation
B: Agricultural and forest fires
C: Refuse disposal
D: Industrial and Manufacturing processes
E: Fuel combustion: stationary sources
(power generation, meating, buildings)

To answer A, forms of transportation contribute 43% of the pollutants in the air; stationary fuel-consuming sources contribute 21% of the pollutants.

$$\frac{0.43}{0.21} \approx 2$$

The amount of air pollution attributable to transportation is approximately two times greater than the amount of pollution attributable to stationary fuel-consuming sources.

To answer B, notice that the only category that is not primarily a "man-made" source of air pollution is agricultural and forest fires. Since all sources shown must total up to 100%,

$$100\% - 17\% = 1.00 - 0.17 = 0.83 \text{ or } 83\%$$

Using graphical information: 

The purpose of this graph is to compare the number of television sets sold by an independent dealer with the number of sets sold by a major department store. The vertical scale unit is 3 sets; the horizontal scale is in 1-month units.

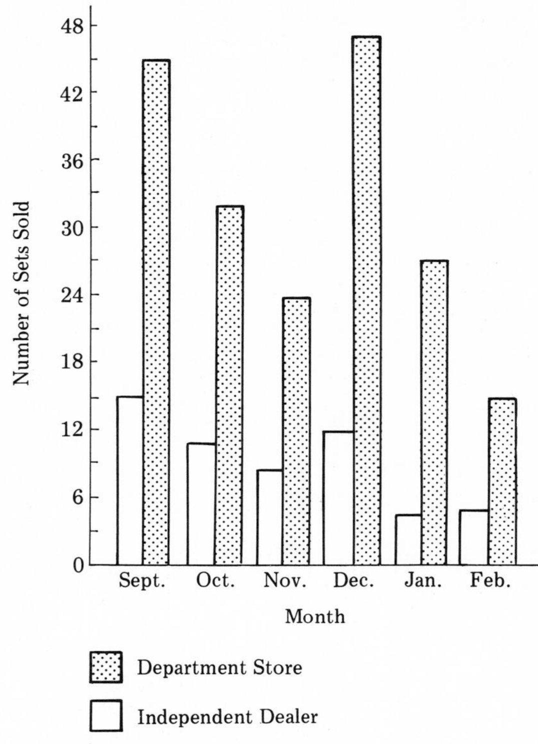

What is the average monthly difference between the number of sets sold by the independent dealer and the number of sets sold by the department store?

Complete the following:

From the graph:

The difference between the number of sets sold in

September is 45 - 15 = 30 sets
October is 32 - 11 = 21 sets
November is □ - 8 = 16 sets
December is 47 - ○ = 35 sets
January is 27 - 4 = △ sets
February is 16 - 5 = ⬡ sets

Calculating the average:

$$\text{Average} = \frac{30 + 21 + 16 + 35 + 23 + 11}{\triangledown}$$

$$= \frac{\Diamond}{\triangledown}$$

$\approx$ 22.7 or 23 sets per month

Using graphical information: (19)

The purpose of this pictograph is to compare the number of single-family dwellings built per year in a particular community. Each symbol represents 200 dwellings completed during the indicated year.

Constructionof Single-family Dwellings
1969 to 1974

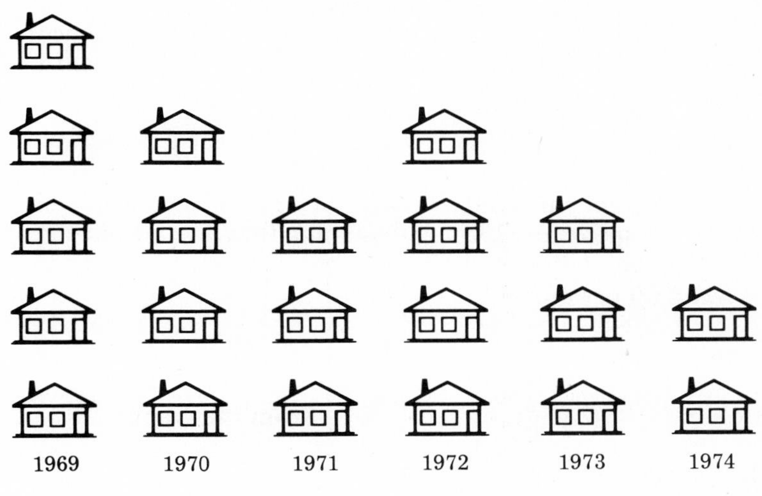

represents 200 single-family dwellings completed

What is the percent of decrease in construction of single-family dwellings from the year when the greatest number of dwellings was completed to the year when the least number of dwellings was completed?

Complete the following:

The year in which the greatest number of single-family dwellings was completed is □; 1,000 dwellings completed.

The year in which the least number of single-family dwellings was completed is 1974; ○ dwellings completed.

$$\textbf{Percent of decrease} = \frac{1{,}000 - \bigcirc}{\hexagon}$$

$$= 0.6$$

$$= \triangle\,\%$$

Using graphical information: 

The purpose of this circle graph is to show the relationships between the budgetary categories in the Allen's budget.

Allen Family's Budget

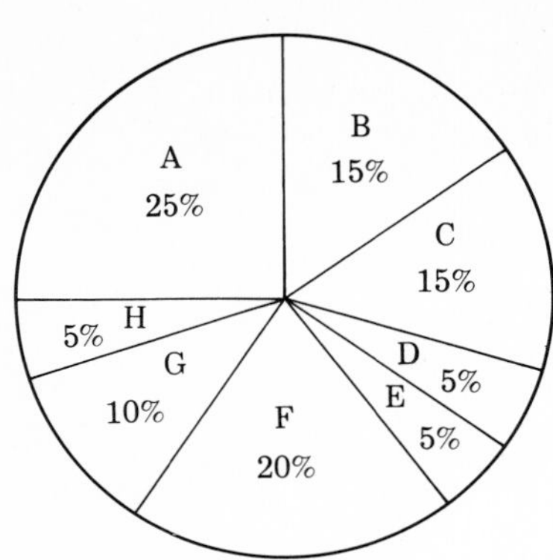

Key

A: Housing (mortage payment and utility bills)
B: Food
C: Insurance and savings
D: Medical expenses
E: Clothing
F: Automobiles (including installment payment and maintenance)
G: Installment payments
H: Miscellaneous

If the Allen family's monthly income is \$1,200.00, what is the difference between the amount spent for housing and the amount spent for clothing (assuming they live within their budget)? At the end of 1 year, what percent of their total yearly income has been used to make installment payments if their car payment is \$138.00 per month?

Complete the following:

From the graph:

The percent of their monthly income spent for housing is ○ %; the percent spent for clothing is △ %.

The amount spent for housing is

○ % of 1,200 or □

The amount spent for clothing is

△ % of 1,200 or ⬡

The amount of difference is

□ - ⬡ or \$240.00

From the graph:

The percent of their monthly income used to make installment payments (other than the car payment) is 10%.

10% of 1,200 is ▽

The total amount spent each month for installment payments is

\$138 + ▽ or \$258

The total yearly amount spent to make installment payments is

\$258 × ◇ or \$3,096

Their total yearly income is

\$1,200 × ◇ or \$14,400

Installment payments account for the following percent of their yearly expense:

$$\left(\frac{\$3,096}{\$14,400} \times 100\right) \% \text{ or } 21\frac{1}{2}\%$$

Practice Exercises

(21)

1. Use the following double bar graph to answer questions (a), (b), and (c):

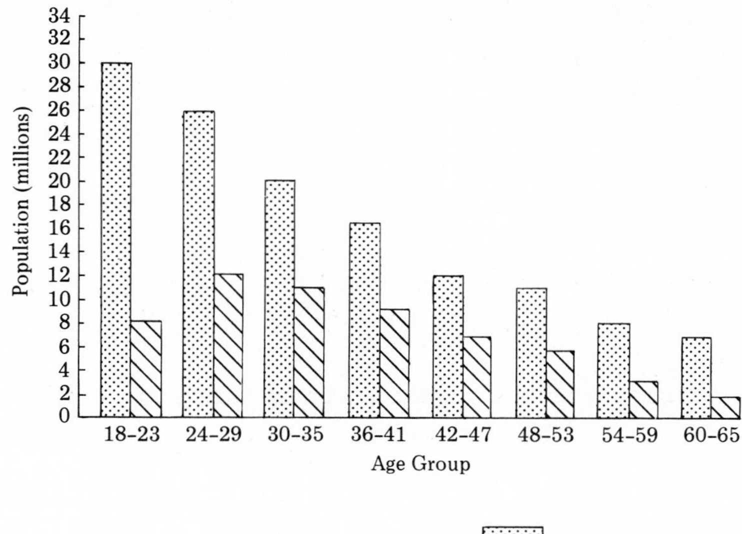

(a) In which bracket is the difference between the number of smokers and the number of non-smokers the greatest?
(b) The non-smokers are what percent of the smokers in the 18–23 years age bracket?
(c) What is the difference between the number of smokers in the age bracket with the most smokers and the age bracket with the least smokers?

2. Use the following line graph to answer questions (a), (b), and (c):

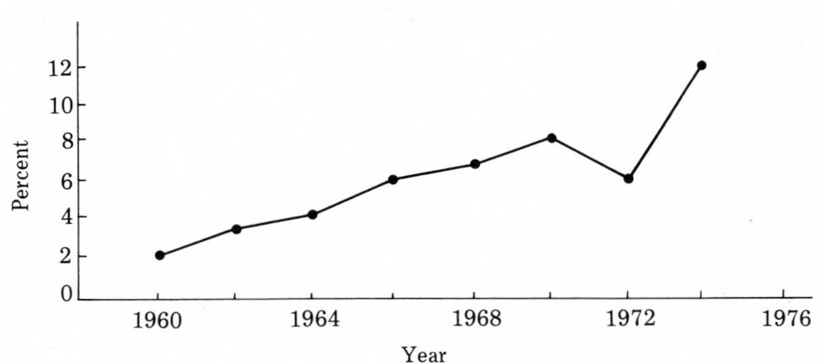

(a) The horizontal scale is in units of how many years?
(b) In which year does the greatest increase in the rate of inflation occur?
(c) What is the percent of increase in the rate of inflation from 1960 to 1974?

3. Use the following circle graph to answer questions (a), (b), and (c):

Population by Ages

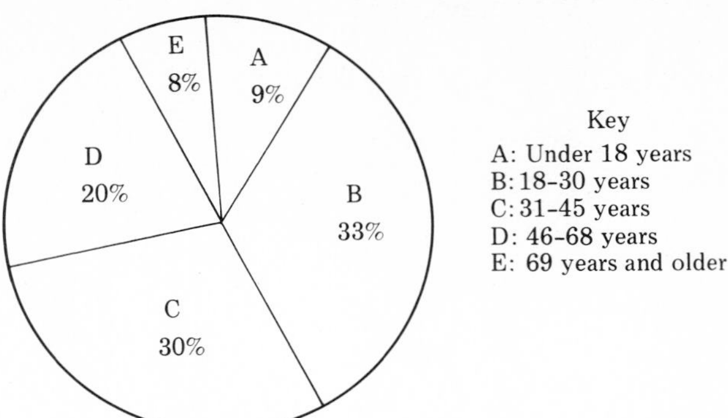

(a) The people under 31 years of age represent what fractional part of the total population?
(b) The 46–68 years group is what percent of the 69 years and older group?
(c) If the total population is 165,000 people, how many of these people are *not* 31–45 years of age?

In the last lesson the first thing that was done with any distribution was to organize the numerical information by using a table. Tables provide an efficient way for organizing any numerical values, especially if the table entries are numbers that must be used frequently. One table familiar to most people is a *sales tax table*, such as the sample shown below. 22

Amount	*Tax*	*Amount*	*Tax*	*Amount*	*Tax*
$1.90–2.09	$0.10	$4.50–4.69	$0.23	$7.10–7.29	$0.36
2.10–2.29	0.11	4.70–4.89	0.24	7.30–7.49	0.37
2.30–2.49	0.12	4.90–5.09	0.25	7.50–7.69	0.38
2.50–2.69	0.13	5.10–5.29	0.26	7.70–7.89	0.39
2.70–2.89	0.14	5.30–5.49	0.27	7.90–8.09	0.40
2.90–3.09	0.15	5.50–5.69	0.28	8.10–8.29	0.41
3.10–3.29	0.16	5.70–5.89	0.29	8.30–8.49	0.42
3.30–3.49	0.17	5.90–6.09	0.30	8.50–8.69	0.43
3.50–3.69	0.18	6.10–6.29	0.31	8.70–8.89	0.44
3.70–3.89	0.19	6.30–6.49	0.32	8.90–9.09	0.45
3.90–4.09	0.20	6.50–6.69	0.33	9.10–9.29	0.46
4.10–4.29	0.21	6.70–6.89	0.34	9.30–9.49	0.47
4.30–4.49	0.22	6.90–7.09	0.35	9.50–9.69	0.48

To see how this table is used, suppose that a sales clerk sells an item for $7.98. What is the sales tax on this amount? 

To use the sales tax table, the clerk reads down the columns labeled "Amount" until the correct category is located. The "correct category" will be one that begins with a number less than $7.98 and ends with a number greater than this amount. The sales tax is then read from the next column; the entry in the same row is the correct tax.

Amount	Tax	Amount	Tax	Amount	Tax
$1.90–2.09	$0.10	$4.50–4.69	$0.23	$7.10–7.29	$0.36
2.10–2.29	0.11	4.70–4.89	0.24	7.30–7.49	0.37
2.30–2.49	0.12	4.90–5.09	0.25	7.50–7.69	0.38
2.50–2.69	0.13	5.10–5.29	0.26	7.70–7.89	0.39
2.70–2.89	0.14	5.30–5.49	0.27	7.90–8.09	0.40
2.90–3.09	0.15	5.50–5.69	0.28	8.10–8.29	0.41
3.10–3.29	0.16	5.70–5.89	0.29	8.30–8.49	0.42
3.30–3.49	0.17	5.90–6.09	0.30	8.50–8.69	0.43
3.50–3.69	0.18	6.10–6.29	0.31	8.70–8.89	0.44
3.70–3.89	0.19	6.30–6.49	0.32	8.90–9.09	0.45
3.90–4.09	0.20	6.50–6.69	0.33	9.10–9.29	0.46
4.10–4.29	0.21	6.70–6.89	0.34	9.30–9.49	0.47
4.30–4.49	0.22	6.90–7.09	0.35	9.50–9.69	0.48

The tax on $7.98 is $0.40; the customer must pay $7.98 + $0.40 or $8.38.

Using a sales tax table: 

Noemi bought a binder for $3.98, paper for $1.98, and two ballpoint pens for $2.19. The sales clerk totaled her purchases before adding the 5% sales tax. What did Noemi have to pay? Would the amount she had to pay have been any different if the items had been taxed separately?

Amount	Tax	Amount	Tax	Amount	Tax
$1.90–2.09	$0.10	$4.50–4.69	$0.23	$7.10–7.29	$0.36
2.10–2.29	0.11	4.70–4.89	0.24	7.30–7.49	0.37
2.30–2.49	0.12	4.90–5.09	0.25	7.50–7.69	0.38
2.50–2.69	0.13	5.10–5.29	0.26	7.70–7.89	0.39
2.70–2.89	0.14	5.30–5.49	0.27	7.90–8.09 →	0.40
2.90–3.09	0.15	5.50–5.69	0.28	8.10–8.29	0.41
3.10–3.29	0.16	5.70–5.89	0.29	8.30–8.49	0.42
3.30–3.49	0.17	5.90–6.09	0.30	8.50–8.69	0.43
3.50–3.69	0.18	6.10–6.29	0.31	8.70–8.89	0.44
3.70–3.89	0.19	6.30–6.49	0.32	8.90–9.09	0.45
3.90–4.09	0.20	6.50–6.69	0.33	9.10–9.29	0.46
4.10–4.29	0.21	6.70–6.89	0.34	9.30–9.49	0.47
4.30–4.49	0.22	6.90–7.09	0.35	9.50–9.69	0.48

Complete the following by using the sample 5% sales tax table:

Calculating what Noemi had to pay:

The total taxable amount is $3.98 + $1.98 + $2.19 or ___.

From the table, the tax on this amount is ___.

Noemi had to pay ___ + ___ or $8.56.

Calculating the total cost if each item is first taxed separately:

The tax on \$3.98 is ◯ ; \$3.98 + ◯ = \$4.18

The tax on \$1.98 is ☐ ; \$1.98 + ☐ = \$2.08

The tax on \$2.19 is ◇ ; \$2.19 + ◇ = \$2.30

If each item is taxed separately, the total amount is

\$4.18 + \$2.08 + \$2.30 or \$8.56

There is no difference in the amounts in this example if the purchases are taxed separately or together.

Other tables all too familiar to any taxpayer are those such as the following: 

Tax Rate Schedule for:

Joint Return of Married Couple
(Provides for split-income benefits)

TAXABLE INCOME	*AMOUNTS AND RATES OF TAX*
\$0 to \$4,000	1%
4,000 to 7,000	\$40 plus 2% of amount over \$4,000
7,000 to 10,000	100 plus 3% of amount over 7,000
10,000 to 13,000	190 plus 4% of amount over 10,000
13,000 to 16,000	310 plus 5% of amount over 13,000
16,000 to 19,000	460 plus 6% of amount over 16,000
19,000 to 22,000	640 plus 7% of amount over 19,000
→ 22,000 to 25,000 →	850 plus 8% of amount over 22,000
25,000 to 28,000	1,090 plus 9% of amount over 25,000
28,000 and over	1,360 plus 10% of amount over 28,000

Tax Rate Schedule for:

Single Person
Separate Return of a Married Person

TAXABLE INCOME	*AMOUNTS AND RATES OF TAX*
\$0 to \$2,000	1%
2,000 to 3,500	\$20 plus 2% of amount over \$2,000
3,500 to 5,000	50 plus 3% of amount over 3,500
5,000 to 6,500	95 plus 4% of amount over 5,000
6,500 to 8,000	155 plus 5% of amount over 6,500
8,000 to 9,500	230 plus 6% of amount over 8,000
9,500 to 11,000	320 plus 7% of amount over 9,500
11,000 to 12,500	425 plus 8% of amount over 11,000
12,500 to 14,000	545 plus 9% of amount over 12,500
14,000 and over	680 plus 10% of amount over 14,000

To see how these tables are used, consider the following example:

> Mr. and Mrs. Hernandez are filing a joint income tax return on their taxable income of $23,700. What is the amount of federal income tax that they must pay? ("Taxable income" is the portion of the total year's income that is taxed after allowable deductions have been subtracted.)

Use the table on the left; read down the first column to determine the correct category for an income of $23,700.

The correct category is $22,000 to 25,000. Reading to the right in this category, we see that the tax is computed by adding $850 to the amount determined by subtracting $22,000 from their income and taking 8% of this difference:

$$\begin{array}{r} \$23{,}700 \\ -22{,}000 \\ \hline \$1{,}700 \end{array} \quad \Rightarrow \quad \begin{aligned} 8\% \text{ of } 1{,}700 &= 0.08 \times 1{,}700 \\ &= \$136 \end{aligned}$$

The amount of tax that Mr. and Mrs. Hernandez must pay is $850 + $136 or $986.

Using a Federal Tax Rate Schedule: 

Earl Williams was married on December 15 last year. If his total earnings for last year was $15,250 and his taxable income was 80% of this amount, how much did Earl save on his taxes by being able to file a joint return?

Tax Rate Schedule for:

Joint Return of Married Couple
(Provides for split-income benefits)

TAXABLE INCOME	*AMOUNTS AND RATES OF TAX*
$0 to $4,000	1%
4,000 to 7,000	$40 plus 2% of amount over $4,000
7,000 to 10,000	100 plus 3% of amount over 7,000
10,000 to 13,000	190 plus 4% of amount over 10,000
13,000 to 16,000	310 plus 5% of amount over 13,000
16,000 to 19,000	460 plus 6% of amount over 16,000
19,000 to 22,000	640 plus 7% of amount over 19,000
22,000 to 25,000	850 plus 8% of amount over 22,000
25,000 to 28,000	1,090 plus 9% of amount over 25,000
28,000 and over	1,360 plus 10% of amount over 28,000

Tax Rate Schedule for:

Single Person
Separate Return of a Married Person

TAXABLE INCOME	*AMOUNTS AND RATES OF TAX*
$0 to $2,000	1%
2,000 to 3,500	$20 plus 2% of amount over $2,000
3,500 to 5,000	50 plus 3% of amount over 3,500
5,000 to 6,500	95 plus 4% of amount over 5,000
6,500 to 8,000	155 plus 5% of amount over 6,500
8,000 to 9,500	230 plus 6% of amount over 8,000
9,500 to 11,000	320 plus 7% of amount over 9,500
11,000 to 12,500	425 plus 8% of amount over 11,000
12,500 to 14,000	545 plus 9% of amount over 12,500
14,000 and over	680 plus 10% of amount over 14,000

Complete the following by using the Federal Tax Rate Schedule:

In order to compare the two amounts, it will be necessary to calculate the taxes for both a single and a joint return. In either case, the first thing that must be determined is the amount of Earl's taxable income.

His taxable income is 80% of $15,250 or $12,200.

Calculating the tax due from the Joint Return Schedule:

The amount over $10,000 is $12,200 - $10,000 = $2,200.

△ % of $2,200 = □

Tax due if a joint return is filed is □ + ○.

Calculating the tax due from the Single Person Schedule:

The amount over $11,000 is $12,200 - $11,000 = $1,200.

⬡ % of $1,200 = ◇

Tax due if a single return is filed is ◇ + ▽.

By filing a joint return, Earl saved $521 - $278 or $243.

The following is a sample of a compound interest table, a table that few people are familiar with. It is included because there seems to be some rather widespread confusion about the difference between *compound* and *simple interest*. 27

Amount of $1.00 Compounded Annually

Number of Years	*1%*	*2%*	*3%*	*4%*	*5%*	*6%*
1	1.01	1.02	1.03	1.04	1.05	1.06
2	1.02	1.04	1.06	1.08	1.10	1.12
3	1.03	1.06	1.09	1.12	1.16	1.19
4	1.04	1.08	1.13	1.17	1.22	1.26
5	1.05	1.10	1.16	1.22	1.28	1.34
6	1.06	1.13	1.19	1.27	1.34	1.42
7	1.07	1.15	1.23	1.32	1.41	1.50
8	1.08	1.17	1.27	1.37	1.48	1.59
9	1.09	1.20	1.30	1.42	1.55	1.69
10	1.10	1.22	1.34	1.48	1.63	1.79

Compound interest is an amount of interest that is paid, into a savings account, for example, at the end of regular intervals of time. The amount of interest for the next interval of time is then computed on both the principal (amount of money originally put in the account) *and* the interest that has been previously added to the account. *Simple interest* is paid only once, when the money is withdrawn.

For example, if $200 is placed in each of two savings accounts, one paying 4% simple interest and the other paying 4% interest compounded annually, how much money can be withdrawn from the two accounts at the end of 5 years?

To determine the amount of money in the account paying 4% simple interest, the following formula is used:

Interest = Principal × Rate × Time (in years)

$$I = \$200 \times 4\% \times 5 \text{ years}$$
$$= (200)(0.04)(5)$$
$$= \$40.00$$

The total amount of money that can be withdrawn from the account paying simple interest is $240.

To determine the amount of money in the account paying 4% interest compounded annually, the compound interest table is used. The amount given by this table is the amount of money in the account, at the end of the number of years shown in the first column, for each $1.00 originally deposited. Read down the first column to 5 years and then across the row to the entry in the 4% column: each $1.00 in the account has grown to $1.22 at the end of this time. Thus, the total amount in the account at the end of 5 years is

200 × $1.22 or $244.00

Using a compound interest table: 

Grant is faced with the following decision. He has $1,000 in a savings account that pays 6% interest that is compounded annually. He is trying to decide if he should withdraw $750 to pay off some debts. If he pays off his debts, he will save $175 interest charges in the next 3 years. Which will be worth more to him: leaving the money in the savings account and letting it draw interest or withdrawing the money and saving the finance charges?

Amount of $1.00 Compounded Annually

Number of Years	*1%*	*2%*	*3%*	*4%*	*5%*	*6%*
1	1.01	1.02	1.03	1.04	1.05	1.06
2	1.02	1.04	1.06	1.08	1.10	1.12
3	1.03	1.06	1.09	1.12	1.16	1.19
4	1.04	1.08	1.13	1.17	1.22	1.26
5	1.05	1.10	1.16	1.22	1.28	1.34
6	1.06	1.13	1.19	1.27	1.34	1.42
7	1.07	1.15	1.23	1.32	1.41	1.50
8	1.08	1.17	1.27	1.37	1.48	1.59
9	1.09	1.20	1.30	1.42	1.55	1.69
10	1.10	1.22	1.34	1.48	1.63	1.79

Use the compound interest table to complete the following:

Calculating the amount of interest that $1,000 will earn at 6% interest in 3 years:

From the table it will be seen that each $1.00 in the account now will have grown to □ at the end of 3 years.

□ X 1,000 = ○

○ - $1,000 = $190, the amount of interest in 3 years

If he withdraws $750 from the account, $1,000 - $750 = $250 will continue to draw interest for the 3 years:

□ X 250 = △

△ - $250 = $47.50, the amount of interest in 3 years

If he leaves the money in the bank, it will be worth $190 at the end of the 3-year period. If he pays off his bills early, it will be worth the money he saves on finance charges plus the amount of interest that the $250 balance will earn in that same time: $175 + $47.50 = $222.50. It will save him $32.50 if he withdraws the money and pays off his debts.

Practice Exercises

1. Use the sample sales tax table in **Frame 24** to answer the following: Nanette purchases a used Algebra book for $6.89 plus tax. If she gives the clerk a ten-dollar bill to pay for the book, how much change should she be given?
2. Use the income tax schedule in **Frame 25** to answer the following: Out of his $11,300 total earnings for the year, $700 was withheld from Dan Brosnahan's wages for federal income tax. If his taxable income is 90% of his total earnings, what will be the amount of the refund if he files as a single person?
3. Use the sample compound interest table in **Frame 27** to answer the following: How much more interest is earned by investing $560 for 4 years at 6% interest, compounded annually, than is earned by investing the same amount for the same length of time at 4% interest, compounded annually?

POST-TEST

1. Use the following graph to answer the questions (a) through (e):

Projected U.S. Population
3 children per family vs. 2 children per family

Population (millions): 0, 100, 200, 300, 400, 500, 600, 700, 800

Year: 1850, 1870, 1910, 1930, 1950, 1970, 1990, 2010, 2030, 2050, 2070

- - - - - 3-children-per-family population projection
•••••••••• 2-children-per-family population projection

(a) What is the name of this type of graph?
(b) What purpose is best served by this type of graph?
(c) If the 3 children per family growth rate continues, the 2070 population will be how many times the 1970 population?
(d) The 1970 population has approximately what percent increase compared to the 1920 population?
(e) What is the difference between the projected populations in 2070?

2. Use the following excerpt from the sample 5% sales tax table to answer questions (a) and (b):

Amount	*Tax*
$4.50–4.69	$0.23
4.70–4.89	0.24
4.90–5.09	0.25
5.10–5.29	0.26
5.30–5.49	0.27
5.50–5.69	0.28

(a) Mistie purchased a book prices $4.97. How much payment, including sales tax, is due?

(b) Steven purchased a tie priced $5.35. If he hands the clerk $6.00 for his purchase, how much change should he be given?

DRILL EXERCISES: UNIT 8

1. An employee reported the following number of overtime hours worked per day for a 10-day period: 2, 0, 1, 3, 2, 3, 4, 0, 2, 1. Determine each of the following correct to the nearest whole number:
   (a) Range
   (b) Mode (or modes, if any)
   (c) Median
   (d) Mean
   (e) Standard deviation
   (f) Graph the frequency polygon for the distribution.
   (g) Graph the histogram for the distribution.
2. A baseball player has made the following number of hits per game while on a 12-game exhibition tour: 5, 2, 0, 1, 4, 3, 0, 4, 2, 1, 3, 1. Determine each of the following correct to the nearest whole number:
   (a) Range
   (b) Mode (or modes, if any)
   (c) Median
   (d) Mean
   (e) Standard deviation
   (f) Graph the frequency polygon for the distribution.
   (g) Graph the histogram for the distribution.
3. A biology student has weighed 15 white mice. Their weights, in grams, are as follows: 53, 50, 51, 47, 53, 47, 49, 51, 50, 49, 54, 53, 51, 55, 50. Determine each of the following correct to the nearest whole number:
   (a) Range
   (b) Mode (or modes, if any)
   (c) Median
   (d) Mean
   (e) Standard deviation
   (f) Graph the frequency polygon for the distribution.
   (g) Graph the histogram for the distribution.
4. The library tax rates, in cents per year, for the last 10 years have been as follows: 22, 23, 20, 22, 26, 23, 25, 22, 25, 23. Determine each of the following correct to the nearest whole number:
   (a) Range
   (b) Mode (or modes, if any)
   (c) Median
   (d) Mean
   (e) Standard deviation
   (f) Graph the frequency polygon for the distribution.
   (g) Graph the histogram for the distribution.
5. A consumer affairs analyist has conducted a survey to determine the selling price of a regular-size box of Brand X detergent at 12 local stores. The selling prices, in cents, are: 69, 72, 70, 70, 75, 75, 75, 70, 74, 70, 70, 70. Determine each of the following correct to the nearest whole number:
   (a) Range
   (b) Mode (or modes, if any)
   (c) Median
   (d) Mean
   (e) Standard deviation
   (f) Graph the frequency polygon for the distribution.
   (g) Graph the histogram for the distribution.

6. The lifetimes, in months, for a sample of 25 similar automobile batteries are as follows: 24, 30, 27, 25, 20, 28, 30, 24, 29, 29, 27, 21, 26, 24, 20, 27, 25, 25, 28, 30, 24, 27, 27, 24, 30. Determine each of the following correct to the nearest whole number:
   (a) Range
   (b) Mode (or modes, if any)
   (c) Median
   (d) Mean
   (e) Standard deviation
   (f) Graph the frequency polygon for the distribution.
   (g) Graph the histogram for the distribution.

Each of the following histograms represents the distribution of scores for a particular class of 21 students on a midterm exam.

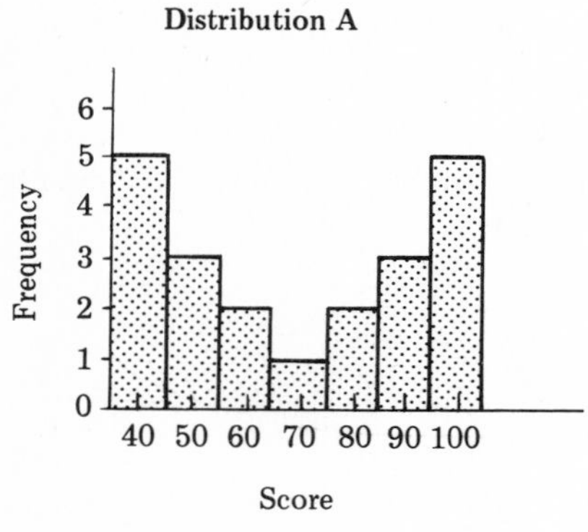

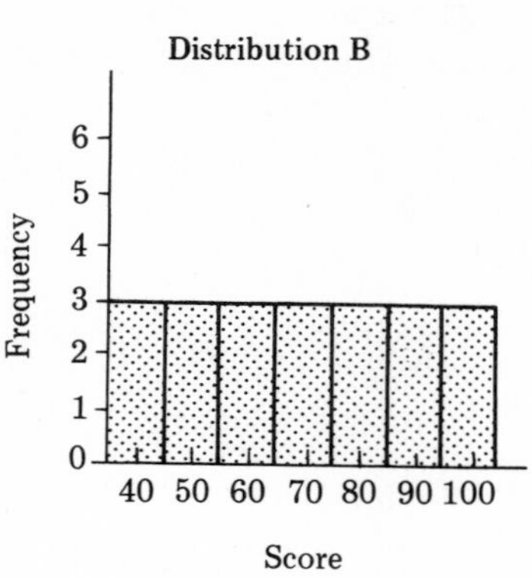

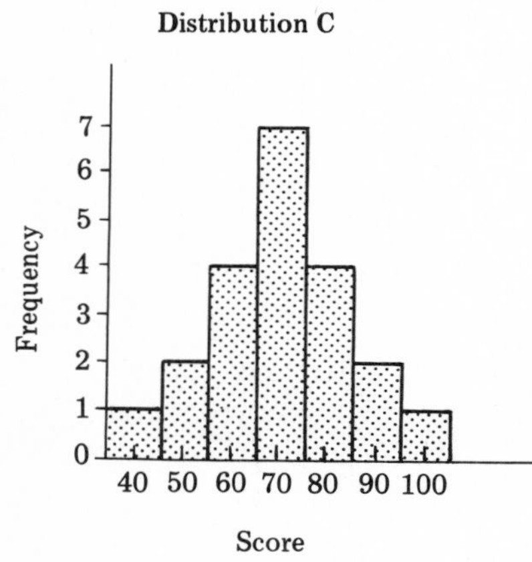

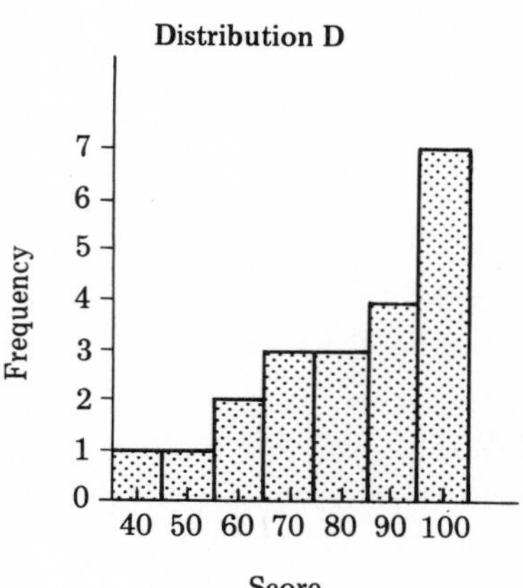

These histograms will be used to answer exercises 7 through 18.

7. Estimate the mode (or modes, if any) for
   (a) Distribution A
   (b) Distribution B
   (c) Distribution C
   (d) Distribution D

8. Estimate the mean for
   (a) Distribution A
   (b) Distribution B
   (c) Distribution C
9. Estimate the median for
   (a) Distribution A
   (b) Distribution B
   (c) Distribution C
10. Estimate the range for distributions A through D.
11. Identify the distribution that has the highest mean score.
12. Identify the distribution for which the number of students who scored the mean is greatest.
13. Identify the distribution that has the lowest mean score.
14. Identify the distribution for which the number of students who scored the mean score is least.

Use the following graph to answer exercises 15 through 19.

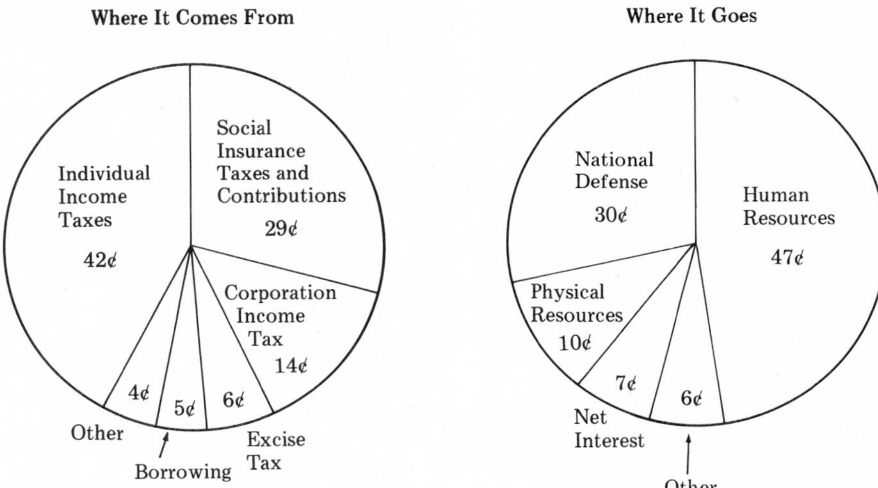

15. Name the purpose for which this type of graph is best suited.
16. Suppose that the federal government collected \$63,500,000,000 in taxes. How much of this money came from corporation income taxes?
17. If the federal government spent a total of \$81,500,000,000, how much money was expended for physical resources, net interest, and the "other" categories combined?
18. The difference between the human resources category and the national defense category is what percent of the expenditures.
19. If \$26,670,000,000 was collected from individual income taxes, how much was collected from all sources?

Use the following graph to answer exercises 20 through 26.

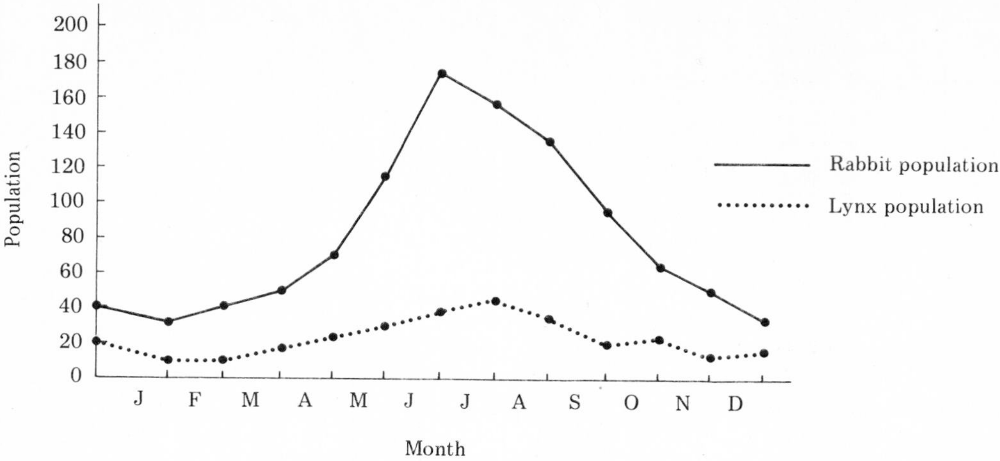

20. Name the purpose for which this type of graph is best suited.
21. The rabbit population increases by what amount from the beginning of February to the end of June?
22. The lynx population decreases by what amount from the beginning of August to the end of November?
23. The rabbit population is how much greater than the lynx population at the end of July?
24. The lynx population is how much less than the rabbit population at the beginning of January?
25. The rabbit population has undergone what percent decrease from the beginning of August to the end of October?
26. The lynx population has undergone what percent increase from the beginning of April to the end of July?

Use the following graph to answer exercises 27 through 31.

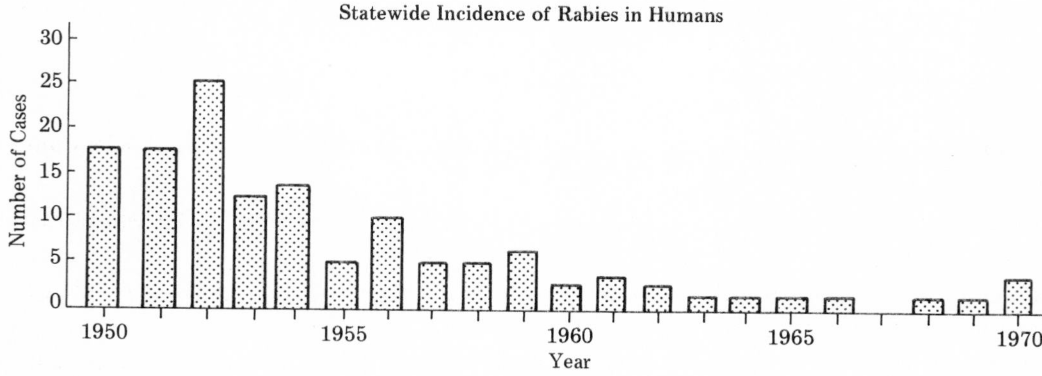

27. Name the purpose for which this type of graph is best suited.
28. Between which two consecutive years has the decrease in the number of rabies cases been the greatest?

29. Between which two consecutive years has the increase in the number of rabies cases been the greatest?

30. What is the percent of increase in the number of rabies cases from 1955 to 1956?

31. What is the percent of decrease in the number of rabies cases from 1956 to 1957?

Use the following table to answer exercises 32 through 39.

Tax Rate Schedule for:

Joint Return of Married Couple
(Provides for split-income benefits)

Taxable Income	*Amounts and Rates of Tax*
$0 to $4,000	1%
4,000 to 7,000	$40 plus 2% of amount over $4,000
7,000 to 10,000	100 plus 3% of amount over 7,000
10,000 to 13,000	190 plus 4% of amount over 10,000
13,000 to 16,000	310 plus 5% of amount over 13,000
16,000 to 19,000	460 plus 6% of amount over 16,000
19,000 to 22,000	640 plus 7% of amount over 19,000
22,000 to 25,000	850 plus 8% of amount over 22,000
25,000 to 28,000	1,090 plus 9% of amount over 25,000
28,000 and over	1,360 plus 10% of amount over 28,000

Tax Rate Schedule for:

Single Person
Separate Return of a Married Person

Taxable Income	*Amounts and Rates of Tax*
$0 to $2,000	1%
2,000 to 3,500	$20 plus 2% of amount over $2,000
3,500 to 5,000	50 plus 3% of amount over 3,500
5,000 to 6,500	95 plus 4% of amount over 5,000
6,500 to 8,000	155 plus 5% of amount over 6,500
8,000 to 9,500	230 plus 6% of amount over 8,000
9,500 to 11,000	320 plus 7% of amount over 9,500
11,000 to 12,500	425 plus 8% of amount over 11,000
12,500 to 14,000	545 plus 9% of amount over 12,500
14,000 and over	680 plus 10% of amount over 14,000

32. An unmarried person has a taxable income of $8,950. Calculate the amount of income tax to be paid.

33. Ralph and Kathleen Conner have a taxable income of $15,200 and will file a joint return. Determine the amount of income tax due.

34. Mitch Goodan and his wife have a taxable income of $21,300. How much tax is due if they file a joint return?

35. Trung Nguyen, a single student, has a taxable income of $6,300. How much tax must he pay?

36. Al Romero has a taxable income of $7,350. If he files a single person's return, how much tax will he pay?

37. Don and Sonnie Walker filed a joint income tax return. If their taxable income is $14,700 and if $534 has been withheld from Don's salary for federal income taxes, how much money, if any, will they get back?

38. The Graves earned $19,300 last year. If their taxable income is 85% of their earnings, how much income tax is due?

39. Debbie Pappas earned $13,250 last year, 90% of which is taxable. If she files a single person's return, how much tax does she pay?

Use the following 5% sales tax table to answer exercises 40 through 43.

Amount	*Tax*	*Amount*	*Tax*	*Amount*	*Tax*
$1.90–2.09	$0.10	$4.50–4.69	$0.23	$7.10–7.29	$0.36
2.10–2.29	0.11	4.70–4.89	0.24	7.30–7.49	0.37
2.30–2.49	0.12	4.90–5.09	0.25	7.50–7.69	0.38
2.50–2.69	0.13	5.10–5.29	0.26	7.70–7.89	0.39
2.70–2.89	0.14	5.30–5.49	0.27	7.90–8.09	0.40
2.90–3.09	0.15	5.50–5.69	0.28	8.10–8.29	0.41
3.10–3.29	0.16	5.70–5.89	0.29	8.30–8.49	0.42
3.30–3.49	0.17	5.90–6.09	0.30	8.50–8.69	0.43
3.50–3.69	0.18	6.10–6.29	0.31	8.70–8.89	0.44
3.70–3.89	0.19	6.30–6.49	0.32	8.90–9.09	0.45
3.90–4.09	0.20	6.50–6.69	0.33	9.10–9.29	0.46
4.10–4.29	0.21	6.70–6.89	0.34	9.30–9.49	0.47
4.30–4.49	0.22	6.90–7.09	0.35	9.50–9.69	0.48

40. Garth made the following purchases: 2 shirts at $2.95 each and 3 pairs of socks at $0.85 each. How much sales tax will have to be paid on the total amount of his purchases?

41. Shelly returned an item that was priced $6.39 plus tax. If she receives a refund which includes the sales tax paid, how much of her money will she get back?

42. Ginnie purchased a blouse priced $7.37. How much change should she receive if she pays for the blouse with a ten-dollar bill?

43. Dena Jay has only $5.45 in her purse. Determine the amount of the largest purchase for which she can pay both the selling price and the sales tax.

Use the following table to answer exercises 44 through 47.

Amount of $1.00 Compounded Annually

Number of Years	*1%*	*2%*	*3%*	*4%*	*5%*	*6%*
1	1.01	1.02	1.03	1.04	1.05	1.06
2	1.02	1.04	1.06	1.08	1.10	1.12
3	1.03	1.06	1.09	1.12	1.16	1.19
4	1.04	1.08	1.13	1.17	1.22	1.26
5	1.05	1.10	1.16	1.22	1.28	1.34
6	1.06	1.13	1.19	1.27	1.34	1.42
7	1.07	1.15	1.23	1.32	1.41	1.50
8	1.08	1.17	1.27	1.37	1.48	1.59
9	1.09	1.20	1.30	1.42	1.55	1.69
10	1.10	1.22	1.34	1.48	1.63	1.79

44. Donald Ellwood invested $400 compounded annually at 6%. Determine the amount of his investment after 3 years.

45. Jerry Beaumount invested $2,000 compounded annually at 4%. Determine the amount of interest earned after 5 years.

46. Harvey, a retired author, has a monthly income of $420. If the rate of inflation is 5%, compounded annually, determine the loss in purchasing power over a 2-year period if prices are assumed to inflate according to the amounts shown on this table.

47. If Linda earns $4.60 per hour now, determine the amount of hourly pay needed to maintain the same purchasing power after 4 years of a 6% rate of inflation (compounded annually).

Use the following frequency distribution to answer questions 1 through 5 correct to the nearest whole number:

Score	*Frequency*	*Cumulative Frequency*	*Frequency × Value*	*Deviation*	*Deviation Squared*	*Frequency × Deviation Squared*
10	2					
9	4					
8	2					
7	0					
6	1					
5	2					
4	1					

1. Determine the range.
   (a) 10
   (b) 6
   (c) 4
   (d) 9
   (e) None of the above
2. Determine the mode.
   (a) 10
   (b) 4
   (c) 7
   (d) 9
   (e) None of the above
3. Determine the median correct to the nearest whole number.
   (a) 9
   (b) 8
   (c) 7
   (d) 6
   (e) None of the above
4. Determine the mean correct to the nearest whole number.
   (a) 6
   (b) 7
   (c) 9
   (d) 6
   (e) None of the above
5. Determine the standard deviation correct to the nearest whole number.
   (a) 1
   (b) 2

(c) 3
(d) 4
(e) None of the above

Use the following histogram to answer questions 6 and 7:

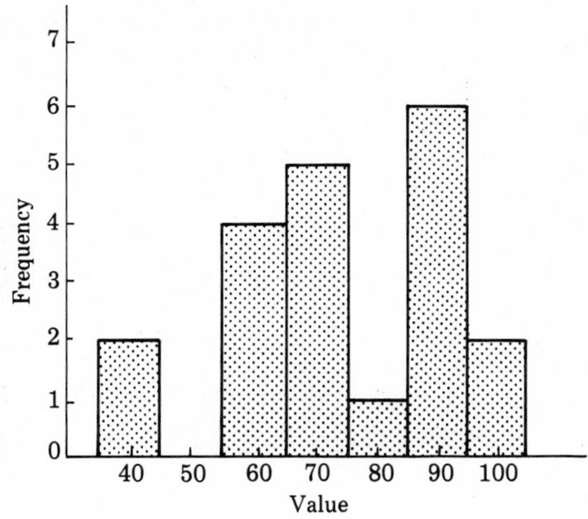

6. Which one of the following statements is false?
   (a) The range is 60.
   (b) The mode is 90.
   (c) The median is 70.
   (d) The frequency of 50 is 2.
   (e) None of the above
7. Which one of the following statements is true?
   (a) The frequency of 80 is less than the frequency of 50.
   (b) The range is 100.
   (c) The mode is 70.
   (d) The median is 60.
   (e) None of the above

Use the following graph to answer questions 8 through 10:

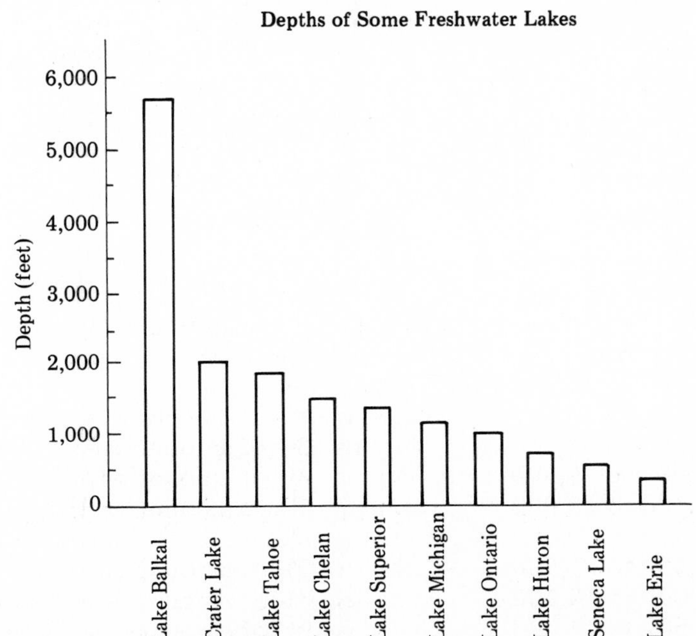

8. The "depth" graph best illustrates which one of the following purposes?
(a) A trend
(b) A comparison
(c) A change
(d) Parts of the whole
(e) None of the above

9. The depth of Seneca Lake is approximately what fractional part of the depth of Crater Lake?
(a) $\frac{1}{4}$
(b) $\frac{2}{5}$
(c) $\frac{1}{2}$
(d) $\frac{4}{5}$
(e) None of the above

10. The difference in depth of Lake Huron and Lake Tahoe is approximately how many feet?
(a) 500 feet
(b) 100 feet
(c) 1,000 feet
(d) 1,500 feet
(e) None of the above

Use the following tables to answer questions 11 through 13:

Tax Rate for:

Joint Return of Married Couple

Taxable Income	*Amounts and Rates of Tax*
$0 to $4,000	1%
4,000 to 7,000	$40 plus 2% of amount over $4,000
7,000 to 10,000	100 plus 3% of amount over 7,000
10,000 to 13,000	190 plus 4% of amount over 10,000
13,000 to 16,000	310 plus 5% of amount over 13,000
16,000 to 19,000	460 plus 6% of amount over 16,000
19,000 to 22,000	640 plus 7% of amount over 19,000
22,000 to 25,000	850 plus 8% of amount over 22,000
25,000 to 28,000	1,090 plus 9% of amount over 25,000
28,000 and over	1,360 plus 10% of amount over 28,000

Tax Rate for:

Single Person's Return

Taxable Income	*Amounts and Rates of Tax*
$0 to $2,000	1%
2,000 to 3,500	$20 plus 2% of amount over $2,000
3,500 to 5,000	50 plus 3% of amount over 3,500
5,000 to 6,500	95 plus 4% of amount over 5,000
6,500 to 8,000	155 plus 5% of amount over 6,500
8,000 to 9,500	230 plus 6% of amount over 8,000
9,500 to 11,000	320 plus 7% of amount over 9,500
11,000 to 12,500	425 plus 8% of amount over 11,000
12,500 to 14,000	545 plus 9% of amount over 12,500
14,000 and over	680 plus 10% of amount over 14,000

11. Mr. and Mrs. Frank Levin had a taxable income of \$23,400. If they prepare a joint return, how much tax will they pay?
(a) \$850
(b) \$858
(c) \$112
(d) \$962
(e) None of the above

12. Linda Thurston, an unmarried woman, earned \$13,700. If her taxable income is 90% of her earnings, how much tax will she pay?
(a) \$108
(b) \$106.40
(c) \$653
(d) \$531.40
(e) None of the above

13. Dave and Eunice Paquette earned \$18,400. If they can deduct 20% of their earnings to obtain their taxable income, what amount of tax refund will be received if \$460 has been withheld?
(a) \$86
(b) \$396
(c) \$64
(d) \$310
(e) None of the above

Use the following graph to answer questions 14 through 16:

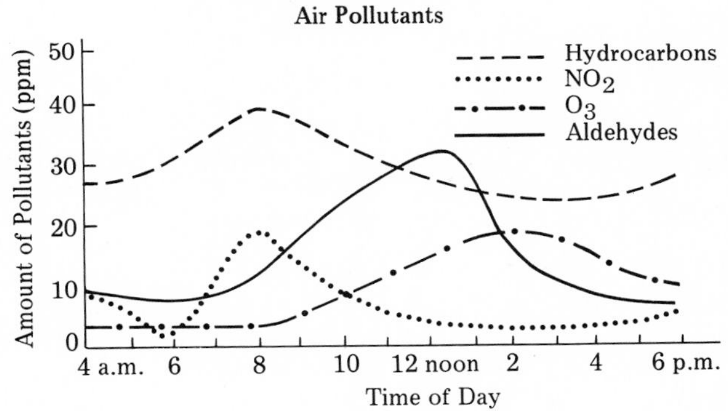

14. During which hour of the day do the hydrocarbons reach a maximum amount?
(a) 3:00
(b) 12:00
(c) 2:00
(d) 6:00 a.m.
(e) None of the above

15. During which hour does the nitrogen oxide, NO_2, have a minimum amount?
(a) 6:00 a.m.
(b) 8:00
(c) 6:00 p.m.
(d) 2:00
(e) None of the above

16. What is approximately the amount of difference in parts per million (ppm) between the highest and lowest aldehyde value?
(a) 25 ppm
(b) 35 ppm
(c) 10 ppm
(d) 20 ppm
(e) None of the above

Use the following table to answer questions 17 through 20:

Amount of $1.00 Compounded Annually

Number of Years	*1%*	*2%*	*3%*	*4%*	*5%*	*6%*
1	1.01	1.02	1.03	1.04	1.05	1.06
2	1.02	1.04	1.06	1.08	1.10	1.12
3	1.03	1.06	1.09	1.12	1.16	1.19
4	1.04	1.08	1.13	1.17	1.22	1.26
5	1.05	1.10	1.16	1.22	1.28	1.34

17. If $150 is invested at 5% compounded annually, what is the amount of the investment after 3 years?
    (a) $24
    (b) $174
    (c) $15
    (d) $165
    (e) None of the above
18. If $500 is invested at 6% compounded annually, how much interest is earned after 4 years?
    (a) $595
    (b) $95
    (c) $630
    (d) $130
    (e) None of the above
19. How much more interest is earned if $200 is invested at 4% compound interest for 5 years than $200 invested at 4% simple interest for 5 years?
    (a) $240
    (b) $244
    (c) $44
    (d) $40
    (e) None of the above
20. Darlene currently earns a weekly salary of $180. If 5% inflation, compounded annually, occurs over the next 2 years, what should her weekly salary be if her purchasing power is to remain the same?
    (a) $185
    (b) $190
    (c) $198
    (d) $204
    (e) None of the above

LESSON 29:

Pre-test

1. (a) Range is 6 (b) Mode is 2 (c) Median is 3
   (d) Mean $\approx$ 3 (e) Standard deviation $\approx$ 2

2.

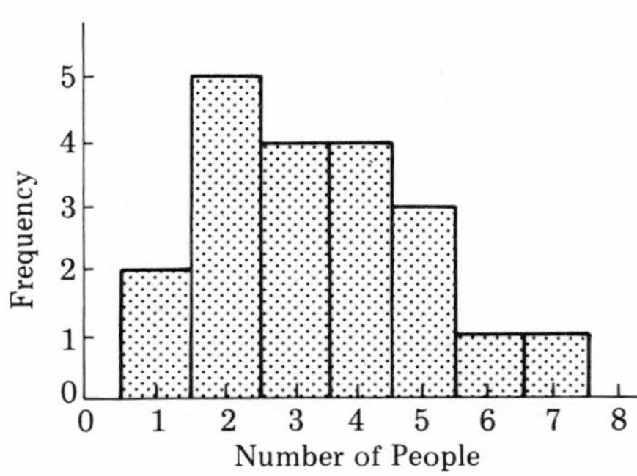

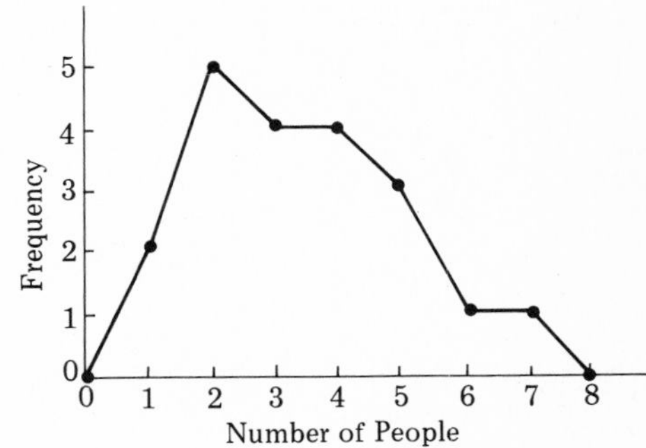

Practice Exercises

Frame 19

1. Range is 91 - 83 = 8
   Mode is 89
   Median is $\dfrac{87 + 88}{2} = 87.5 \approx 88$
   Mean is $\dfrac{1{,}044}{12} = 87$
   Standard deviation $\approx \sqrt{5.3} \approx 2$
2. Range is 20 - 10 = 10
   Modes are 16 and 18
   Median is $\dfrac{16 + 16}{2} = 16$
   Mean is $\dfrac{481}{30} \approx 16$
   Standard deviation $= \sqrt{8.5} \approx 3$
3. Range is 8
   Mode is 8
   Median is 8
   Mean is $\dfrac{196}{25} \approx 8$
   Standard deviation $\approx \sqrt{5.48} \approx 2$

Frame 25

1.

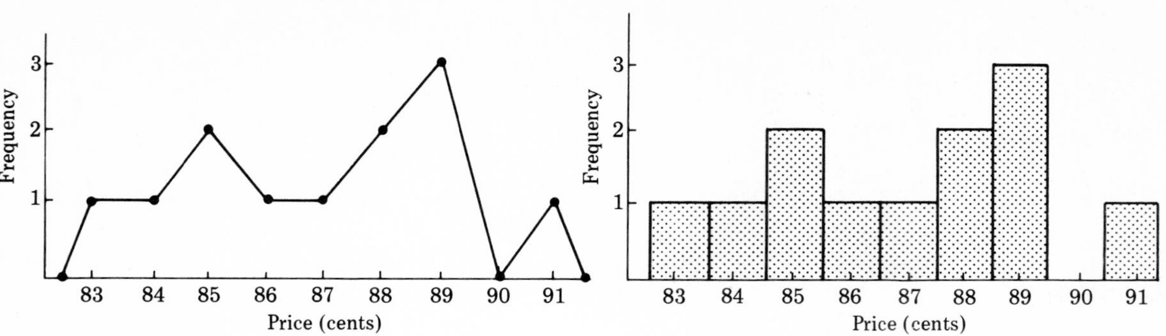

2.

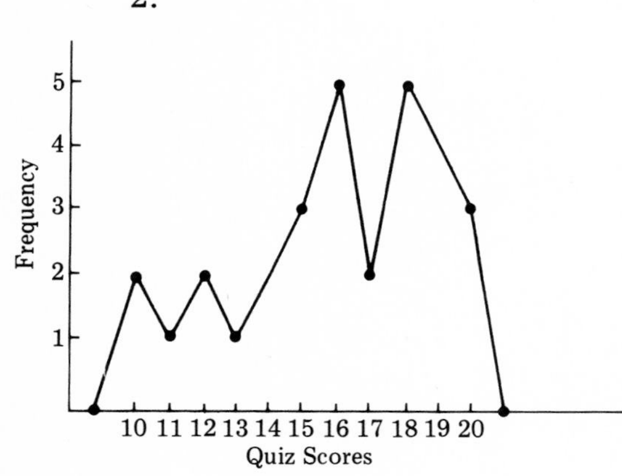

Frequency

10 11 12 13 14 15 16 17 18 19 20

Quiz Scores

3.

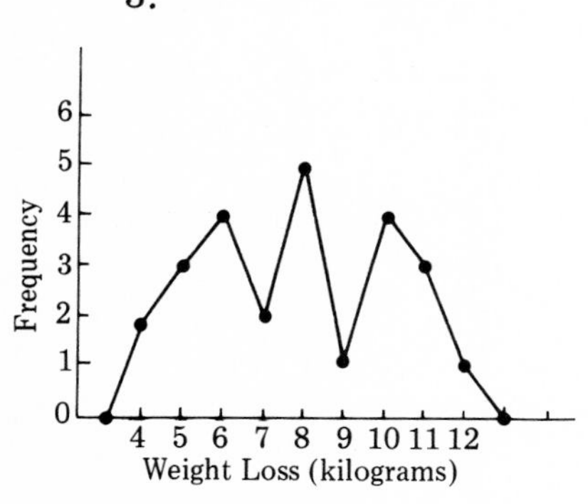

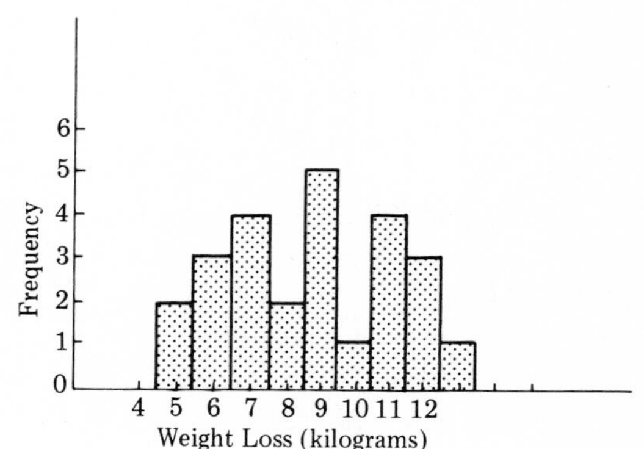

Post-test

1. (a) Range is 7
   (b) Mode is 20
   (c) Median is 19
   (d) Mean is $\frac{287}{15} \approx 19$
   (e) Standard deviation is approximately $\sqrt{3.3} \approx 2$

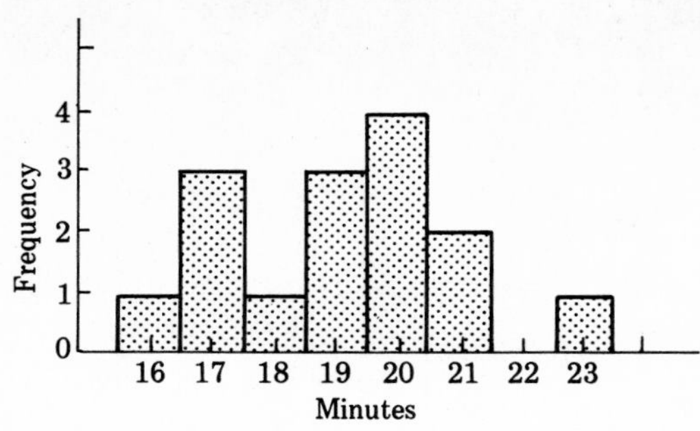

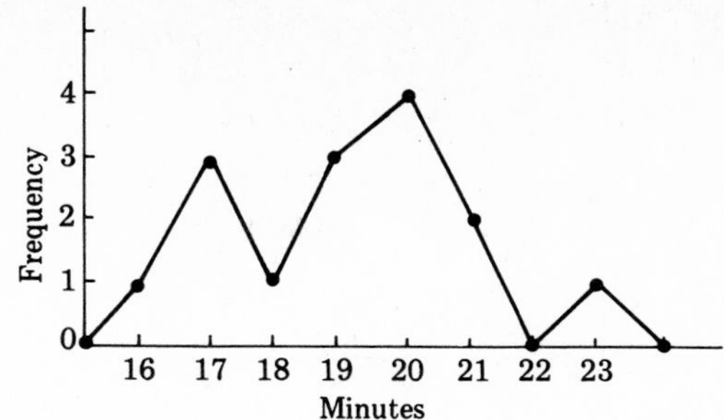

LESSON 30:

Pre-test

1. (a) Line graph
   (b) To show a trend
   (c) 10 miles per gallon
   (d) 10 miles per gallon
   (e) 75% decrease
2. (a) \$928 (b) \$142

Practice Exercises

Frame 12

1. Comparisons 2. Line 3. Bar
4. Line, bar 5. Parts, whole

Frame 21

1. (a) 18–23 years (b) 375% (c) 10 million
2. (a) 4 years (b) 1974 (c) 500% increase
3. (a) 42% or $\frac{21}{50}$ (b) 250% (c) 115,500

Frame 29

1. \$2.77 change 2. \$333.10 refund 3. \$50.40 more interest

Post-test

1. (a) Line graph (b) To show a trend (c) 4 times
   (d) 100% increase (e) 500,000,000 persons
2. (a) \$5.22 (b) \$0.38

Answers to Drill Exercises: Unit 8

1. (a) 4 (b) 2 (c) 2 (d) 2 (e) $\sqrt{1.6} \approx 1$

(f)

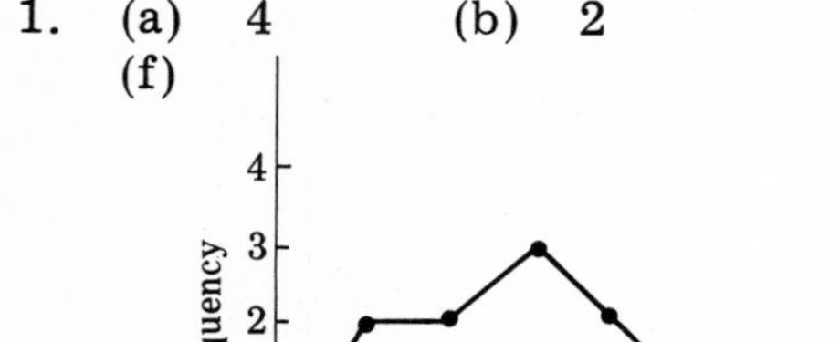

(g)

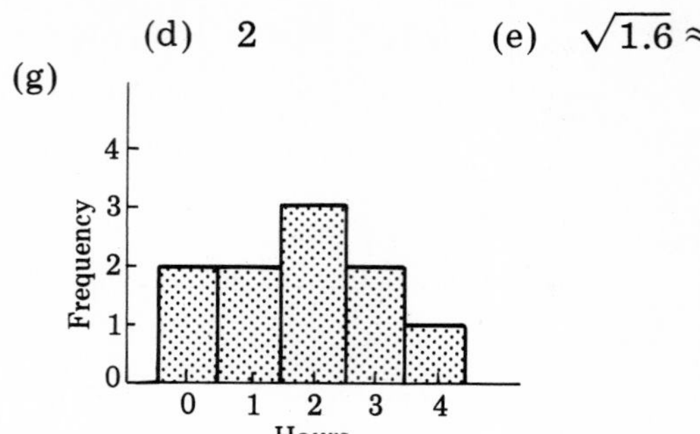

2. (a) 5 (b) 1 (c) 2 (d) 2 (e) 2

(f)

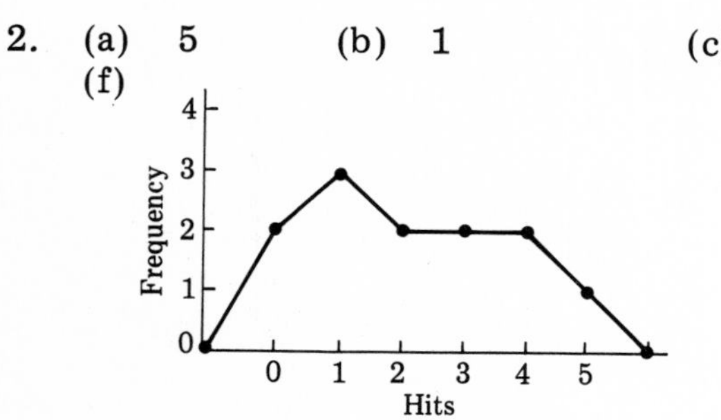

(g)

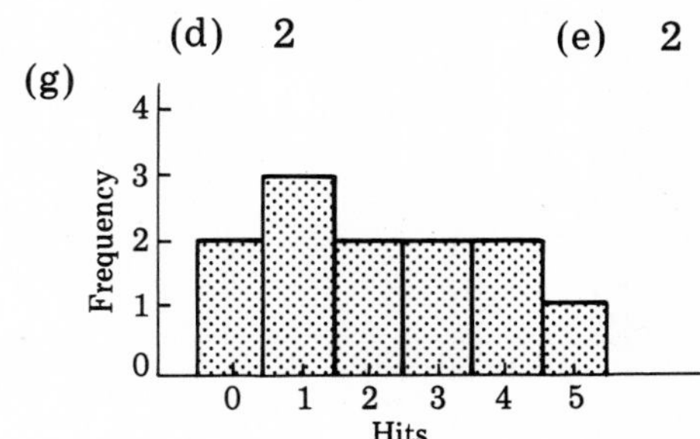

3. (a) 8 (b) 50, 51, and 53 (c) 51 (d) 51 (e) 2

(f)

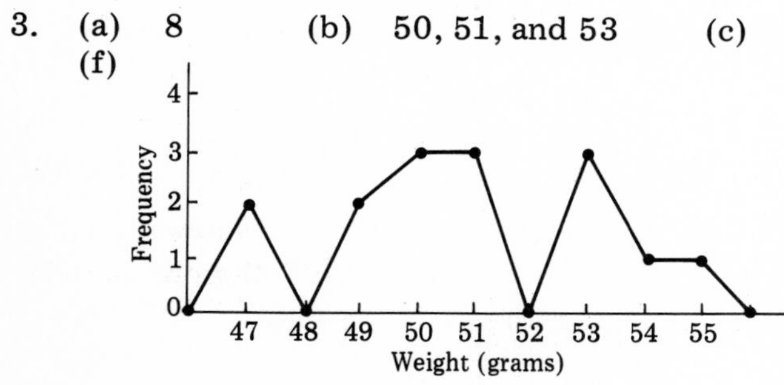

(g)

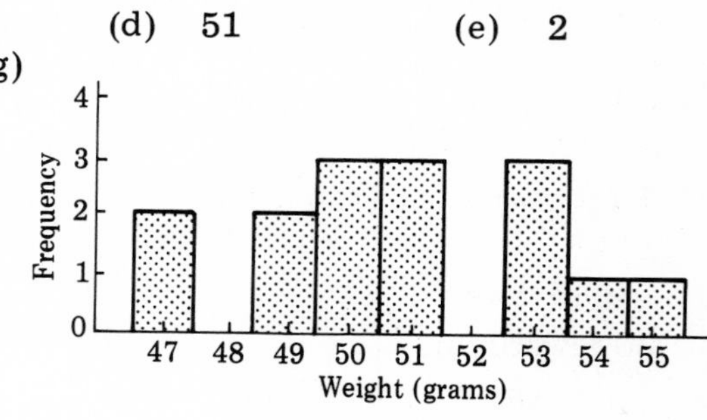

4. (a) 6 (b) 22 and 23 (c) 23 (d) 23 (e) 2

(f)

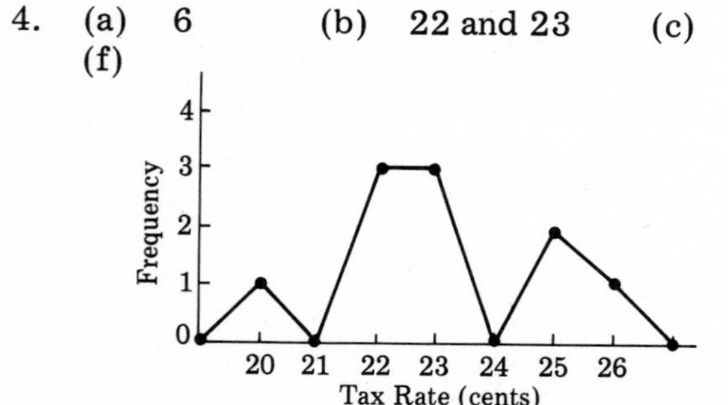

(g)

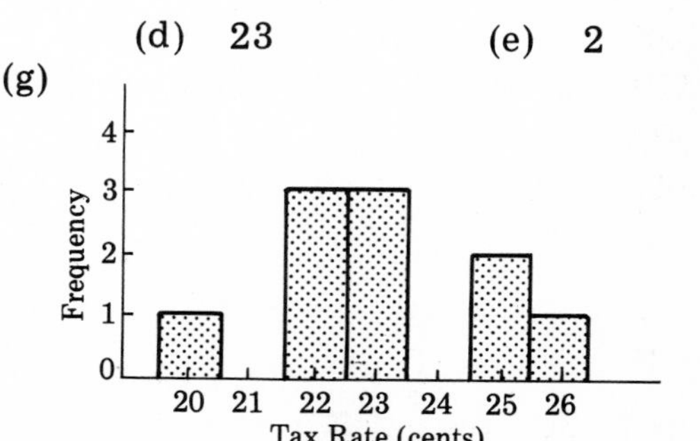

5. (a) 6 (b) 70 (c) 70 (d) 72 (e) 2

(f) (g)

6. (a) 10 (b) 24 and 27 (c) 27 (d) 26 (e) 3

(f) (g)

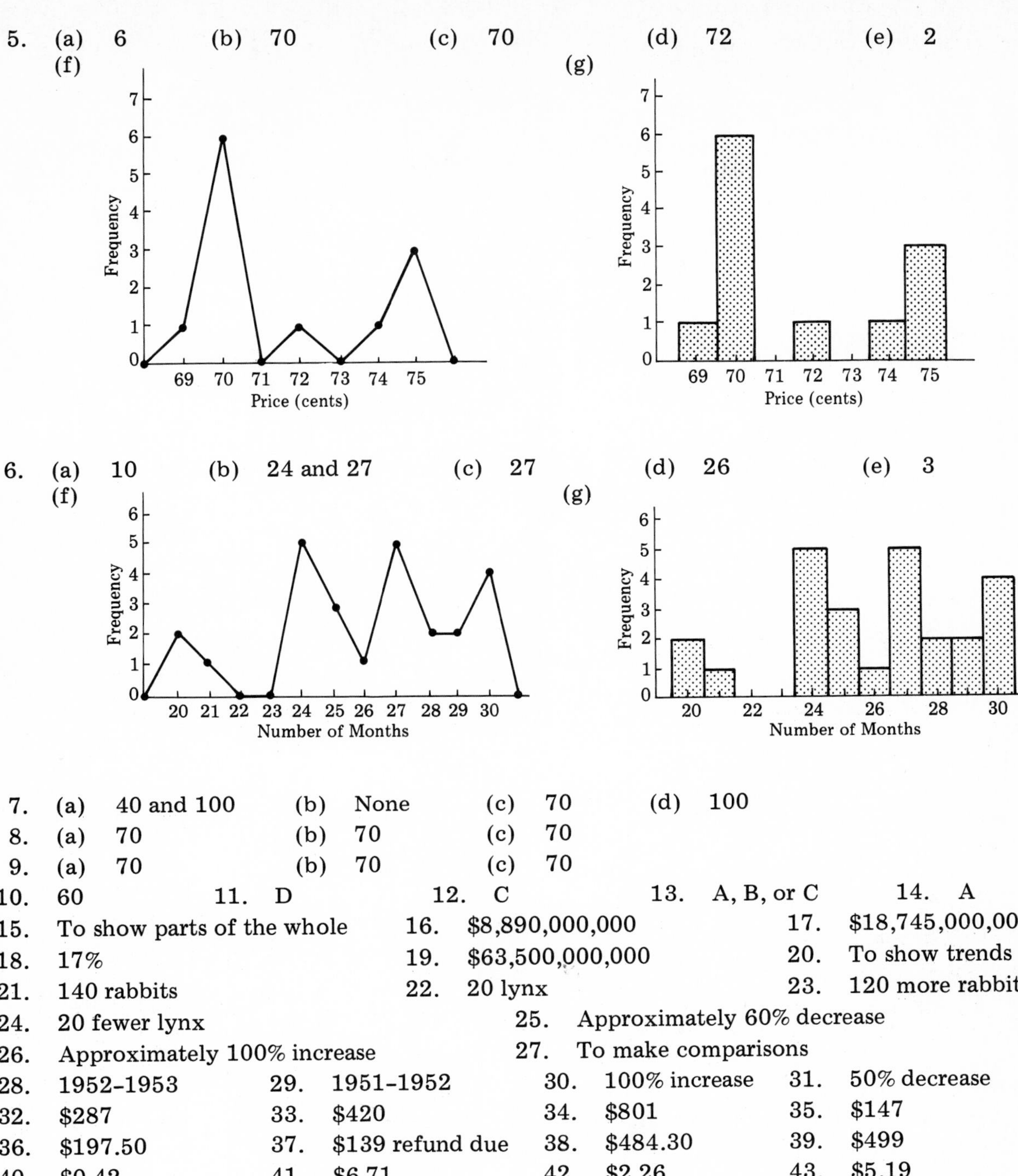

7. (a) 40 and 100 (b) None (c) 70 (d) 100
8. (a) 70 (b) 70 (c) 70
9. (a) 70 (b) 70 (c) 70
10. 60 11. D 12. C 13. A, B, or C 14. A
15. To show parts of the whole 16. \$8,890,000,000 17. \$18,745,000,000
18. 17% 19. \$63,500,000,000 20. To show trends
21. 140 rabbits 22. 20 lynx 23. 120 more rabbits
24. 20 fewer lynx 25. Approximately 60% decrease
26. Approximately 100% increase 27. To make comparisons
28. 1952–1953 29. 1951–1952 30. 100% increase 31. 50% decrease
32. \$287 33. \$420 34. \$801 35. \$147
36. \$197.50 37. \$139 refund due 38. \$484.30 39. \$499
40. \$0.42 41. \$6.71 42. \$2.26 43. \$5.19
44. \$476 45. \$440 46. \$42 loss 47. \$5.80

Answers To Self-Test: Unit 8

1. (b) 2. (d) 3. (a) 4. (e) 5. (b) 6. (d) 7. (e)
8. (b) 9. (a) 10. (c) 11. (d) 12. (d) 13. (c) 14. (e)
15. (a) 16. (a) 17. (b) 18. (d) 19. (e) 20. (c)

Appendix

Table 1:	ADDITION AND MULTIPLICATION COMBINATIONS
Table 2:	SQUARES AND SQUARE ROOTS
Table 3:	ENGLISH SYSTEM WEIGHTS AND MEASURES
Table 4:	METRIC PREFIXES
Table 5:	METRIC SYSTEM WEIGHTS AND MEASURES
Table 6:	METRIC-ENGLISH EQUIVALENTS
Table 7:	MEASURES OF TIME
Table 8:	GEOMETRIC FIGURES
Table 9:	PERCENT, DECIMAL, AND COMMON FRACTION EQUIVALENTS
Table 10:	MISCELLANEOUS FORMULAS

TABLE 1

ADDITION COMBINATIONS

Addend (columns), *Addend* (rows)

+	0	1	2	3	4	5	6	7	8	9
0	0	1	2	3	4	5	6	7	8	9
1	1	2	3	4	5	6	7	8	9	10
2	2	3	4	5	6	7	8	9	10	11
3	3	4	5	6	7	8	9	10	11	12
4	4	5	6	7	8	9	10	11	12	13
5	5	6	7	8	9	10	11	12	13	14
6	6	7	8	9	10	11	12	13	14	15
7	7	8	9	10	11	12	13	14	15	16
8	8	9	10	11	12	13	14	15	16	17
9	9	10	11	12	13	14	15	16	17	18

MULTIPLICATION COMBINATIONS

Factor (columns), *Factor* (rows)

×	0	1	2	3	4	5	6	7	8	9
0	0	0	0	0	0	0	0	0	0	0
1	0	1	2	3	4	5	6	7	8	9
2	0	2	4	6	8	10	12	14	16	18
3	0	3	6	9	12	15	18	21	24	27
4	0	4	8	12	16	20	24	28	32	36
5	0	5	10	15	20	25	30	35	40	45
6	0	6	12	18	24	30	36	42	48	54
7	0	7	14	21	28	35	42	49	56	63
8	0	8	16	24	32	40	48	56	64	72
9	0	9	18	27	36	45	54	63	72	81

TABLE 2

SQUARES AND SQUARE ROOTS

n	n^2	$\sqrt{n}$	$\sqrt{10n}$
1.0	1.00	1.000	3.162
1.1	1.21	1.049	3.317
1.2	1.44	1.095	3.464
1.3	1.69	1.140	3.606
1.4	1.96	1.183	3.742
1.5	2.25	1.225	3.873
1.6	2.56	1.265	4.000
1.7	2.89	1.304	4.123
1.8	3.24	1.342	4.243
1.9	3.61	1.378	4.359
2.0	4.00	1.414	4.472
2.1	4.41	1.449	4.583
2.2	4.84	1.483	4.690
2.3	5.29	1.517	4.796
2.4	5.76	1.549	4.899
2.5	6.25	1.581	5.000
2.6	6.76	1.612	5.099
2.7	7.29	1.643	5.196
2.8	7.84	1.673	5.292
2.9	8.41	1.703	5.385
3.0	9.00	1.732	5.477
3.1	9.61	1.761	5.568
3.2	10.24	1.789	5.657
3.3	10.89	1.817	5.745
3.4	11.56	1.844	5.831
3.5	12.25	1.871	5.916
3.6	12.96	1.897	6.000
3.7	13.69	1.924	6.083
3.8	14.44	1.949	6.164
3.9	15.21	1.975	6.245
4.0	16.00	2.000	6.325
4.1	16.81	2.025	6.403
4.2	17.64	2.049	6.481
4.3	18.49	2.074	6.557
4.4	19.36	2.098	6.633
4.5	20.25	2.121	6.708
4.6	21.16	2.145	6.782
4.7	22.09	2.168	6.856
4.8	23.04	2.191	6.928
4.9	24.01	2.214	7.000
5.0	25.00	2.236	7.071
5.1	26.01	2.258	7.141
5.2	27.04	2.280	7.211
5.3	28.09	2.302	7.280
5.4	29.16	2.324	7.348

n	n^2	$\sqrt{n}$	$\sqrt{10n}$
5.5	30.25	2.345	7.416
5.6	31.36	2.366	7.483
5.7	32.49	2.387	7.550
5.8	33.64	2.408	7.616
5.9	34.81	2.429	7.681
6.0	36.00	2.449	7.746
6.1	37.21	2.470	7.810
6.2	38.44	2.490	7.874
6.3	39.69	2.510	7.937
6.4	40.96	2.530	8.000
6.5	42.25	2.550	8.062
6.6	43.56	2.569	8.124
6.7	44.89	2.588	8.185
6.8	46.24	2.608	8.246
6.9	47.61	2.627	8.307
7.0	49.00	2.646	8.367
7.1	50.41	2.665	8.426
7.2	51.84	2.683	8.485
7.3	53.29	2.702	8.544
7.4	54.76	2.720	8.602
7.5	56.25	2.739	8.660
7.6	57.76	2.757	8.718
7.7	59.29	2.775	8.775
7.8	60.84	2.793	8.832
7.9	62.41	2.811	8.888
8.0	64.00	2.828	8.944
8.1	65.61	2.846	9.000
8.2	67.24	2.864	9.055
8.3	68.89	2.881	9.110
8.4	70.56	2.898	9.165
8.5	72.25	2,915	9.220
8.6	73.96	2.933	9.274
8.7	75.69	2.950	9.327
8.8	77.44	2.966	9.381
8.9	79.21	2.983	9.434
9.0	81.00	3.000	9.487
9.1	82.81	3.017	9.539
9.2	84.64	3.033	9.592
9.3	86.49	3.050	9.644
9.4	88.36	3.066	9.695
9.5	90.25	3.082	9.747
9.6	92.16	3.098	9.798
9.7	94.09	3.114	9.849
9.8	96.04	3.130	9.899
9.9	98.01	3.146	9.950

TABLE 3

ENGLISH SYSTEM WEIGHTS AND MEASURES

Measure of Length

12 inches (in.)	= 1 foot (ft)
3 feet (ft)	= 1 yard (yd)
$16\frac{1}{2}$ feet (ft)	= 1 rod (rd)
5,280 feet (ft)	= 1 mile (mi)
	= 1,760 yards (yd)
	= 0.87 nautical mile
$5\frac{1}{2}$ yards (yd)	= 1 rod (rd)

Measure of Area

144 square inches (in.^{2})	= 1 square foot (ft^2)
9 square feet (ft^2)	= 1 square yard (yd^2)
30.25 square yards (yd^2)	= 1 square rod (rd^2)
160 square rods (yd^2)	= 1 acre
640 acres	= 1 square mile (mi^2)

Measure of Volume

1728 cubic inches (in.^{3})	= 1 cubic foot (ft^3)
27 cubic feet (ft^3)	= 1 cubic yard (yd^3)

Measure of Capacity

Dry Measure

2 pints (pt)	= 1 quart (qt)
8 quarts (qt)	= 1 peck (pk)
4 pecks (pk)	= 1 bushel (bu)

Liquid Measure

16 fluid ounces (fl oz)	= 1 pint (pt)
	= 2 cups (c)
2 pints (pt)	= 1 quart (qt)
4 quarts (qt)	= 1 gallon (gal)

Measure of Weight — Avoirdupois

2,000 pounds (lb)	= 1 ton (T.)
2,240 pounds (lb)	= 1 long ton (l. ton)
16 ounces (oz)	= 1 pound (lb)
437.5 grains (gr)	= 1 ounce (oz)

TABLE 3

ENGLISH SYSTEM WEIGHTS AND MEASURES *(continued)*

Equivalents — Weight, Volume, Capacity

231 cubic inches ($in.^3$)	= 1 gallon (gal)
7.5 gallons (gal)	= 1 cubic foot (ft^3)

1 cubic foot (ft^3) of fresh water weighs 62.5 pounds (lb)

1 cubic foot (ft^3) of salt water weighs 64 pounds (lb)

TABLE 4

METRIC PREFIXES—NAMES, SYMBOLS, DECIMAL EQUIVALENTS

Prefix	*Symbol*	*Decimal Equivalent*	
tera-	T	1000 000 000 000	$= 10^{12}$
giga-	G	1 000 000 000	$= 10^9$
mega-	M	1 000 000	$= 10^6$
kilo-	k	1 000	$= 10^3$
hecto	h	100	$= 10^2$
deka	da	10	$= 10^1$
deci-	d	0.1	$= \frac{1}{10^1}$
centi-	c	0.01	$= \frac{1}{10^2}$
milli-	m	0.001	$= \frac{1}{10^3}$
micro-	u	0.000 001	$= \frac{1}{10^6}$
nano-	n	0.000 000 001	$= \frac{1}{10^9}$
pico-	p	0.000 000 000 001	$= \frac{1}{10^{12}}$

TABLE 5

METRIC SYSTEM WEIGHTS AND MEASURES

Measure of Length

1 kilometer (km)	= 1 000 meters (m)
1 hectometer (hm)	= 100 meters (m)
1 dekameter (dam)	= 10 meters (m)
1 Meter	
1 decimeter (dm)	= 0.1 meter (m)
1 centimeter (cm)	= 0.01 meter (m)
1 millimeter (mm)	= 0.001 meter (m)

TABLE 5
METRIC SYSTEM WEIGHTS AND MEASURES *(continued)*

Measure of Area

1 square kilometer (km^2)	= 1 000 square meters (m^2)
1 square hectometer (hm^2)	= 10 000 square meters (m^2)
1 square dekameter (dam^2)	= 100 square meters (m^2)
1 Square Meter	
1 square decimeter (dm^2)	= 0.01 square meter (m^2)
1 square centimeter (cm^2)	= 0.0001 square meter (m^2)
1 square millimeter (mm^2)	= 0.000 0001 square meter (m^2)

Measure of Volume

1 cubic decimeter (dm^3)	= 0.001 cubic meter (m^3)
1 cubic centimeter (cm^3)	= 0.000 001 cubic meter (m^3)
1 cubic millimeter (mm^3)	= 0.000 000 001 cubic meter (m^3)

Measure of Capacity

1 kiloliter (kl)	= 1 000 liters (l)
1 hectoliter (hl)	= 100 liters (l)
1 dekaliter (dal)	= 10 liters (l)
1 Liter	
1 deciliter (dl)	= 0.1 liter (l)
1 centiliter (cl)	= 0.01 liter (l)
1 milliliter (ml)	= 0.001 liter (l)

Measure of Weight

1 metric ton (t)	= 1 000 kilograms (kg)
1 kilogram (kg)	= 1 000 grams (g)
1 hectogram (hg)	= 100 grams (g)
1 dekagram (dag)	= 10 grams (g)
1 Gram	
1 decigram (dg)	= 0.1 gram (g)
1 centigram (cg)	= 0.01 gram (g)
1 milligram (mg)	= 0.001 gram (g)

Equivalents—Weight, Capacity, Volume

1 liter (l)	= 1 000 cubic centimeters (cm^3)
1 milliliter (ml)	= 1 cubic centimeter (cm^3)

1 liter of water at 4°C weighs 1 kilogram (kg)

1 milliliter of water at 4°C weighs 1 gram (g)

TABLE 6

METRIC-ENGLISH EQUIVALENTS

Measure of Length

1 inch (in.)	= 2.54 centimeters (cm)
	= 0.0254 meter (m)
1 foot (ft)	= 30.48 centimeters (cm)
	= 0.3048 meter (m)
1 yard (yd)	= 91.44 centimeters (cm)
	= 0.9144 meter (m)
1 mile (mi)	= 1 609 meters (m)
	= 1.609 kilometers (km)

Measure of Weight

1 pound (lb)	= 453.6 grams (g)
	= 0.4536 kilogram (kg)
1 ounce (oz)	= 28.35 grams (g)
(avoirdupois)	= 0.0284 kilogram (kg)

Measure of Volume

1 cubic inch (in.^{3})	= 16.39 cubic centimeters (cm^3)
1 cubic foot (ft^3)	= 0.03 cubic meter (m^3)
	= 28.32 liters (l)
1 cubic yard (yd^3)	= 0.76 cubic meter (m^3)
1 quart	= 0.95 liter
1 gallon	= 3.79 liters

TABLE 7

MEASURES OF TIME

1 millennium	= 1,000 years (yr)
1 century	= 100 years (yr)
1 decade	= 10 years (yr)
1 year (yr)	= 365 days (da)
	= 52 weeks (wk)
	= 12 months (mo)
1 week (wk)	= 7 days (da)
1 day (da)	= 24 hours (hr)
1 hour (hr)	= 60 minutes (min)
1 minute (m)	= 60 seconds (sec)

TABLE 8
GEOMETRIC FIGURES

Name	*Sketch*	*Perimeter*	*Area*
Square	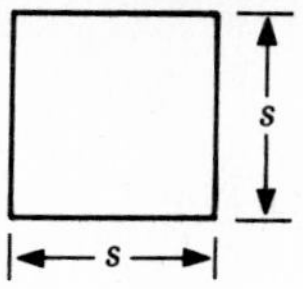	$P = 4s$	$A = s^2$
Rectangle	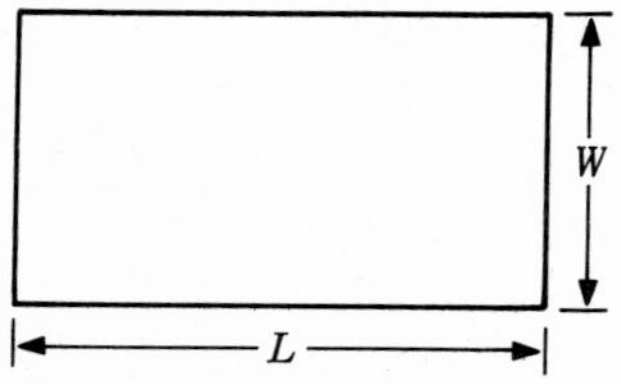	$P = 2L + 2W$	$A = L \times W$
Triangle	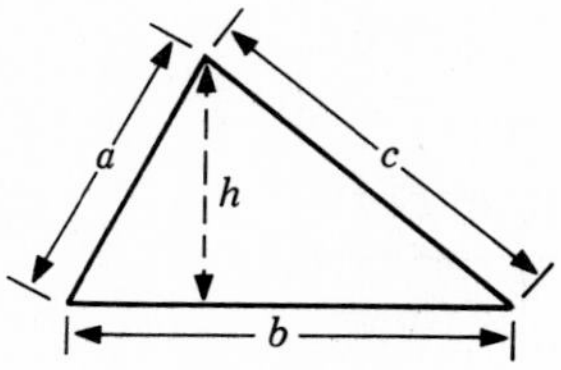	$P = a + b + c$	$A = \sqrt{s(s-a)(s-b)(s-c)}$ where $s = \frac{1}{2}(a + b + c)$
Parallelogram	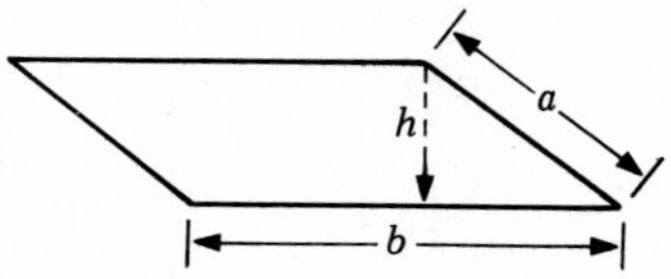	$P = 2b + 2a$	$A = bh$
Circle	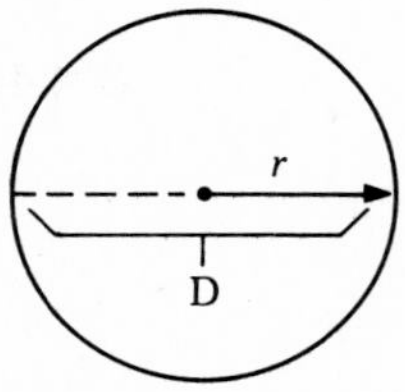	$C = 2\pi r$ $= \pi D$ where $\pi = 3.14159\ldots$	$A = \pi r^2$
Trapezoid	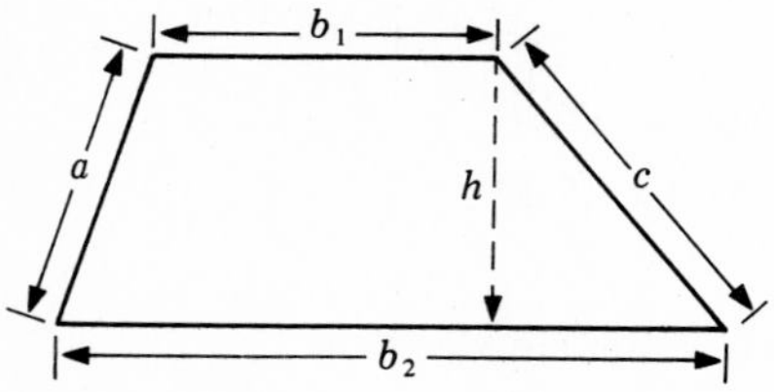	$P = a + b_1 + c + b_2$	$A = \frac{1}{2}h\,(b_1 + b_2)$

TABLE 8
GEOMETRIC FIGURES (*continued*)

Name	*Sketch*	*Volume*	*Surface Area*
Cube	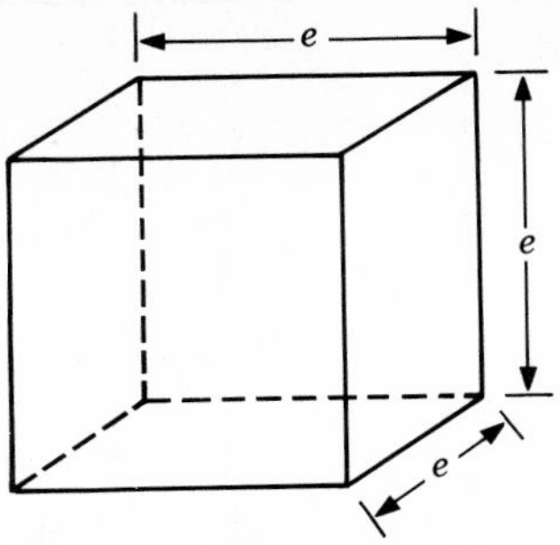	$V = e^3$	$S = 6e^2$
Box	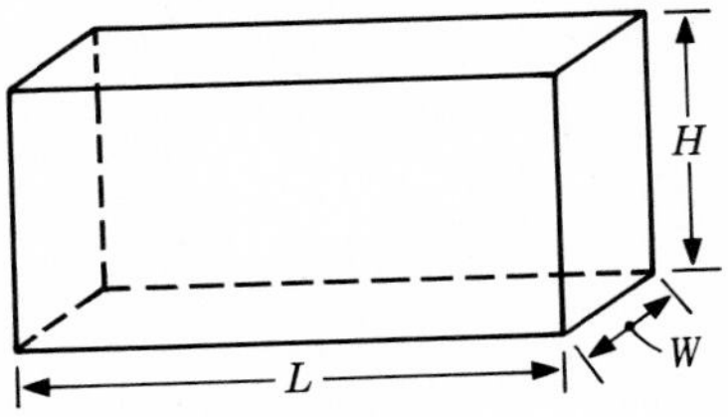	$V = L \cdot W \cdot H$	$S = 2(LW + WH + LH)$
Cylinder	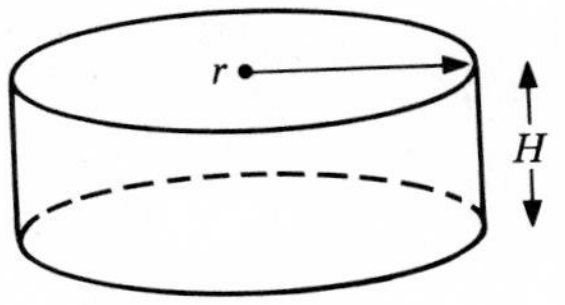	$V = \pi r^2 H$	$S = 2\pi r^2 + 2\pi rH$
Sphere	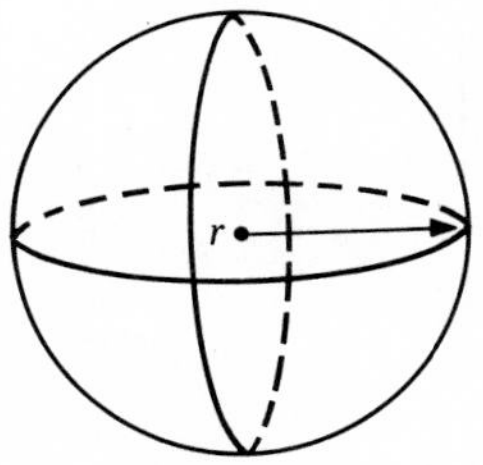	$V = \frac{4}{3}\pi r^3$	$S = 4\pi r^2$

TABLE 9

PERCENT, DECIMAL, AND COMMON FRACTION EQUIVALENTS

Percent	*Decimal*	*Common Fraction*
5%	0.05	$\frac{1}{20}$
$6\frac{1}{4}\%$	$0.06\frac{1}{4}$ or 0.0625	$\frac{1}{16}$
$8\frac{1}{3}\%$	$0.08\frac{1}{3}$ or $0.08\overline{3}$	$\frac{1}{12}$
10%	0.10 or 0.1	$\frac{1}{10}$
$12\frac{1}{2}\%$	$0.12\frac{1}{2}$ or 0.125	$\frac{1}{8}$
$16\frac{2}{3}\%$	$0.16\frac{2}{3}$ or $0.16\overline{6}$	$\frac{1}{6}$
20%	0.20 or 0.2	$\frac{1}{5}$
25%	0.25	$\frac{1}{4}$
30%	0.30 or 0.3	$\frac{3}{10}$
$33\frac{1}{3}\%$	$0.33\frac{1}{3}$ or $0.33\overline{3}$	$\frac{1}{3}$
$37\frac{1}{2}\%$	$0.37\frac{1}{2}$ or 0.375	$\frac{3}{8}$
40%	0.40 or 0.4	$\frac{2}{5}$
50%	0.50 or 0.5	$\frac{1}{2}$
60%	0.60 or 0.6	$\frac{3}{5}$
$62\frac{1}{2}\%$	$0.62\frac{1}{2}$ or 0.625	$\frac{5}{8}$
$66\frac{2}{3}\%$	$0.66\frac{2}{3}$ or $0.66\overline{6}$	$\frac{2}{3}$
70%	0.70 or 0.7	$\frac{7}{10}$
75%	0.75	$\frac{3}{4}$
80%	0.80 or 0.8	$\frac{4}{5}$
$83\frac{1}{3}\%$	$0.83\frac{1}{3}$ or $0.83\overline{3}$	$\frac{5}{6}$
$87\frac{1}{2}\%$	$0.87\frac{1}{2}$ or 0.875	$\frac{7}{8}$
90%	0.90 or 0.9	$\frac{9}{10}$
100%	1.00 or 1	

TABLE 10
MISCELLANEOUS FORMULAS

1. $D = RT$	Distance = Rate × Time
2. $P = RB$	Percentage = Rate × Base
3. $I = PRT$	Interest = Principal × Rate × Time
4. $°C = \frac{5}{9}(°F - 32)$	Fahrenheit to Celsius
5. $°F = \frac{9}{5}°C + 32$	Celsius to Fahrenheit

6. Distance Formula:

$$d = \sqrt{(x^2 + y^2)}$$

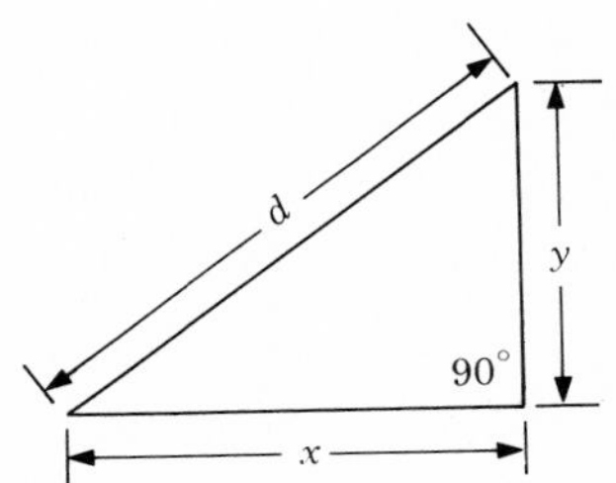

Index

A

Accuracy, 338–39, 342
 in computations, 344
Addend, 9
Addition, 9
 of decimals, 200–01
 of denominate numbers, 299–300, 344
 of fractions, 101, 103–05
 of mixed numbers, 119, 120
 of whole numbers, 9–13
Addition Combinations, 9, 432 (*table*)
Algebra, 154
Approximate numbers, 336
 computations with (*see* Accuracy; Precision)
Area, 321–27
 of a circle, 348, 438
 of a rectangle, 323, 438
 surface, 348–49, 352–53
 of triangles, 351, 438
Associative property:
 of Addition, 10
 of Multiplication, 19
Average, arithmetic, 374

B

Bar graph:
 horizontal, 393–94, 397–98
 vertical, 390–92 (*see also* Histogram)
Base:
 in exponential form, 32
 in percentage formula, 256
Borrowing, 14, 202, 300 (*see* Subtraction)

C

Cancellation, 78, 296
Capacity (*see* Volume)
Carrying:
  in addition, 11, 200
  in denominate number operations, 219
  in multiplication, 22
Celsius (*see* Temperature)
Circle, 345
  area, 348, 438
  circumference, 346, 438
  diameter, 346, 438
  radius, 348, 438
Circle graph, 395–96, 399
Circumference, 346, 438
Commission, 272–73
Common multiple, 64
Common prime factor, 65
Commutative property:
  of addition, 9
  of multiplication, 19
Complex fraction, 112–13
Composite number, 59
Compound interest, 409–10
Conversion factor, 296 (*see also* Dimension analysis)
Conversions:
  English-to-English, 296–99, 434 (*table*)
  English-to-metric, 316–17, 437 (*table*)
  metric-to-metric, 312–16, 435–36 (*table*)
  temperature, 333–34, 441
Cross product, 170
Cube number, 34
Cube root, 37–38
Cumulative frequency, 372

D

Decimal:
  repeating, 219–23
  terminating, 219–20
Decimal place-value names, 196
Decimal point, 193
Decimals, 193
  adding, 200–01
  dividing, 208–13, 215–17
  fractional equivalents of, 194, 218, 220–22
  multiplying, 203–07
  percent equivalents of, 255–56
  powers of, 204
  rounding, 213–16
  subtracting, 201–02
  word forms of, 197

Decimal-shift method:
of division, 216–18, 229
of multiplication, 206–07, 229
Denominate numbers, 296
adding, 299–300, 344
converting, 296–99, 312–17
dividing, 302, 344
multiplying, 301, 302, 344
subtracting, 300–01, 344
Denominator, 69
Least common, 104–06
Depth, 327
Descriptive statistics, 371–81
Deviation, 377 (*see also* Standard deviation)
Diameter, 346
Difference, 13
Digit, 4
significant, 340–42
test (*see* rounding)
Dimension analysis, 296–99
Discount, 275
Distribution, 372
Distributive Property:
of addition, 20, 165
of subtraction, 21, 165
Dividend, 25
Divisible, 42
Divisibility, rules of, 42–43
Division, 25
of decimals by decimals, 211–12
of decimals by whole numbers, 208–11
of denominate numbers, 302, 344
of fractions by fractions, 111–13
of fractions by whole numbers, 111–12
of mixed numbers, 124
by one, 29, 71
by powers of ten, 216–17 (*see* Decimal-shift method)
of whole numbers by decimals, 212
of whole numbers by fractions, 110
of whole numbers by whole numbers, 25–29
by zero, 29, 155

E

English-metric conversions, 316–17, 437 (*table*)
English system of measures, 293, 434–35 (*table*)
Equation, 155
Equivalent fractions, 74–75, 116–17 (*see also* Decimals; Percents)
Equivalent phrases, 155
Evaluating:
formulas, 176 (*see* Formulas)
by using order of operations, 39–42
Even number, 42
Exact number, 336

Expanded notation:
 decimals, 195
 whole numbers, 5, 34
Exponent, 32
Exponential form, 32
Extremes, 170

F

Factor, 18
 common prime, 65
Factoring, 60–61
Fahrenheit (*see* Temperature)
Formulas, 176–80
 evaluating, 176
 for geometric figures, 438 (*table*)
 miscellaneous, 441 (*table*)
 percentage, 266
 simple interest, 277
Fractional parts, 127–32
 in probability, 133–37
Fractions, 69
 adding, 101, 103–05
 complex, 112–13
 decimal equivalents of, 194, 215, 220, 222
 dividing, 110–13
 as indicated divisions, 69, 110
 equivalent, 74
 improper, 116
 multiplying, 72–74
 as numbers, 69
 as parts of a whole, 69, 127
 percent equivalents of, 252, 253
 powers of, 73
 as ratios, 69
 reducing, 76–79
 subtracting, 102, 106
Frequency, 372
 cumulative, 372
Frequency distribution, 372
Frequency polygon, 375–76

G

Graph:
 bar, 390–92, 393–94, 397–98
  histogram, 383–86
 circle, 395–96, 399
 line, 390–91, 393, 394, 398
  frequency polygon, 375–76
 picto-, 394–95, 401
Greatest common factor (GCF), 65–66
Greatest possible error (GPE), 337–39

H

Hectare, 326
Histogram, 383–86

I

Improper fraction, 116
Income tax, 406–08
Interest:
  compound, 408–10
  simple, 277–78
International System of Units (SI), 305
Inverse operations, 15, 30, 34, 37
  in solving equations, 162

K

Kelvin (*see* Temperature)

L

Least common denominator (LCD), 104
Least common multiple (LCM), 64–65, 104
Line graph, 390–91, 393, 394, 398 (*see also* Frequency polygon)

M

Mean, 374
Median, 373–74
Metric measures:
  of area, 321, 323, 436 (*table*)
    hectare, 326
    square meter, 321
  of length, 309, 435 (*table*)
    meter, 305
  of temperature, 332
  of volume, 328, 329, 436 (*table*)
    cubic centimeter, 330
    cubic meter, 328
    liter, 310
  of weight, 310, 436 (*table*)
    gram, 309
    ton, 310
Metric prefixes, 313, 441 (*table*)
  centi-, 306
  deci-, 306
  deka-, 307
  hecto-, 308
  kilo-, 308
  milli-, 307
Metric system, 305
Mill, 234

Minuend, 13
Mixed numbers, 116
  adding, 119
    alternate method, 119, 120
  dividing, 124
  decimal equivalents of, 194
  fractional equivalents of, 116
  multiplying, 123
  percent equivalents of, 252, 257
  powers of, 124
  reciprocals of, 126
  subtracting, 122
    alternate method, 122–23
Mode, 372
Multiple, 62–64
Multiplication, 18
  of decimals, 203–07
  by decimal-shift method, 206–07, 229
  of denominate numbers, 301, 302, 344
  of fractions, 72–74
  of mixed numbers, 123
  by one, 20, 75
  by powers of ten, 206
  of whole numbers, 17, 22–25
  by zero, 20
Multiplication combinations, 19, 432 (*table*)

N

Numeral, 4
Numeration system, 4
Numerator, 69

O

Open sentences, 157–59
Order of operations, 39–40

P

Parenthetical expression, 40
Percent, 251
  decimal equivalents of, 255, 256
  of discount, 275–76
  fractional equivalents of, 252, 253
  of increase or decrease, 273–75
  mixed number equivalents of, 252, 257
  of profit or loss, 276–77
Percentage, 266
Percentage formula, 266
Percent error, 338–39 (*see also* Relative error)
Percent-fraction-decimal triangle, 258–60
Perimeter, 349

Phrases:
equivalent, 155
mathematical, 154
word, 155
Pi, 346
computations using approximations of, 346
Pictograph, 395, 401
Placeholder, 205
Place-value names, 6, 196
Power, 32
Powers:
of decimals, 204
of fractions, 73
of mixed numbers, 124
of ten, 33, 193, 194, 226, 312
of whole numbers, 32 (*see also* Square numbers; Cube numbers)
Precision, 336–40
in computation, 342–44
Prime factor, 65
Prime factorization, 60 (*see also* Factoring)
Prime number, 58
Principal, 277
Probability, 132–37
Product, 18
Proportions, 169–74
cross product, 170
extremes, 170
means, 170

Q

Quotient, 25

R

Radius, 348
Range, 372
Rate, 266
Ratio, 69
in proportions, 169
Reciprocal, 108
Reciprocals:
of fractions, 108–09
of mixed numbers, 126
used in solving equations, 162
of whole numbers, 109
Rectangle, 321
areas of, 322, 438
perimeter of, 349, 438
Reducing (to lowest terms), 76–79
Relative Error, 338–39 (*see also* Percent error)
Remainder, 26
Repeating decimal, 219–23

Rounding, 213–16
Rules of Divisibility, 42–43

S

Sales tax, 404–06
Scientific notation, 225–28
Sieve of Eratosthenes, 59
Significant digits, 340–42
Solution (of an open sentence), 157
Solving equations, 161–67
Square, 345, 438
Square number, 34, 433 (*table*)
Square root, 34–36, 433 (*table*)
Standard deviation, 376–78
Statistics, descriptive, 371–81
Subtraction, 13
  of decimals, 201–02
  of denominate numbers, 300–01, 344
  of fractions, 102, 106
  of mixed numbers, 122–23
  of whole numbers, 13–15
Subtrahend, 13
Sum, 9

T

Table, 404
  compound interest, 409
  income tax, 407
  sales tax, 404
Tally, 372
Taxes, 234–35, 271–72, 404–08
  table of, 404, 407
Temperature, 332–334
  Celsius, 332
  conversions, 333–34
  Fahrenheit, 332
  Kelvin, 332
Terminating decimal, 219–20
Triangle, 351, 438

U

Unit scale:
  of area, 321
  of length, 320
  of volume, 328

V

Variable, 157
Volume, 328
  cylinder, 347, 439
  sphere, 353, 439

W

Weight, 310
Whole numbers, 5
 adding, 9–13
 dividing, 25–29
 expanded notation for, 5, 34
 fractional form of, 70
 multiplying, 17, 22–25
 powers of, 32 (*see also* Square number; Cube number)
 rounding, 214
 subtracting, 13
 word forms of, 5
Word problems, 80–85
 fractional parts, 131–32
 percent, 263–68
 proportions, 171–74
 tax rate, 234–35
 using graphical information, 396–402
 using tables, 404–11
 writing open sentences for, 156–59